Zoophysiology   Volume 13

Coordinating Editor: D. S. Farner

Editors:

D. S. Farner B. Heinrich K. Johansen
H. Langer G. Neuweiler D. J. Randall

Stefan Nilsson

Autonomic Nerve Function in the Vertebrates

With 83 Figures

Springer-Verlag
Berlin Heidelberg New York 1983

Dr. STEFAN NILSSON
Department of Zoophysiology
Göteborg University
S-400 31 Göteborg, Sweden

QP
368
.N54
1983

The cover picture symbolizes an autonomic neuron with an abundance of varicose nerve terminals.

ISBN 3-540-12124-2 Springer-Verlag Berlin Heidelberg New York
ISBN 0-387-12124-2 Springer-Verlag New York Heidelberg Berlin

Library of Congress Cataloging in Publication Data. Nilsson, Stefan, 1946–. Autonomic nerve function in the vertebrates. (Zoophysiology; v. 13). Includes bibliographical references. 1. Nervous system, Autonomic. 2. Nervous system – Vertebrates. I. Title. II. Series. (DNLM: 1. Autonomic nervous system – Physiology. 3. Vertebrates – Physiology. W1 Z0615M v. 13). QP368.N54. 1983. 596'.0188 82-19659.

This work is subject to copyright. All rights are reserved, whether the whole or part of the material is concerned, specifically those of translation, reprinting, re-use of illustrations, broadcasting, reproduction by photocopying machine or similar means, and storage in data banks. Under § 54 of the German Copyright Law where copies are made for other than private use, a fee is payable to "Verwertungsgesellschaft Wort", Munich.

© by Springer-Verlag Berlin Heidelberg 1983
Printed in Germany

The use of registered names, trademarks, etc. in this publication does not imply, even in the absence of a specific statement, that such names are exempt from the relevant protective laws and regulations and therefore free for general use.

Typesetting, printing and bookbinding: Brühlsche Universitätsdruckerei, Gießen.
2131/3130-543210

*This book is dedicated to
Susanne and Jenny*

Preface

The intention of this book is to offer a comprehensive description and discussion of autonomic nerve function in the vertebrates from several points of view. Sections on anatomy, biochemistry of the transmitter substances and the structure, physiology and pharmacology of the different types of autonomic neurons have been included, together with chapters dealing with the autonomic nervous control of some organs and organ systems in the different vertebrate groups. Although knowledge in several of these areas is based primarily on studies of mammals, a certain emphasis has been placed on the autonomic nerve functions in the non-mammalian vertebrates to describe, from a comparative physiological point of view, the adaptations and possible "phylogenetic trends" in the development of the autonomic nerve functions in the vertebrates.

It is very obvious that the literature created by the vigorous research activities within the fields of autonomic nervous anatomy, histochemistry, biochemistry, pharmacology and physiology is vast indeed, and not all aspects of the subject may have received fair treatment in the present volume. With an analogy from astronomy, it is hoped that the mass compressed into this book has reached the level of an energy-emitting neutron star, rather than the black hole which would be the result of compressing too large a mass.

In the preparation of this book I have had excellent assistance from a number of people. I am particularly indebted to Barbro Egnér, Aino Falck-Wahlström, Susanne Holmgren, Inger Holmqvist, Carin Smederöd, Lena Utter and Birgitta Vallander for help with figures, photographic work and preparation of the reference list. Bits and pieces of the manuscript have been read by Ragnar Fänge, Ian Gibbins, Matts Henning and Judy Morris, to whom I am very grateful for suggestions. My very special thanks go to David Grove, Susanne Holmgren and Robert Santer for reading the entire manuscript and giving numerous essential comments and suggestions, and to the editor responsible for this volume, Professor Kjell Johansen, Viking and Physiologist, who not only talked me into writing this book, but also provided many critical suggestions, much encouragement and sound advice to help get it written. Several linguistic and other comments from Professor Donald S. Farner, coordinating editor of the series, are gratefully appreciated.

My thanks also to the publishers and authors who allowed me to use their published (or sometimes unpublished) material, and finally to the staff of the Bio-Medical Library in Göteborg for always doing their very best to help.

Göteborg, January 1983   STEFAN NILSSON

Contents

Chapter 1
Introduction 1

1.1 The Comparative Approach 2
1.2 The Vertebrate Pedigree 4
1.3 Useful Reviews on the Autonomic Nervous System . 5

Chapter 2
Anatomy of the Vertebrate Autonomic Nervous Systems . 6

2.1 Mammals 7
    2.1.1 The Mammalian Spinal Autonomic System . . 8
    2.1.2 The Mammalian Cranial Autonomic System . 13
    2.1.3 The Mammalian Enteric Nervous System . . . 14
2.2 Cyclostomes 15
    2.2.1 The Cyclostome Spinal Autonomic System . . 16
    2.2.2 The Cyclostome Cranial Autonomic System . . 16
    2.2.3 The Cyclostome Enteric Nervous System . . . 17
2.3 Elasmobranchs 17
    2.3.1 The Elasmobranch Spinal Autonomic System . 17
    2.3.2 The Elasmobranch Cranial Autonomic System . 19
    2.3.3 The Elasmobranch Enteric Nervous System . . 19
2.4 Teleosts 20
    2.4.1 The Teleost Spinal Autonomic System 20
    2.4.2 The Teleost Cranial Autonomic System . . . 22
    2.4.3 The Teleost Enteric Nervous System 22
2.5 Dipnoans 23
2.6 Amphibians 23
    2.6.1 The Amphibian Spinal Autonomic System . . 23
    2.6.2 The Amphibian Cranial Autonomic System . . 26
    2.6.3 The Amphibian Enteric Nervous System . . . 26
2.7 Reptiles 26
    2.7.1 The Reptilian Spinal Autonomic System . . . 26
    2.7.2 The Reptilian Cranial Autonomic System . . . 28
    2.7.3 The Reptilian Enteric Nervous System 30
2.8 Birds 30
    2.8.1 The Avian Spinal Autonomic System 30

2.8.2　The Avian Cranial Autonomic System 34
　　2.8.3　The Avian Enteric Nervous System 34
2.9　Conclusions 36

Chapter 3
Neurotransmission 41

3.1　Transmitter Substances 42
3.2　The Autonomic Neuron 43
　　3.2.1　General Arrangement 43
　　3.2.2　Release of Transmitter 45
　　3.2.3　Axonal Transport 46
　　3.2.4　Presynaptic Receptor Systems
　　　　　 (Auto-Receptors) 47
　　3.2.5　Neurons with More Than One Transmitter . 48
　　3.2.6　Adrenergic Neurons 49
　　　　　 3.2.6.1　Synthesis of Catecholamines 49
　　　　　 3.2.6.2　Neuronal Uptake of Catecholamines . 53
　　　　　 3.2.6.3　Metabolic Degradation of
　　　　　　　　　 Catecholamines 54
　　　　　 3.2.6.4　Histochemical Demonstration of
　　　　　　　　　 Adrenergic Neurons 56
　　3.2.7　Cholinergic Neurons 57
　　　　　 3.2.7.1　Histochemical Demonstration of
　　　　　　　　　 Cholinergic Neurons 59
　　3.2.8　Purinergic Neurons 59
　　　　　 3.2.8.1　Histochemical Demonstration of
　　　　　　　　　 Purinergic Neurons 62
　　3.2.9　Serotonergic Neurons 62
　　3.2.10 Peptidergic Neurons 64
3.3　Conclusions 66

Chapter 4
Receptors for Transmitter Substances 68

4.1　Drug–Receptor Interaction 68
　　4.1.1　Specificity and Selectivity 68
　　4.1.2　The "Dose-Response" Concept 69
　　4.1.3　The Theoretical Model for Drug-Receptor
　　　　　 Interaction 70
　　4.1.4　Receptor Reserve 73
　　4.1.5　Affinity and Relative Intrinsic Activity . 74
　　4.1.6　Competitive Antagonism 74
　　4.1.7　Non-Competitive Antagonism 75
　　4.1.8　Functional Interaction 76

	4.1.9 Radio-Ligand Binding Studies	77
	4.1.10 Stimulus–Effect Coupling	77
4.2	Classification of Receptors for Neurotransmitters	79
	4.2.1 Adrenoceptors	79
	4.2.1.1 Subdivision of α- and β-Adrenoceptors	81
	4.2.1.2 The Adrenoceptor Interconversion Hypothesis	82
	4.2.2 Cholinoceptors	83
	4.2.3 Purinoceptors	83
	4.2.4 Receptors for 5-Hydroxytryptamine (Serotonergic Receptors)	84

Chapter 5
Chemical Tools 85

5.1	Drugs That Affect Ion Permeability of Cell Membranes	86
5.2	Drugs That Affect Adrenergic Transmission	86
	5.2.1 Synthesis	86
	5.2.2 Storage	88
	5.2.3 Chemical Sympathectomy	88
	5.2.4 Release	89
	5.2.5 Neuronal Uptake	90
	5.2.6 Degradation	90
	5.2.7 Adrenoceptor Agonists	91
	5.2.8 Adrenoceptor Antagonists	91
5.3	Drugs That Affect Cholinergic Transmission	92
	5.3.1 Release and Neuronal Uptake	93
	5.3.2 Degradation	94
	5.3.3 Cholinoceptor Agonists	94
	5.3.4 Cholinoceptor Antagonists	96
5.4	Drugs That Affect Purinergic Transmission	97
	5.4.1 Uptake and Degradation	97
	5.4.2 Purinoceptor Agonists and Antagonists	97
5.5	Drugs That Affect Serotonergic Transmission	98
	5.5.1 Uptake and Storage	98
	5.5.2 5-HT Receptor Agonists and Antagonists	98
5.6	Drugs That Affect Peptidergic Transmission	99

Chapter 6
Chromaffin Tissue 100

6.1	Histochemical Demonstration of Chromaffin Cells	102
6.2	Chromaffin Tissue in Mammals	103
	6.2.1 Adrenal Medulla	103
	6.2.2 Extra-Adrenal Chromaffin Cells	104
6.3	Chromaffin Tissue in Cyclostomes	105

6.4	Chromaffin Tissue in Elasmobranchs	106
6.5	Chromaffin Tissue in Teleosts	107
6.6	Chromaffin Tissue in Ganoids	108
6.7	Chromaffin Tissue in Dipnoans	109
6.8	Chromaffin Tissue in Amphibians	109
6.9	Chromaffin Tissue in Reptiles	110
6.10	Chromaffin Tissue in Birds	110
6.11	Conclusions	110

Chapter 7
The Circulatory System 112

7.1	Cyclostomes	115
	7.1.1 The Cyclostome Heart	115
	7.1.2 The Cyclostome Vasculature	116
7.2	Elasmobranchs	117
	7.2.1 The Elasmobranch Heart	117
	7.2.2 The Elasmobranch Branchial Vasculature	118
	7.2.3 The Elasmobranch Systemic Vasculature	118
7.3	Teleosts	119
	7.3.1 The Teleost Heart	119
	7.3.2 The Teleost Branchial Vasculature	120
	7.3.3 The Teleost Systemic Vasculature	122
7.4	Ganoids	124
7.5	Dipnoans	125
7.6	Amphibians	126
	7.6.1 The Amphibian Heart	126
	7.6.2 The Amphibian Pulmonary Vasculature	127
	7.6.3 The Amphibian Systemic Vasculature	129
7.7	Reptiles	129
	7.7.1 The Reptilian Heart	129
	7.7.2 The Reptilian Pulmonary Vasculature	132
	7.7.3 The Reptilian Systemic Vasculature	133
7.8	Birds	133
	7.8.1 The Avian Heart	133
	7.8.2 The Avian Vasculature	134
7.9	Mammals	136
	7.9.1 The Mammalian Heart	137
	7.9.2 The Mammalian Vasculature	138
7.10	Conclusions	139

Chapter 8
Spleen 141

| 8.1 | Adrenergic Control | 142 |
| 8.2 | Cholinergic Control | 144 |

Chapter 9
The Alimentary Canal 146

9.1 Mammals 147
    9.1.1 Structure of the Mammalian Alimentary Canal . 147
    9.1.2 Gastro-Intestinal Reflexes in Mammals 148
        9.1.2.1 Gastric Receptive Relaxation 148
        9.1.2.2 The Peristaltic Reflex 149
        9.1.2.3 The Vasodilatory Reflex 151
    9.1.3 "Rebound Excitation" 151
    9.1.4 Neurotransmitters in the Mammalian Enteric Nervous System 152
        9.1.4.1 Noradrenaline 154
        9.1.4.2 Acetylcholine 155
        9.1.4.3 Adenosine Triphosphate (ATP) 155
        9.1.4.4 5-Hydroxytryptamine (Serotonin) . . . 156
        9.1.4.5 Vasoactive Intestinal Polypeptide (VIP) . 156
        9.1.4.6 Substance P 157
        9.1.4.7 Somatostatin 158
        9.1.4.8 Enkephalin 159
        9.1.4.9 Other Peptides 159
9.2 Cyclostomes 159
9.3 Elasmobranchs 161
    9.3.1 The Elasmobranch Stomach 161
    9.3.2 The Elasmobranch Intestine 162
9.4 Teleosts 163
    9.4.1 The Teleost Stomach 164
    9.4.2 The Teleost Intestine 167
9.5 Amphibians 168
    9.5.1 The Amphibian Stomach 168
    9.5.2 The Amphibian Intestine 169
9.6 Reptiles 170
    9.6.1 The Reptilian Stomach 170
    9.6.2 The Reptilian Intestine 170
9.7 Birds 171
    9.7.1 The Avian Oesophagus, Crop and Gizzard . . 171
    9.7.2 The Avian Intestine 172
9.8 Conclusions 174

Chapter 10
Swimbladder and Lung 176

10.1 The Teleost Swimbladder 176
    10.1.1 Inflation of the Swimbladder 178
    10.1.2 Deflation of the Swimbladder 180
10.2 Fish Lungs 182

10.3 Tetrapod Lungs 182
    10.3.1 The Amphibian Lung 183
    10.3.2 The Reptilian Lung 185
    10.3.3 The Avian Lung 185
    10.3.4 The Mammalian Lung 186

Chapter 11
Urinary Bladder 187

11.1 The Teleost Urinary Bladder 188
11.2 The Amphibian Urinary Bladder 189
11.3 The Reptilian Urinary Bladder 189
11.4 The Mammalian Urinary Bladder 190

Chapter 12
Iris 192

12.1 The Elasmobranch Iris 193
12.2 The Teleost Iris 193
12.3 The Amphibian Iris 194
12.4 The Reptilian Iris 194
12.5 The Avian Iris 196
12.6 The Mammalian Iris 197

Chapter 13
Chromatophores 198

13.1 Teleost Chromatophores 198
13.2 Reptilian Chromatophores 201

Chapter 14
Concluding Remarks 202

References 204
Subject Index 249

Chapter 1

Introduction

> "I propose the term 'autonomic nervous system' for the sympathetic system and the allied nervous system of the cranial and sacral nerves, and for the local nervous system of the gut"
>
> Langley (1898)

The first observations of the anatomy of the autonomic nervous system were made by Galenos (A.D. 130–200) (see Ackerknecht 1974) during dissection of pigs. Galenos regarded the nerves as hollow tubes, through which the animal spirit could pass from one organ to the other, creating a coordination of the organs called sympathy. Although Galenos' conclusions were somewhat at variance with the modern views of nerve function, the term sympathy (or rather sympathetic) has been retained till this day. More precisely, the term "les grands nerfs sympathiques" was used by Winslow (1732) to describe the human paravertebral ganglionic chains (sympathetic chains), which are a part of the autonomic nervous system.

The terminology of Langley (1898, 1921) replaced earlier, often confusing terms such as "involuntary nervous system" or "vegetative nervous system", and provided an anatomical subdivision of the autonomic nervous system into three parts: the *sympathetic*, with central nervous connections in the thoracic and lumbar segments of the spinal cord; the *parasympathetic*, with cranial and sacral connections; and finally the *enteric* nervous system which comprises the intrinsic neurons of the gut. This terminology is still valid on an anatomical basis for the description of the mammalian autonomic nervous system but, as will be argued in Chapter 2, its usefulness in non-mammalian vertebrates is limited.

The autonomic nervous system is composed of all efferent nervous pathways that have a ganglionic synapse outside the central nervous system (CNS) (Campbell 1970a). This includes the nervous supply to all smooth muscle, cardiac muscle, striated muscle in the iris of reptiles and birds, possibly striated muscle in the gut of some fish, certain glands including the adrenal and extra-adrenal chromaffin cells as well as enterochromaffin cells of the gut, adipose tissue, particularly brown fat in mammals, and melanophores in some groups and photophores in fish. All neurons intrinsic to the gut, i.e., including the sensory neurons and interneurons, will be regarded as enteric autonomic neurons, since a subdivision into afferent and efferent components is at present difficult if not impossible to make. The extrinsic input to the enteric neurons is very limited, and the enteric nervous system is capable of integrated activity in the absence of all central nervous connections. It may thus be regarded as even more "autonomic" than the other divisions.

Autonomic nerves reach almost every part of the vertebrate body, and the involvement of autonomic nerve functions in the maintenance of "the fixity of the internal environment" (Bernard 1878) or "homeostasis" (Cannon 1929) was early recognized. Particular attention has been paid to the role of the "sympathico-adrenal" system in the "defence reaction" of the animal, i.e., the circulatory and other adjustments brought about by an increased adrenergic activity in "fight or flight" emergencies (Cannon 1929). A double antagonistic control of organs via "sympathetic" and "parasympathetic" nerves is often present, and in the early view of autonomic nerve function the "sympathetic" system was held synonymous with an adrenergic nervous system, whilst the "parasympathetic" system was thought to be composed of cholinergic nerves. Although this gross oversimplification of matters still prevails in some textbooks, it should now be quite clear that the autonomic nervous system offers a complex variety of neuron types with more than a dozen different putative transmitter substances.

1.1 The Comparative Approach

The aim of this book is to describe the function of the autonomic nervous system in the various vertebrate groups, with a special emphasis on the non-mammalian vertebrates. Since, however, the knowledge of mammalian autonomic nerve function in many respects is far superior to that for the other vertebrate groups, the mammals form a necessary and convenient frame of reference for many of the discussions.

The comparative approach to the physiology of the autonomic nervous system is useful for several reasons. First of all it offers an insight into the role of a fundamentally important control system in the adaptation of physiological functions of animals with environmental conditions and prerequisites that may be vastly different from those of man and his favourite laboratory mammals. Secondly, it offers a new and fresh angle of attack on fundamental problems regarding the mechanisms of autonomic nerve functions. It seems more than likely that a sampling of the wide variety of vertebrate organisms will provide information of great value for the basic understanding of nerve function, and the evolution of transmitter mechanisms. As will be seen in the following chapters, there are in fact several such examples.

It should be stressed immediately that our knowledge of the physiological role of the autonomic innervation of a certain organ or organ system, particularly in the non-mammalian species, is very often restricted to the *potential* role of the nerves concluded from experiments with a variety of in vitro preparations. So far there is a conspicuous lack of understanding of the *actual* functions of the autonomic innervation in the in vivo situation. The act of bringing the animal into the laboratory is likely to induce more or less severe disturbances in the delicate balance of autonomic (and endocrine) functions, and quite possibly triggers a "defence reaction" with an increased adrenergic influence on the organ functions. It is to be expected that a full understanding of the various components of the autonomic innervation is hampered by such a dramatic adrenergic background.

In Chapters 6–13 of this book, the autonomic innervation of different organs and organ systems is reviewed, and in some cases summarized in generalized diagrams of the innervation patterns. The idea of these diagrams comes from the review by Burnstock (1969), and it is hoped that they, although generalized and in continuous need of revision thanks to new information, will provide a general view of the complexity (or simplicity) of the innervation patterns. It should be noticed that the complexity presented of the innervation patterns not only reflects an actual situation, but also to a great extent the amount of research dedicated to a certain organ (e.g., amphibian lung).

It is also hoped that the later chapters of this book will provide some basis for speculations and deductions about the phylogenetic evolution of autonomic nerve functions and innervation patterns, despite the fact that the available information from a certain vertebrate group very often rests on experiments with a single or a

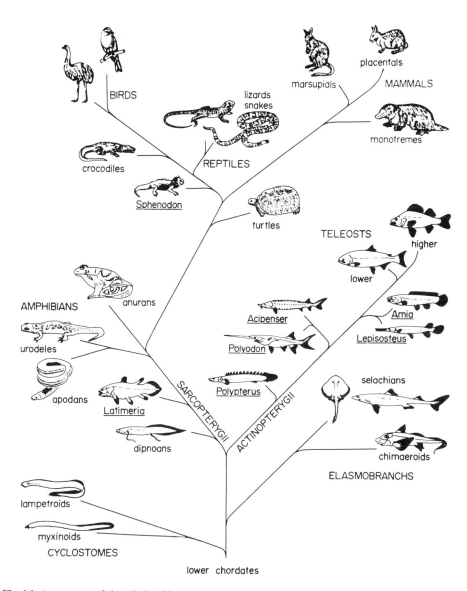

Fig. 1.1. A summary of the relationships among the major vertebrate groups: cyclostomes, elasmobranchs, teleosts, amphibians, reptiles, birds and mammals and some of the phylogenetically interesting but smaller groups

few species. To provide a taxonomic framework for such phylogenetic speculations, a brief summary of the vertebrate pedigree is offered below and in Fig. 1.1.

1.2 The Vertebrate Pedigree

The first primitive vertebrates appeared during the Ordovician period, some 500 million years ago, as descendants of the ancestral chordates. The fossil record offers some clues about the evolution of the different groups, although the exact relationships among the fishes and reptiles, for instance, are disputed. The present brief summary is *not* a contribution to this dispute, but simply an attempt to give some kind of basis for the phylogenetic speculations about the evolution of functions of the autonomic nervous system. The relationships are based mainly on the conclusions reached by Romer (1962).

It is customary, but not necessarily correct in all respects, to regard the steps of development of the major vertebrate groups in the order: cyclostomes, elasmobranchs, teleosts, amphibians, reptiles, birds and mammals. It is important, however, to note that the teleosts are not the direct ancestors of the amphibians, and that the birds and mammals form two independent lines of evolution from the stem reptiles (Fig. 1.1).

Among the fish, the cyclostomes are the most primitive extant forms. These are jawless fish (class Agnatha), descending from the most primitive vertebrate forms, the now extinct ostracoderms. The cyclostomes form two loosely related groups: the myxinoids (hagfishes) and the lampetroids (petromyzontids, lampreys).

From the jawless fish sprang a group of jawed, strange-looking fish, the placoderms (class Placodermi), which are now extinct but which gave rise to the two major classes of modern fish: the cartilaginous fish (class Chondrichthyes) and the bony fish (class Osteichthyes). The modern chondrichthyans are the elasmobranchs, which form two groups: the selachians (sharks and rays) and the chimaeroids (holocephali).

The bony fish split into two major lines of evolution early on in their history. The first line (subclass Actinopterygii) gave rise to three major groups: chondrosteans, holosteans and teleosts. The first two of these groups are often together labelled ganoids, and comprise the bichir (*Polypterus*) and sturgeons (*Acipenser, Polyodon*) (chondrosteans) and the gar and bowfin (*Lepisosteus, Amia*) (holosteans). The teleosts form the tip of the actinopterygian branch in which it is possible to distinguish more primitive ("lower") teleosts such as salmon, trout and herring, from the more advanced forms ("higher teleosts") such as cod, flounder, labrid and perch.

The second line of evolution of the Osteichthyes is the Sarcopterygii, from which the tetrapods descended. The surviving groups of sarcopterygian fish are the dipnoans (lungfish) and the coelacanth (*Latimeria*).

The remaining vertebrate groups each comprise a separate class. The amphibians (class Amphibia) are represented by three orders: urodeles (newts and salamanders), anurans (frogs and toads) and a group of rare tropical worm-like animals, the apodans.

The living reptiles (class Reptilia) are represented by four major groups: the turtles (order Chelonia), the tuatara (*Sphenodon;* order Rhynchocephala), the lizards and snakes (order Squamata) and the crocodiles (order Crocodilia).

Finally, the birds and mammals sprang as two separate lines from the ancestral stem reptiles (Fig. 1.1).

1.3 Useful Reviews on the Autonomic Nervous System

The present book covers a wide variety of aspects of autonomic nerve function, anatomy and pharmacology. Several excellent monographs which deal in depth with such individual aspects are available, of which the following suggestions represent a sample of this vast literature. Further reference to less extensive reviews is given in the following chapters.

A general treatment of the anatomy autonomic nervous system is given by Gabella (1976) and Pick (1970) who also include the non-mammalian vertebrates. The pharmacology of the autonomic nervous system is dealt with in most general textbooks of pharmacology, and extensively by Goodman and Gilman (1970 and later editions), Carrier (1972) and Day (1979).

Functional aspects can be found in Bennett (1972), who also includes a thorough account of the electrophysiology of neurons and smooth muscle, and Kalsner (1979). The function of adrenergic neurons is specifically treated by Iversen (1967), Burnstock and Costa (1975) and Paton (1979).

The literature on autonomic nerve function of the non-mammalian species is not extensive, but several excellent reviews on individual vertebrate groups are available, e.g. Bennett (1974) and Bolton (1971a, b), birds; Taxi (1976), amphibians; Santer (1977) adrenergic neurons in fish; Berger and Burnstock (1979), reptiles Rovainen (1979), lampetroids.

A very useful comprehensive review on the evolution of vertebrate autonomic innervation is that of Burnstock (1969), and the older literature on autonomic innervation in fish and amphibians is reviewed by Nicol (1952).

Chapter 2

Anatomy of the Vertebrate Autonomic Nervous Systems

This chapter provides an introduction to gross anatomy and basic terminology of the autonomic nervous systems in the different vertebrate groups. Since the literature is replete with interesting anatomical peculiarities of individual species, emphasis has been placed on generalized "type animals" from each group to avoid bewilderment by details. Where appropriate, supplementary descriptions of the more striking differences among orders and species have been included.

The mammals are by far the most-studied vertebrates, which makes it practical to present first a description of the general arrangement of the mammalian autonomic nervous system, before proceeding to the other vertebrate groups.

As already stated in Chap. 1, the autonomic nervous system was subdivided by Langley (1898, 1921) into a sympathetic, a parasympathetic and an enteric portion. The reason for this subdivision was primarily anatomical, but the functional differences between the sympathetic and the parasympathetic portions, which are often antagonistic, have contributed to establishment of the terminology. In mammals, the numerous deviations from the over-simplified picture of sympathetic = adrenergic and parasympathetic = cholinergic were pointed out by Cambell (1970a). It is now clear that a *functional* basis for the separation of the three portions of the autonomic nervous system in mammals does not exist. The *anatomical* basis for a subdivision of the autonomic nervous system according to Langley's terminology is somewhat stronger. The point of outflow of the autonomic nervous pathways from the central nervous system gives a basis for distinguishing between a sympathetic (thoraco-lumbar outflow) and a parasympathetic (cranio-sacral outflow) system, but the mingling of fibers in the pelvic plexus (see below) and the occurrence of both "sympathetic" and "parasympathetic" ganglion cells in this plexus (Gabella 1979), make it difficult to trace the anatomical origin of the individual autonomic pathways as they reach their effector organs.

The situation in the non-mammalian vertebrates is even more confusing, and Young (1933c, 1936) and Lutz (1931) concluded that there is little evidence on which to base a separation of a sympathetic and a parasympathetic system in elasmobranch and teleost fish. Later, doubts about the anatomical distinction of a sacral parasympathetic system in amphibians were expressed (Burnstock 1969, Pick 1970), and Campbell (1970b) as well as Bennett (1974) abandoned the use of the terms sympathetic and parasympathetic in their reviews on the autonomic nervous system of fish and birds, respectively.

To provide a practical anatomical classification of the sub-units of the autonomic nervous system, which is also suitable for the present comparative treatment, a modified terminology slightly different from that of Langley (1921) is adopted (Table 2.1). Thus the term *cranial autonomic system* is used to describe the

Table 2.1. Suggested modification of the terminology used for the anatomical classification of the different portions of the autonomic nervous system of the vertebrates, compared with the terminology of Langley (1921)

Type of autonomic pathway	Langley (1921)	Modified terminology
Cranial	Parasympathetic	Cranial autonomic
Thoracic	Sympathetic	Spinal autonomic
Lumbar		
Sacral	Parasympathetic	
Intrinsic neurons of the gut	Enteric	Enteric

autonomic pathways that leave the CNS in the cranial nerves. This system is identical with the cranial parasympathetic system in Langley's terminology.

The second recognizable portion of the autonomic nervous system is the *spinal autonomic system*, which consists of all autonomic pathways leaving the CNS in the spinal segments. This includes the sympathetic system (thoraco-lumbar outflow), and the sacral parasympathetic system in Langley's terminology. The term *sympathetic chains*, derived from the very early descriptions of the nervous system (Winslow 1732, Gaskell 1886), has been retained in the present account to describe the paired paravertebral ganglionic chains. It is a well-established term which can hardly lead to misunderstanding even when discussing non-mammalian vertebrates.

The third portion of the autonomic nervous system, the *enteric system*, although thoroughly neglected in textbooks for the past 50 years, has recently regained a well-deserved interest (Gershon et al. 1979, Gershon 1981). This term will be used here as originally proposed by Langley (1898) to describe the local nervous system of the gut including sensory neurons, interneurons and motor neurons (excitatory or inhibitory) (Furness and Costa 1980, Gershon 1981). A summary of the basic arrangement of the different autonomic nervous pathways is presented in Fig. 2.1.

2.1 Mammals

Although an anatomical basis for subdividing the autonomic nervous system as originally proposed by Langley exists in mammals, the slightly modified terminology described above will be used here also for this group. The aim is to provide a unified frame of reference for comparison with the anatomy of the autonomic nervous systems of the non-mammalian vertebrates.

More extensive treatment of mammalian autonomic nervous anatomy can be found in the classical work of Langley (1921), and in the more recent comprehensive monographs by Pick (1970) and Gabella (1976).

2.1.1 The Mammalian Spinal Autonomic System

There are three main types of ganglia in the spinal autonomic pathways. The first type consists of the interconnected ganglia of the paired paravertebral chains (sympathetic chains), which are characterized by short preganglionic neurons and long postganglionic neurons (Figs. 2.1 and 2.2). The second type is the prevertebral (collateral) ganglia, such as the coeliac and mesenteric ganglia, and the ganglia of the pelvic plexus. The third type are the ganglia within the wall of the effector organs. The preganglionic neurons of the two last types are relatively longer, while the postganglionic neurons vary in length (Figs. 2.1 and 2.2).

The longitudinal connections between the sympathetic chain ganglia are composed of pre- and postganglionic fibres which enter the paravertebral ganglion of one segment and run to another (anterior or posterior) segment before synapsing with a postganglionic neuron or leaving the chain respectively. This type of arrangement allows the sympathetic chains to continue into both cervical and sacral regions, but there are no preganglionic nerves from the spinal cord that join the chains in these segments (Fig. 2.2).

The cell bodies of the preganglionic neurons are situated in the intermediolateral nucleus (lateral horn) and to a lesser extent in the intermediomedial nucleus of the grey matter in the spinal cord (Pick 1970, Gabella 1976). The preganglionic fibres to the sympathetic chain ganglia leave the spinal cord via the ventral roots of the spinal nerves of the thoracic and lumbar vertebrae, and run from the spinal nerves to the ganglia in connectives called white *rami communicantes* (sing. *ramus communicans*). Postganglionic fibres leave the sympathetic chains either as separate nerves (e.g., the cardiac and pulmonary nerves or the nerves forming the carotid plexus to the head), or rejoin the spinal nerves via grey *rami communicantes* (Fig. 2.2). The reason for the difference in colour between the white and the grey *rami communicantes* is the presence of light-reflecting myelin sheaths around the axons of the preganglionic neurons, which are absent in the postganglionic nerves. The postganglionic fibres that rejoin the spinal nerves via the grey *rami communicantes* are distributed to the blood vessels of the skeletal muscle and skin, pilomotor muscles and skin glands.

The cervical part of each sympathetic chain consists of three ganglia, which receive their preganglionic input from the white *rami communicantes* entering the chains in the thoracic segments. Thus the only direct connections between the cer-

Fig. 2.1. Diagrammatic representation of the components of cranial, spinal and enteric autonomic pathways. The cranial, classically "parasympathetic" pathways, are characterized by long preganglionic and short postganglionic neurons, with the cell bodies of the postganglionic neurons close to or within the effector organ. The cell bodies of the postganglionic neurons in the spinal autonomic pathways are situated in the paravertebral ganglia (sympathetic chains) (*upper example*), in prevertebral (collateral) classically "sympathetic" ganglia (*middle example*) or within the effector organ as in the classically "parasympathetic" sacral pathways (*lower example*). Neurons of the enteric nervous system may or may not receive an extrinsic input, and all types of neurons that are intrinsic to the gut (sensory neurons, interneurons and motorneurons) belong to the enteric nervous system

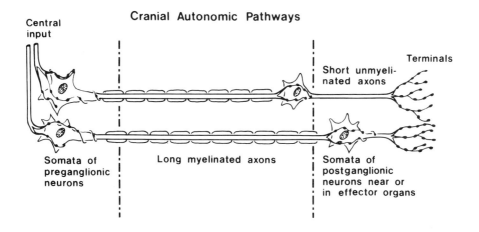
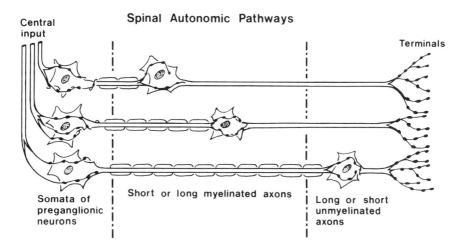
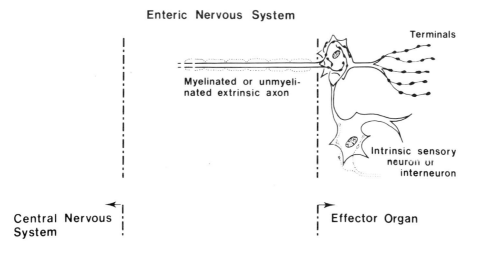

vical chain ganglia and the spinal nerves are the grey *rami communicantes*. The most anterior of the cervical ganglia is the large *superior cervical ganglion*, from which fibres to the head (carotid plexus), heart and lungs emerge (Fig. 2.3). In some mammalian species a small *middle cervical ganglion* may give off additional fibres to the heart and lungs. The third cervical ganglion is the *inferior cervical ganglion* which often fuses with the first thoracic ganglion to form the *stellate ganglion* or stellate complex (Fig. 2.3). This ganglion gives off the major cardiac (*nervi accelerantes*) and pulmonary branches.

The postganglionic fibres of the anterior thoracic ganglia run mainly forward in the chains to join the spinal autonomic nerves to the heart and lungs from the stellate ganglia. The middle and posterior part of the thoracic region of the chains contain small ganglia, which give off comparatively few postganglionic fibres that join the spinal nerves. Most of the preganglionic fibres at this level pass the chains without synapsing, and continue in the greater and lesser *splanchnic nerves* to the *coeliac ganglion* (coeliac complex) from which postganglionic fibres run along abdominal blood vessels to the stomach, small intestine, anterior part of the large intestine, liver, spleen and kidney. Some preganglionic fibres also pass the coeliac ganglion and run directly to the adrenal glands, where they innervate the catecholamine-secreting (adreno-medullary) chromaffin cells (Fig. 2.3).

Most of the preganglionic fibres from the posterior thoracic and the lumbar region also pass the sympathetic chains without synapsing, and run to the *superior* and *inferior mesenteric ganglia*. These prevertebral ganglia give off postganglionic fibres to the large intestine, rectum and urogenital organs. Part of the fibres from the inferior mesenteric ganglion form the *hypogastric nerves*, which enter the extensive *pelvic plexus* and are distributed to the lower abdominal viscera (rectum, large intestine and urogenital organs).

The spinal autonomic outflow from the sacral segments, classically referred to as the "sacral parasympathetic system", consists of preganglionic fibres which form the *pelvic nerves (nervi erigentes)*. These mix with the hypogastric nerves to form the pelvic plexus, which contains numerous ganglion cells mainly of sacral, but also of lumbar autonomic pathways (Figs. 2.2 and 2.3).

Fig. 2.2. Arrangement of the thoracolumbar and sacral autonomic pathways of a mammal. In the thoracolumbar segments, preganglionic fibres from the spinal cord enter the paravertebral ganglia of the sympathetic chains via the ventral roots of the spinal nerves and the white *rami communicantes*. Preganglionic fibres may synapse with postganglionic cell bodies within the same sympathetic chain ganglion, or may continue to other chain ganglia or prevertebral ganglia before synapsing with a postganglionic neuron. Postganglionic fibres from neurons in the sympathetic chain ganglia may also run in the chain to other segments. The postganglionic fibres leave the chain either in separate nerves to the viscera, or in the grey *rami communicantes* to join the spinal nerves and run with these. The preganglionic neurons of the classically "parasympathetic" pathways in the sacral segments of the spine leave in the ventral roots of the spinal nerves and synapse with postganglionic neurons situated in the ganglia of the pelvic plexus or within the effector organ. Abbreviations: *coel g*, coeliac ganglion; *postg n*, postganglionic neuron; *preg n*, preganglionic neuron; *r comm*, ramus communicans; *sc*, sympathetic chain; *spl n*, splanchnic nerve

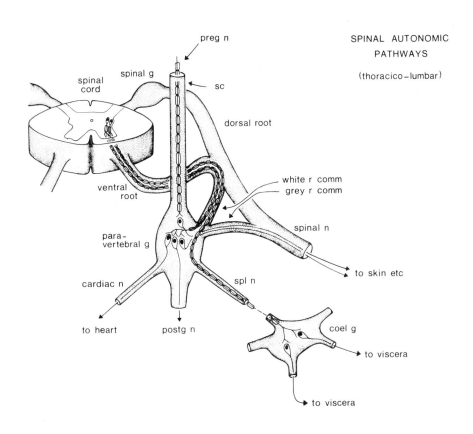
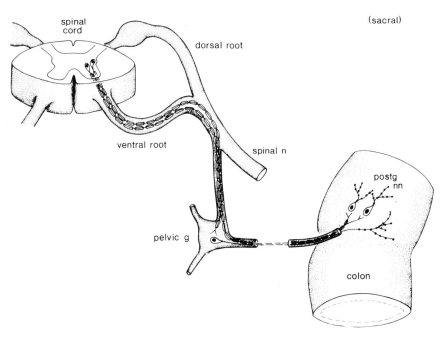

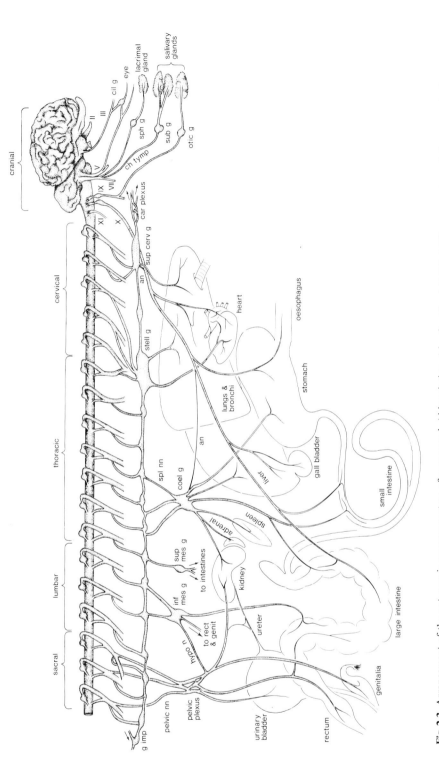

Fig. 2.3. Arrangement of the autonomic nervous system of a mammal. Abbreviations (also for Figs. 2.6–2.8, 2.10): *an*, anastomosis between spinal autonomic and cranial nerves; *ant spl n*, anterior splanchnic nerve; *car plexus*, carotid plexus; *ceph sc*, cephalic sympathetic chain; *ch tymp*, chorda tympani (internal mandibular nerve); *cil g*, ciliary ganglion; *coel g*, coeliac ganglion; *comm*, commissure between right and left sympathetic chain; *deep ceph symp*, deep cephalic sympathetic; *deep cerv symp*, deep cervical sympathetic; *g imp*, ganglion impar; *hypo n*, hypogastric nerve; *inf mes g*, inferior mesenteric ganglion; *mid spl n*, middle splanchnic nerve; *nod g*, nodose (vagal) ganglion; *pet g*, petrous (glossopharyngeal)ganglion; *post spl nn*, posterior splanchnic nerves; *r comm*, ramus communicans; *sph g*, sphenopalatine ganglion; *sub g*, submandibular ganglion; *sup ceph symp*, superior cervical sympathetic; *sup cerv g*, superior cervical ganglion; *sup cerv symp*, superior cervical sympathetic; *sup mes g*, superior mesenteric ganglion; *stell g*, stellate ganglion. *Roman numbers* refer to the cranial nerves

2.1.2 The Mammalian Cranial Autonomic System

The cranial part of the autonomic nervous system consists of the autonomic pathways that run in the paired cranial nerves. The ganglia of the cranial autonomic pathways are located either close to (e.g., salivary glands) or embedded in (e.g., heart) the effector organ (Fig. 2.4).

Autonomic fibres are found in cranial nerves III (oculomotor nerve), VII (facial nerve), IX (glossopharyngeal nerve) and X (vagus). The preganglionic fibres of the oculomotor nerve (III) arise in the Edinger-Westphal nucleus of the midbrain, and run in the oculomotor nerve to the *ciliary ganglion*. This ganglion lies close to the eye on each side, and receives three "roots": the short ciliary root (*radix brevis*) is the autonomic contribution from the oculomotor nerve, the long ciliary root (*radix longa*) contains the sensory fibres from the eye to the ophthalmic branch of the trigeminal nerve, and the third ciliary root (*radix sympathica*) contributes postganglionic fibres from the ipsilateral superior cervical ganglion (Fig. 12.1). Postganglionic fibres in the long and short *ciliary nerves* from the ciliary ganglion innervate the iris and the ciliary muscles of the eye.

The facial nerve (VII) is divided into three main branches: the *palatine, chorda tympani* (internal mandibular) and *hyomandibular* branch. Cranial autonomic preganglionic fibres from the medulla oblongata run in the palatine branch to the *sphenopalatine ganglion*, and in the chorda tympani to the *submandibular ganglion*. Postganglionic fibres from the sphenopalatine ganglion innervate the lachrymal (tear) glands and blood vessels in the nasal mucosa. Postganglionic fibres from the

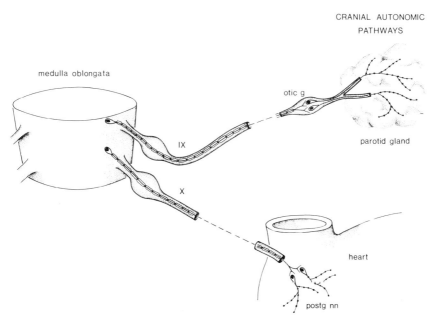

Fig. 2.4. Arrangement of the cranial autonomic pathways. Preganglionic fibres run in the cranial nerves, and the cell bodies of the postganglionic neurons form ganglia near or within the effector organ. Abbreviations as in Fig. 2.2

submandibular ganglion innervate the submandibular and sublingual salivary glands.

Preganglionic fibres that originate in the medulla oblongata also run in the glossopharyngeal (IX) nerve to the *otic ganglion*, which sends off postganglionic fibres to the large parotid salivary glands. It is worth mentioning that the ganglion on the glossopharyngeal nerve near its exit from the cranium, the *petrous ganglion*, is purely sensory in function, and corresponds to the dorsal root ganglia of the spinal nerves. It is thus not involved in the efferent (autonomic) pathways.

The most proximal ganglion in the vagus (X) nerve is likewise only sensory. This ganglion consists of an upper *superior* or *jugular ganglion*, and the adjacent *inferior* or *nodose ganglion*. The autonomic ganglia in the vagal pathways are usually embedded in the wall of the effector organ (Fig. 2.4), and do not form distinct ganglionic swellings on the nerves as do the other cranial autonomic ganglia.

Vagal fibres are distributed to the heart, trachea, lungs, liver and gut as far back as the upper part of the large intestine. Fibres from the superior cervical ganglion may enter the ipsilateral vagus nerve, and there are also connections between the vagi and the coeliac ganglion (Fig. 2.3).

2.1.3 The Mammalian Enteric Nervous System

The mammalian enteric nervous system contains a remarkable number of neurons: between 10 and 100 million neurons in man, which is comparable to the total number of neurons in the entire spinal cord (Furness and Costa 1980). The neurons are organized as two major interconnected plexuses which are present along the whole length of the gut (Fig. 2.5). The outer plexus, the *myenteric plexus* (Auerbach's plexus) is situated between the outer longitudinal and the inner circular smooth muscle layer in the wall of the gut whereas the inner plexus, the *submucous plexus* (Meissner's plexus) lies in the submucosa (Fig. 2.5). A good review of the arrangement of the enteric neurons is given by Gabella (1979).

The enteric nervous system comprises all intrinsic neurons of the gut, i.e., sensory and motor neurons and interneurons (Fig. 2.1). Unlike the cranial and spinal autonomic systems, the majority of the enteric neurons appear to be devoid of an extrinsic input from the central nervous system. The enteric motor neurons (excitatory or inhibitory) may be activated directly by local physical or chemical stimuli, or may be involved in local reflexes initiated by intrinsic sensory neurons (Fig. 2.1, Gershon et al. 1979, Furness and Costa 1980, Gershon 1981).

The enteric nervous system shows several peculiarities that set it apart from the other autonomic divisions. The neurons are uni-, bi- and multipolar, and some of them resemble primary sensory neurons and may be involved in intrinsic reflexes (Schofield 1968, Gershon 1981). Histochemistry (Schofield 1968, Furness and Costa 1971), electron microscopy (Baumgarten et al. 1970, Gabella 1972, Cook and Burnstock 1976a, Furness and Costa 1978, 1980, Wilson et al. 1981b), electrophysiological methods (Furness and Costa 1980, North et al. 1980, Gershon 1981, Wood 1981) and the content of peptides and other possible transmitter substances (Furness and Costa 1980, Larsson 1980, Sundler et al. 1980, Gershon 1981) have been used to describe more than ten different types of enteric neurons.

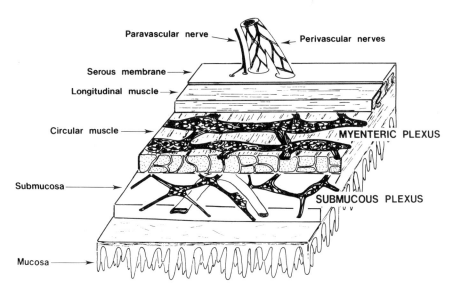

Fig. 2.5. Arrangement of the two major enteric plexuses in the mammalian gut wall. The myenteric plexus (Auerbach's plexus) is located between the circular and the longitudinal smooth muscle layer, while the submucous plexus (Meissner's plexus) is located in the submucosa bordering the circular muscle layer. Extrinsic nerves to the gut run either as distinct nerves along the blood vessels (paravascular nerves), or form a plexus in the wall of blood vessels (perivascular nerves)

The supporting cells of the mammalian enteric plexuses are also peculiar in resembling more the astroglia of the central nervous system than "normal" peripheral supporting Schwann cells. The "enteric glial cells" enclose the enteric ganglia, which are not penetrated by blood capillaries. Since the capillaries bordering the outer surface of the envelope are non-fenestrated, they may be an essential part of a "blood-enteric neuron barrier" comparable to the "blood-brain barrier" of the central nervous system (Gabella 1972, Cook and Burnstock 1976b, Gershon and Bursztajn 1978, Gabella 1981, Gershon 1981, Wilson et al. 1981a).

The extrinsic innervation of the gut is both excitatory and inhibitory (Campbell 1970a, Gershon et al. 1979) and reaches the enteric plexuses of the anterior part of the gut via the vagi and the posterior part of the gut via the pelvic nerves and perivascular plexuses (Figs. 2.3 and 2.5). This means that some of the enteric neurons can also be regarded as postganglionic neurons of the extrinsic autonomic pathways (see Chap. 9).

2.2 Cyclostomes

The autonomic nervous system in lampetroids and, especially, myxinoids is poorly developed, and the knowledge of its functions fragmentary. There are neurons in both cyclostome groups that can be regarded as autonomic, but in many cases it is not possible to distinguish between autonomic and sensory neurons.

2.2.1 The Cyclostome Spinal Autonomic System

Sympathetic chains are absent in both groups of cyclostomes, but there are scattered ganglionic clusters that may represent the autonomic ganglia. The ventral and dorsal roots of the spinal nerves, incompletely united in myxinoids and not at all in lampetroids, continue as separate nerves. The spinal autonomic outflow may take place in both the dorsal and the ventral spinal nerves, but as mentioned above the possible presence of sensory neurons makes the anatomical evaluation difficult (Nicol 1952, Johnels 1956, Romer 1962, Campbell 1970b).

In the hagfish, *Myxine glutinosa*, the ventral spinal nerves give off fine visceral branches, which form the spinal autonomic outflow from the central nervous system. Clusters of ganglion cells connected to the visceral branches are found along the dorsal aorta, and may represent the autonomic ganglia in the spinal pathways of *Myxine* (Marcus 1910, Lindström 1949, Fänge et al. 1963). The visceral branches are probably vasomotor in function, but it is doubtful that there is any innervation of the gut by such fibres (Nicol 1952, Fänge et al. 1963, Campbell 1970b).

In the lamprey, *Lampetra*, visceral branches leave both the dorsal and the ventral spinal nerves and run directly to the kidneys, gonads and blood vessels, and there are also ganglionated branches from the posterior spinal nerves to the hindgut (rectum), cloaca and ureters (Johnels 1956). Ganglionated nerve plexuses are found in the cardinal veins, sinus venosus of the heart, intestine, buccal and branchial mucosa, urogenital organs, blood vessels and associated with chromaffin bodies (Nicol 1952). Some spinal autonomic fibres in lampetroids exhibit catecholamine fluorescence (Chap. 3.2.6.4) and may thus be adrenergic (Leont'eva 1966).

Spinal autonomic fibres also reach a widely distributed subcutaneous nerve plexus in both cyclostome groups. The neurons of this plexus could be involved in autonomic control of colour change (*Petromyzon*), production and release of mucus (myxinoids) and cutaneous blood flow (Bone 1963, Fänge et al. 1963, Campbell 1970b).

2.2.2 The Cyclostome Cranial Autonomic System

In myxinoids, cranial autonomic fibres occur only in the vagus (X), while in lampetroids facial (VII) and glossopharyngeal (IX) autonomic fibres have also been postulated. The eyes of myxinoids are degenerate, and there is no oculomotor nerve or ciliary ganglion. The existence of a ciliary ganglion, and thus autonomic nerve fibres in the oculomotor nerve of lampetroids is also doubtful (Tretjakoff 1927, Nicol 1952).

The facial (VII) nerve in lampetroids mediates a vasomotor influence on the gills. Glossopharyngeal (IX) fibres run to ganglia in the gill region (Lindström 1949).

The left and right vagi of cyclostomes unite to become the *nervus* (or *ramus*) *intestinalis impar*, which forms a plexus near the heart and continues as a single ganglionated strand in the dorsal part of the intestine. In spite of the position of the vagal plexus close to the heart, there are no cardiac vagal fibres in myxinoids, and the physiological role of a vagal innervation of the poorly developed intestinal smooth muscle remains uncertain (Greene 1902, Carlson 1904, Augustinsson et al. 1956, Johnels 1957, Fänge et al. 1963, Campbell 1970b, Holmgren and Fänge 1981). There is, however, a well-developed vagal control of the gall bladder of *Myxine* (Fänge and Johnels 1958).

In lampetroids, there is a vagal innervation of the heart and gut (Carlson 1906, Nicol 1952).

2.2.3 The Cyclostome Enteric Nervous System

A ganglionated nerve plexus is found in the wall of the gut in both cyclostome groups. The exact nature and function of the enteric neurons is poorly understood, but an involvement in the control of peristalsis is possible (see Chap. 9.2) (Brandt 1922, Johnels and Östlund 1958, Milochin 1960, Fänge et al. 1963, Campbell 1970b).

2.3 Elasmobranchs

2.3.1 The Elasmobranch Spinal Autonomic System

The paravertebral ganglia in elasmobranchs are arranged segmentally, except in the most anterior part, and there are one or two ganglia connected to each spinal nerve by white *rami communicantes*. The existence of recurrent grey *rami communicantes* was denied by Young (1933c). The segmentally arranged ganglia are irregularly connected longitudinally and with the contralateral paravertebral ganglia, but there are no distinct sympathetic chains of the type found in the more advanced vertebrate groups.

The most anterior pair of paravertebral ganglia of elasmobranchs are the *axillary bodies* ("Axillar-Herz" of Leydig 1853) which are situated within the posterior cardinal sinuses. These axillary bodies are made up of ganglion cells and also large masses of catecholamine storing chromaffin cells (Fig. 2.28). The axillary body ganglia receive white *rami communicantes* from several of the anterior spinal nerves, and give off the anterior splanchnic nerves (Fig. 2.6). These are composed mainly, if not entirely, of postganglionic fibres. The left anterior splanchnic nerve crosses to join the right, forming a plexus along the coeliac artery to the gut and liver (Stannius 1849, Chevrel 1887, Young 1933c, Nicol 1952). There is often a small cluster of ganglion cells anterior to the axillary bodies (Chevrel 1887, Young 1933c), and there may be a fine nerve of myelinated fibres from the axillary ganglion that joins the vagus (Fig. 2.6). This connection has been interpreted as a vagal contribution to the gastric ganglia (axillary bodies) (Young 1933c, Nicol 1952). An anastomosis between the vagus and the anterior splanchnic nerves also occurs in some elasmobranchs (Young 1980a).

The paravertebral ("sympathetic") ganglia behind the axillary bodies are smaller and lie in the dorsal wall of the cardinal sinuses. The suprarenal chromaffin

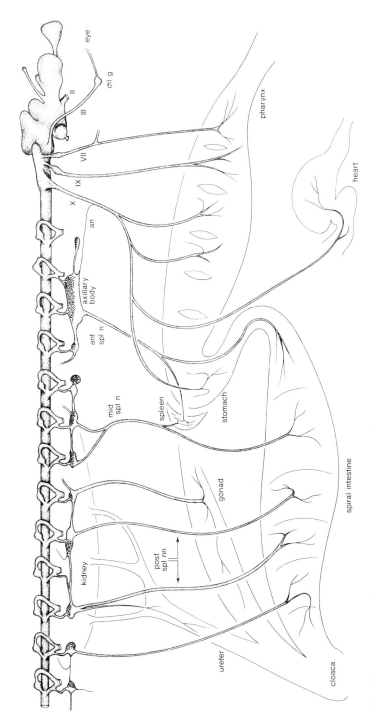

Fig. 2.6. Arrangement of the autonomic nervous system of an elasmobranch. Note chromaffin tissue (*coarse stipple*) associated with paravertebral ganglia. Abbreviations as in Fig. 2.3. (Based mainly on Young 1933c)

tissue is often associated with the ganglia (as in the axillary bodies), but may exist separately (Fig. 2.6). A middle splanchnic nerve arises usually from the middle trunk ganglia on the left side, and innervates the anterior part of the spiral intestine, pancreas and spleen. The more posterior ganglia give off segmental nerves (posterior splanchnic nerves) to the posterior part of the intestine, the urogenital organs and the abdominal blood vessels (Young 1933c). The most posterior spinal autonomic ganglia are located where the dorsal aorta enters the haemal canal (Young 1933c).

The spinal autonomic outflow in elasmobranchs appears to occur chiefly in the ventral roots of the spinal nerves (Young 1933c), but a dorsal outflow to blood vessels is not excluded (Pick 1970). Spinal autonomic fibres do not enter the head, and there are only single observations of fibres from the axillary bodies to the heart (*Mustelus*, Pick 1970).

There are no prevertebral (collateral) ganglia in elasmobranchs.

2.3.2 The Elasmobranch Cranial Autonomic System

Cranial autonomic fibres leave the central nervous system in the oculomotor (III), facial (VII), glossopharyngeal (IX) and vagus (X) nerves in elasmobranchs. A ciliary ganglion is located close to the oculomotor nerve, and postganglionic fibres from this ganglion innervate the dilator muscle of the iris, blood vessels of the eye and the *retractor lentis* muscle. In some species, the ciliary nerves form a plexus together with the ophthalmic (profundus) nerve (Young 1933c).

A cranial autonomic outflow in the hyomandibular branch of the facial (VII) nerve has been concluded by Nicol (1952), although Young (1933c) denied the existence of such an outflow. If they exist, the fibres are probably vasomotor and may also control the pharyngeal smooth muscles. Similarly the post-trematic glossopharyngeal (IX) and vagal (X) fibres innervate the same region, and possibly the gill arches.

There are no salivary or lachrymal glands in fish, but in the absence of a contribution to the head region of fibres from the anterior paravertebral ganglia, the cranial autonomic fibres may play an important role in vasomotor control.

The vagus (X) carries autonomic fibres in its cardiac branches to the heart, where a distinct ganglionic plexus is found in the wall of the sinus venosus. Intestinal vagal branches innervate the anterior part of the gut down to the pylorus and anterior part of the spiral intestine.

2.3.3 The Elasmobranch Enteric Nervous System

Both myenteric and submucous plexuses are present in the stomach and intestine of elasmobranchs. The nature and function of the neurons are not known in any detail, but an involvement in various gastro-intestinal reflexes is likely (Kirtisinghe 1940, Nicol 1952).

2.4 Teleosts

2.4.1 The Teleost Spinal Autonomic System

The sympathetic chain ganglia of teleosts are well developed, and connected longitudinally to form distinct sympathetic chains of the type found in the tetrapods. The sympathetic chains of the teleosts (and some of the ganoids) are unique among the vertebrates in that they continue into the head and carry large ganglia attached to the cranial nerves by grey *rami communicantes* (Fig. 2.7). Both white and grey *rami communicantes* connect the sympathetic chain ganglia of the trunk region with the spinal nerves.

In *Uranoscopus scaber*, the two chains are interconnected along their length (Young 1931 b), but in *Gadus morhua* the connections are restricted to a commissure from the left sympathetic chain to the coeliac ganglion and anterior splanchnic nerve and a fusion between the two chains with a *ganglion impar* at the level of the 20th spinal segment (Nilsson 1976) (Fig. 2.7).

In the head, well-developed sympathetic chain ganglia lie in close contact with the outflow of cranial nerves V + VII, IX and X, and send postganglionic fibres into these cranial nerves. The preganglionic fibres to the cephalic sympathetic chain ganglia leave the spinal cord in the anterior (especially 3rd and 4th) spinal nerves and run forward in the chain into the head. The contribution of spinal autonomic fibres to the vagi (X) is particularly rich, and the branchial, cardiac and intestinal branches of the teleost vagi may therefore be regarded as "vago-sympathetic trunks" of the type found particularly in amphibians and reptiles. In some species, such as *Lampanyctus*, only two pairs of sympathetic chain ganglia are present in the head, and it is not clear to what extent fibres from these enter the cranial nerves (Ray 1950).

In the most anterior end of the sympathetic chains, postganglionic fibres from the sympathetic chain ganglion at the trigeminal (V) and facial (VII) nerve on each side enter the long ciliary nerve and the long ciliary root. In the cod, *Gadus morhua*, some adrenergic ganglion cells are present in the ciliary ganglion, which could mean that the paravertebral sympathetic chains continue anteriorly to include part of the ciliary ganglion in this species (cf. Young 1931 b).

Spinal autonomic contributions to the V + VII cranial nerves are distributed to the skin and somatic structures of the head, and control the dermal melanophores and blood vessels. Fibres from the cephalic sympathetic chain ganglia entering the vagi follow these to the gills, heart, swimbladder, stomach and liver.

There are neither accessory (XI) nor hypoglossal (XII) nerves in teleosts; the first two spinal nerves leave the spinal cord close together. In the cod, fibres from the small elongated sympathetic chain ganglion corresponding to these spinal nerves join the spinal nerves via grey *rami communicantes*, and run to the heart (Stannius 1849, Nilsson 1976, Holmgren 1977).

In teleosts, the large coeliac ganglion on the right side is composed of the fused first 2–3 sympathetic chain ganglia of the trunk region, and is not a prevertebral (collateral) ganglion of the mammalian type. Scattered ganglion cells and small clusters of ganglia are, however, present in the splanchnic nerve in the cod (Abra-

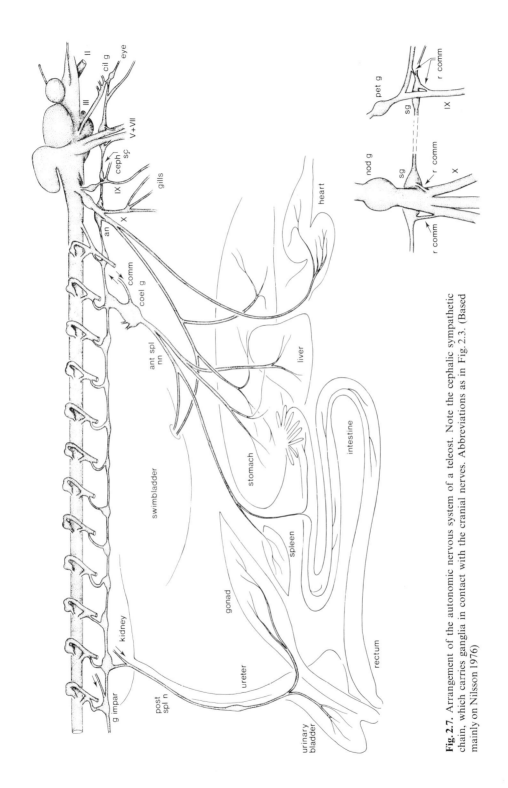

Fig. 2.7. Arrangement of the autonomic nervous system of a teleost. Note the cephalic sympathetic chain, which carries ganglia in contact with the cranial nerves. Abbreviations as in Fig. 2.3. (Based mainly on Nilsson 1976)

21

hamsson 1979a), and the "swimbladder nerve ganglion" (see Chap. 10.1) may be regarded as a prevertebral ganglion in teleosts (Figs. 2.7 and 10.3).

The anterior splanchnic nerves arise from the coeliac ganglion, and run along the coeliac and mesenteric arteries to the viscera. It should be noticed that the splanchnic nerves in teleosts (and elasmobranchs) are thus composed mainly of postganglionic fibres, while the splanchnic nerves of the tetrapods (amphibians, reptiles, birds and mammals) carry preganglionic fibres *to* the prevertebral ganglia (including the coeliac ganglion).

The teleost coeliac ganglion receives a substantial contribution of fibres from the left sympathetic chain via a transverse commisure. In addition to this contribution, the left anterior sympathetic chain ganglia in the cod, especially the small "satellite ganglion" give off unmyelinated fibres which join the intestinal branches of the left vagus (Nilsson 1976). Preganglionic fibres also pass through the "satellite ganglion" to innervate the chromaffin tissue in the wall of the posterior cardinal veins.

In the posterior end of the sympathetic chains, a posterior splanchnic nerve (vesicular nerve) is given off to the urogenital organs and, at least in *Salmo trutta*, the rectum (Young 1931b, Burnstock 1958a, b, 1959, Nilsson 1976). Along this nerve is a ganglion which in the cod contains both adrenergic and non-adrenergic ganglion cells. This ganglion may be homologous with the extensive ganglionated pelvic plexus of the tetrapods (Fig. 2.7).

The arrangement of the sympathetic chains in ganoids is very similar to the arrangement in teleosts, and cephalic sympathetic chain ganglia in contact with at least the vagus and glossopharyngeal have been described in both chondrosteans and holosteans (Allis 1920, Nicol 1952).

2.4.2 The Teleost Cranial Autonomic System

The need for cranial autonomic vasomotor fibres in the head as found in elasmobranchs disappears in teleosts which instead possess a rich supply of spinal autonomic fibres to the cephalic region. The cranial autonomic outflow in teleosts is restricted to the oculomotor (III) and vagus (X) nerves (Fig. 2.7).

The autonomic fibres of the oculomotor nerve enter the ciliary ganglion via the short ciliary root (*radix brevis*), and postganglionic fibres from the ciliary ganglion reach the eyeball in the short ciliary nerve (see also Fig. 12.1).

Autonomic fibres also run in the branchial, cardiac and intestinal branches of the vagus (X). The branchial fibres are vasomotor to the gill vasculature. The cardiac fibres run to the cardiac ganglion at the sino-atrial junction in the heart, from which postganglionic fibres reach the sinus venosus and atrium. The intestinal branch of the vagus innervates the pharynx, oesophagus and stomach down to the pylorus, and the swimbladder. In some stomachless teleosts (e.g., the tench, *Tinca tinca*) the vagus reaches the entire gut.

2.4.3 The Teleost Enteric Nervous System

There are both myenteric and submucous plexuses in the alimentary canal of teleosts, although the submucous plexus contains few or no ganglion cells (Kirtisinghe 1940, Burnstock 1959, Holmgren et al. 1982). The function of the teleost enteric neurons is discussed in Chap. 9.4.

2.5 Dipnoans

Before the work of Giacomini (1906), the existence of sympathetic chains was denied in the lungfish. It is now known that a spinal autonomic system exists in these fish, but it is poorly developed at least in *Protopterus* and *Lepidosiren*. The paravertebral sympathetic chains are extremely thin, and the ganglionic swellings hardly visible. Since the most anterior parts communicate with the vagi, it is possible that spinal autonomic pathways enter the vagi. There are, however, no cephalic sympathetic chains as in teleosts.

The sympathetic chains form loops (annuli) around the intercostal arteries and send off fine fibres to the chromaffin tissue in the wall of the intercostal arteries (Giacomini 1906, Holmes 1950). There is also a fine nerve branch to the left cardinal vein (azygos vein), presumably innervating the chromaffin tissue in the wall of this vein (Fig. 2.27). There are no prevertebral ganglia or splanchnic nerves.

The cranial autonomic outflow is restricted to the vagus (X) in *Protopterus* and *Lepidosiren*. In *Neoceratodus* an oculomotor (III) component may be present in view of the much better-developed eye of this species (Nicol 1952). Autonomic fibres in the vagus reach the gut and lung, with the left half of the lung innervated by the right vagus and vice versa. An interesting feature of the *Protopterus* heart is the storage of large quantities of catecholamines in the atrium, an arrangement resembling that in cyclostomes (Abrahamsson et al. 1979a, Scheuermann 1979). An innervation of these cells, probably from the cardiac branch of the vagus, has been postulated (Scheuermann 1979).

There are no descriptions so far of the enteric nervous system in dipnoans.

2.6 Amphibians

The autonomic nervous systems of anuran amphibians (frogs and toads) have been the subject of numerous and detailed anatomical studies, and have also been used extensively for physiological experiments. Thus both the anatomy and function of the different components in the autonomic nervous system of anuran amphibians are comparatively well understood. The autonomic nervous system of urodeles is less well known (Francis 1934), and that of apodans almost not at all (Nicol 1952).

2.6.1 The Amphibian Spinal Autonomic System

It has been argued that the spinal autonomic outflow in amphibians takes place in both the ventral and dorsal roots of the spinal nerves (see Pick 1970). Since several physiological studies have failed to find evidence for a dorsal root outflow

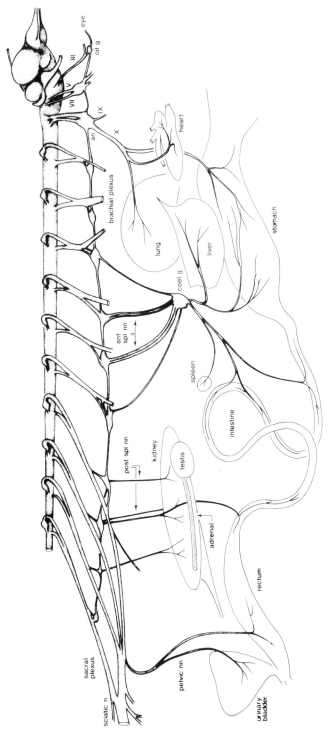

Fig. 2.8. Arrangement of the autonomic nervous system of an anuran amphibian. Abbreviations as in Fig. 2.3

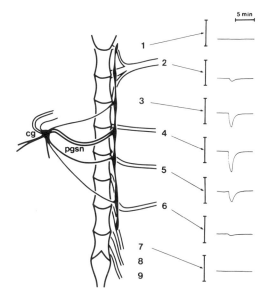

Fig. 2.9. Responses of the perfused toad (*Bufo marinus*) spleen to electrical stimulation of the spinal roots *1* to *9* show that fibres to the coeliac ganglion run in the ventral roots *2* to *6*, but no response to dorsal root stimulation was seen. *Vertical markers* beside the tracings show the perfusion flow range 20–60 drops per min. Abbreviations: *cg*, coeliac ganglion; *pgsn*, preganglionic splanchnic nerves. (Reproduced with permission from Nilsson 1978)

(Langley and Orbeli 1911, Campbell and Duxson 1978, Nilsson 1978) this longstanding controversy is still not resolved.

In anuran amphibians postganglionic fibres from the sympathetic chains extend forward to join the vagus (X) and the glossopharyngeal (IX), and are distributed to the heart, lungs and gut via these nerves. At least in some species, fibres from the sympathetic chain ganglia also enter the facial (VII) and trigeminal (V) nerves. From the ophthalmic branch of the trigeminal nerve, fibres of spinal autonomic origin enter the ciliary ganglion via the long ciliary root (Fig. 2.8).

There are usually 10 pairs of sympathetic chain ganglia in the trunk of the frogs and toads, and the anterior splanchnic nerves leave the left and right sympathetic chain somewhere between the 2nd and 7th pair of ganglia (Fig. 2.9) (Langley and Orbeli 1911, Nicol 1952, Pick 1970). Anterior ganglia give off fine fibres that follow the blood vessels to the head and fore-limbs, and there are also (posterior) splanchnic (renal) nerves to the hindgut and urogenital organs (Fig. 2.8).

The anterior splanchnic nerves extend to the coeliac ganglion (or coeliac plexus), from which postganglionic fibres run to the gut and spleen.

In the urodeles, the abdominal sympathetic chains split to give off ventral trunks, which also carry ganglia.

In the posterior end of the spinal cord, the spinal nerves form a sacral plexus from which pelvic nerves extend to the urinary bladder and rectum. The pelvic nerves have classically been regarded as a "parasympathetic" autonomic component, since early workers noticed a gap in the autonomic outflow at the 8th spinal nerve (Langley and Orbeli 1911) (Fig. 2.8).

The chromaffin tissue is mainly restricted to a ventral band on each kidney in the anurans, but may also be present within the sympathetic chain ganglia. In urodeles, the chromaffin tissue is mainly found within the sympathetic chain ganglia (cf. elasmobranchs) (Nicol 1952).

2.6.2 The Amphibian Cranial Autonomic System

Autonomic fibres are present in the oculomotor nerve (III) from which they enter the ciliary ganglion via the short ciliary root, and postganglionic fibres reach the eye in the short ciliary nerve, as in teleosts and higher vertebrates (Fig. 12.1). The existence of cranial autonomic fibres in the facial (VII) and glossopharyngeal nerves has not been fully established in amphibians, but fibres to the lachrymal and salivary glands are likely to occur in the palatine branch of the facial nerve, and in the anterior branch of the glossopharyngeal. In urodeles, there are ganglia in the palatine branch of the facial nerve, and in the anterior and posterior branches of the glossopharyngeal (Nicol 1952, Pick 1970).

Since the vagi (X) receive large contributions from the anterior sympathetic chain ganglia, they can be regarded as "vago-sympathetic trunks". Vagal branches run to the anterior part of the gut (oesophagus, stomach and intestine), the lungs and the heart.

2.6.3 The Amphibian Enteric Nervous System

There are well-developed myenteric and submucous plexuses in the oesophagus, stomach and intestine of amphibians, but ganglion cells are few or absent in the submucous plexus (Gunn 1951, see also Chap. 9.5).

2.7 Reptiles

The reptilian autonomic nervous system, which is very similar to that of mammals, does not vary fundamentally among the living orders of reptiles. Since the bulk of knowledge of the autonomic nerves is based on investigations on various lizards, this group will be used to represent the general reptilian picture. Detailed descriptions of the reptilian autonomic nervous anatomy in several of the reptilian orders can be found in Hirt (1921), Stiemens (1934), and Adams (1942), summarized in Berger and Burnstock (1979).

2.7.1 The Reptilian Spinal Autonomic System

Preganglionic spinal autonomic fibres leave the spinal cord between the limb plexuses. *Rami communicantes* anterior to the 9th and posterior to the 24th spinal nerve in *Lacerta* are composed solely of non-myelinated, presumably postganglionic, fibres (Hirt 1921, Adams 1942).

From the anterior part of the thoracic sympathetic chains, fibres run forward to the stellate complex at the level of the brachial plexus (Fig. 2.10). From the stellate complex, the main part of each sympathetic chain extends forward as the superficial cervical sympathetic, while a fine branch, the deep cervical sympathetic, runs closer to the vertebral column and sends fibres to the cervical spinal nerves. The superficial cervical trunk sends off a branch to the vagus on the ipsilateral side, close to the vagus trunk ganglion (nodose ganglion), and divides again into a deep

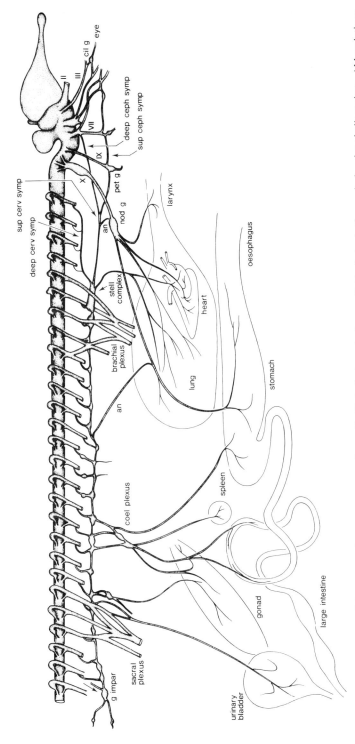

Fig. 2.10. Arrangement of the autonomic nervous system of a reptile. Note the branched sympathetic chains in the cervical and cephalic regions. Abbreviations as in Fig. 2.3. [Based mainly on Hirt (1921) and Adams (1942)]

27

(medial) and a superficial (lateral) cephalic sympathetic. The deep branch crosses the glossopharyngeal (IX) and enters the palatine branch of the facial nerve (VII). The superficial branch enters the glossopharyngeal (IX) near the petrous ganglion, and continues forward to the hyomandibular branch of the facial (VII) and maxillary branch of the trigeminal nerve (V) (Fig. 2.10, Hirt 1921, Stiemens 1934, Adams 1942, Berger and Burnstock 1979). Many of the cephalic fibres from the anterior sympathetic chain ganglia are myelinated (Berger and Burnstock 1979), but may still be postganglionic since it is well known that postganglionic autonomic fibres in birds frequently carry a myelin sheath (Langley 1904).

The prevertebral ganglia are diffuse in reptiles, but postganglionic neurons innervating the gut, spleen and gonads can be found in the coeliac plexus. Cardiac and pulmonary fibres leave the sympathetic chains from the stellate complex, and there are also abdominal connections between the sympathetic chains and the vagi (Fig. 2.10).

In the trunk there are ganglia at each segment throughout the length of the sympathetic chains. In *Lacerta* a *ganglion impar* is formed by the fused 27th pair of chain ganglia, from which fine strands continue into the tail. Pelvic nerves arising in the sacral spinal cord innervate the rectum and urogenital organs (Burnstock and Wood 1967a, Berger and Burnstock 1979).

2.7.2 The Reptilian Cranial Autonomic System

Autonomic fibres leave the central nervous system in the oculomotor (III), facial (VII), glossopharyngeal (IX), and vagus (X) nerves of reptiles. Virtually nothing is known about the function of the facial and glossopharyngeal fibres, but a control of the lacrimal and salivary glands as in mammals seems feasible (Berger and Burnstock 1979).

The arrangement of the oculomotor nerve and the ciliary ganglion is the same as in mammals. Non-myelinated fibres, which may originate from the sympathetic chains, enter the ciliary nerves from the ophthalmic branch of the trigeminus (V), but the role of these fibres in pupillary control is obscure (Iske 1929, Berger and Burnstock 1979, Chap. 12.4).

Fig. 2.11. Ganglion cell in the vagal trunk of the toad, *Bufo marinus*. Note the preganglionic nerve terminal spiralling around the axon-hillock of the ganglion cell (*arrow*). Calibration bar = 50 μm. Silver-staining method

Fig. 2.12. Extra-adrenal chromaffin cells (*arrow*) in the superior cervical ganglion of the chicken, *Gallus domesticus*. Calibration bar = 50 μm. (Courtesy of Dr. Robert Santer)

Fig. 2.13. Extra-adrenal chromaffin cells (SIF cells) (*arrows*) surrounding the axon-hillock region of an adrenergic neuron from a sympathetic chain ganglion of the cod, *Gadus morhua*. Note also the nucleus (*n*) in the adrenergic nerve cell body. Calibration bar = 50 μm. Falck-Hillarp fluorescence histochemistry

Fig. 2.14. Varicose adrenergic nerve terminal from the atrium of a lizard, *Agama caudospinosa*. Note the individual varicosities along the nerve terminal. Calibration bar = 5 μm. Falck-Hillarp fluorescence histochemistry

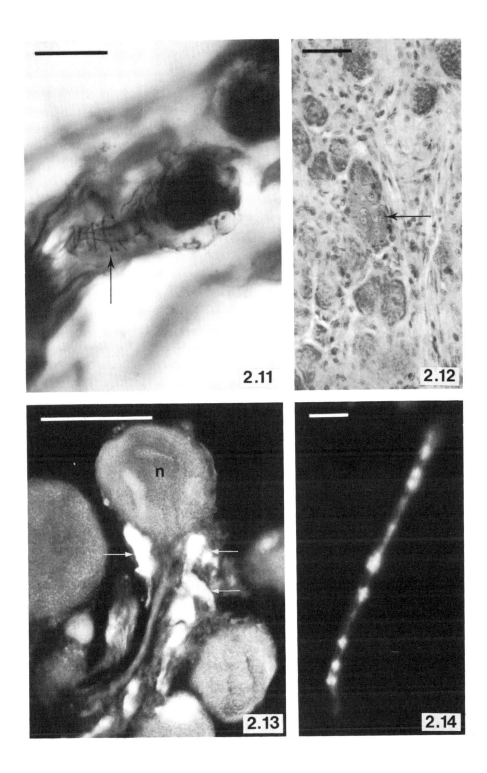

Autonomic fibres run in the palatine and chorda tympani branches of the facial (VII) nerve. Ganglia are present in both branches: the palatine and the (sphen-)ethmoidal ganglia in the palatine branch, and the mandibular ganglion in the chorda tympani branch. Palatine autonomic fibres innervate glands of the mouth and nose, and lachrymal glands. The chorda tympani fibres supply glands of the lower jaw.

Cranial autonomic fibres may also run in the glossopharyngeal to the lingual glands, but there are no otic ganglia in reptiles (Berger and Burnstock 1979).

The anatomy and function of the vagus (X) is well known in reptiles. Vagal fibres innervate the heart, trachea, lung (including the pulmonary vasculature) and gut. An anatomical peculiarity is the vagal trunk ganglion (nodose ganglion) which in reptiles is located in the lower neck region. It is possible that postganglionic cell bodies in the vagal autonomic pathways to the stomach of crocodiles lie along the vagal trunk (Gaskell 1886).

2.7.3 The Reptilian Enteric Nervous System

As in the amphibians, nerve cell bodies are absent from the submucous plexus of the reptilian gut. Nerve cell bodies, including adrenergic ganglion cells, have been described in the myenteric plexus of the large intestine of the lizard (*Trachydosaurus rugosus*) (Read and Burnstock 1968, Chap. 9.6).

2.8 Birds

The avian autonomic nervous system shows many similarities with those of reptiles and mammals. The most striking differences are the long, segmentally ganglionated cervical sympathetic chains, and the partial enclosure of the chains within the cervical and thoracic vertebrae. In addition, the majority of the postganglionic spinal autonomic fibres are myelinated in birds, which means that there are no anatomically distinct grey *rami communicantes*. Finally, the arrangement of Remak's nerve along the gut is uniquely avian (Pick 1970, Bolton 1971a, Bennett 1974).

The arrangement of the autonomic nervous system differs little among the species studied, and the present account is based primarily on descriptions of the domestic fowl (chicken) (*Gallus domesticus*).

2.8.1 The Avian Spinal Autonomic System

In the cervical and thoracic regions of birds, the sympathetic chains are partially enclosed in the vertebrae, with the inter-ganglionic chain segments passing through the transverse foramen of each vertebrae (Pick 1970). As has been observed by

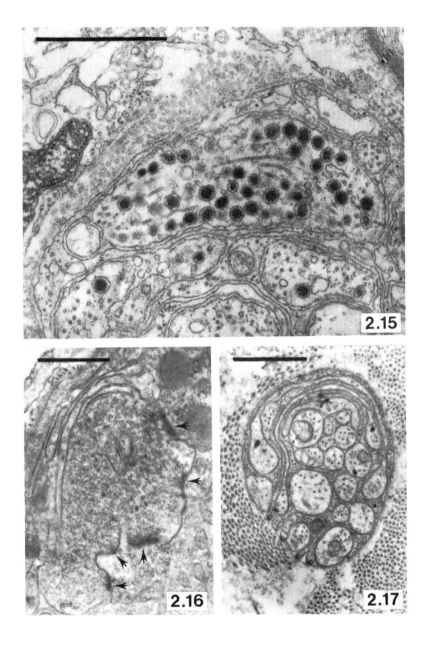

Fig. 2.15. "p-type" nerve profile in a small bundle of axons in the myenteric plexus of the rainbow trout, *Salmo gairdneri*. Calibration bar = 1 µm. (Courtesy of Drs. Robert Santer and Susanne Holmgren)

Fig. 2.16. Synaptic specialization (*arrows*) in a cholinergic axo-somatic junction in the cardiac ganglion of the plaice, *Pleuronectes platessa*. Calibration bar = 1 µm. (Courtesy of Dr. Robert Santer)

Fig. 2.17. Small axon bundle with clearly visible microtubules within the nerve profiles. Trout heart (*Salmo gairdneri*). Calibration bar = 1 µm. (Courtesy of Dr. Robert Santer)

many workers, this arrangement makes dissections of the avian sympathetic chains difficult.

In the chicken, there are usually 37 (35–38) pairs of sympathetic chain ganglia: 14 cervical, 7 thoracic, 13 lumbosacral and 3 coccygeal ganglia (Bolton 1971 a, Bennett 1974). The most anterior outflow of preganglionic fibres takes place in the 14th cervical or 1st thoracic segment (Langley 1904, Yntema and Hammond 1945), from which fibres extend forward in the cervical sympathetic chains to the neck and head. The *rami communicantes* of the 2nd to 12th cervical and the anterior lumbosacral spinal nerves are short or even absent, and the ganglia of the sympathetic chains fuse with the spinal nerves and dorsal root ganglia to a variable degree (Pick 1970, Bolton 1971 a).

A cephalic nerve leaves the superior cervical ganglion and sends off branches to the cranial nerves XII, XI, X, IX, VII and V (and possibly others). Spinal autonomic fibres also follow arteries in the head, and may thus reach all parts of the head (Bennett 1974).

The first thoracic ganglia give off the cardiac and pulmonary nerves, which form several plexuses in the heart and lungs. Vagal fibres also contribute to these plexuses.

The greater splanchnic nerves arise from the 2nd to 5th (or 6th) thoracic ganglia and form, together with vagal fibres, a ganglionated coeliac plexus about the aorta and coeliac artery. The coeliac plexus extends to form the splenic, hepatic, pancreatico-duodenal and gastric plexuses, which send fibres to the spleen, liver and gut. A thoracic oesophageal plexus is formed from parts of the cardiac, pulmonary and coeliac plexuses and vagal fibres, but the role of the spinal autonomic fibres in oesophageal control is obscure (Bolton 1971 a, Bennett 1974).

The lesser splanchnic nerves arise from the 5th thoracic to the 6th lumbosacral sympathetic chain ganglia, and form the aortic plexus about the aorta. This plexus extends to form the richly ganglionated adrenal, anterior mesenteric, genital and renal plexuses (Bolton 1971 a, Bennett 1974).

Fibres from the 6th to 12th lumbo-sacral ganglia form the hypogastric plexus, which extends to the posterior mesenteric, pelvic and cloacal plexuses, and send numerous fibres into Remak's nerve. This heavily ganglionated nerve, which has been compared to the hypogastric and pelvic plexuses in mammals (Bennett 1974), runs from the duodenum along the entire intestine to the caudal end of the rectum. Contributions of fibres from the vagi, coeliac, mesenteric, hypogastric and pelvic plexuses and the sympathetic chains are plentiful along the entire length of Remak's nerve (Bolton 1971 a, Bennett 1974).

Fig. 2.18. Adrenergic type ("a-type") (*lower left*) and cholinergic type ("c-type") (*upper right*) nerve profiles from toad (*Bufo marinus*) atrium. Note the dense chromaffin reaction product in the small granular vesicles of the adrenergic nerve profile, and the presence of a few large granular vesicles in the cholinergic type nerve profile. Calibration bar = 1 µm. (Reproduced with permission from Gibbins 1982)

Fig. 2.19. "p-type" nerve profile from toad (*Bufo marinus*) atrium, with about 50% large granular vesicles. It is possible that this type of nerve profile represents cholinergic neurons, which may have the capacity to release additional transmitters, possibly somatostatin. Calibration bar = 1 µm. (Reproduced with permission from Gibbins 1982)

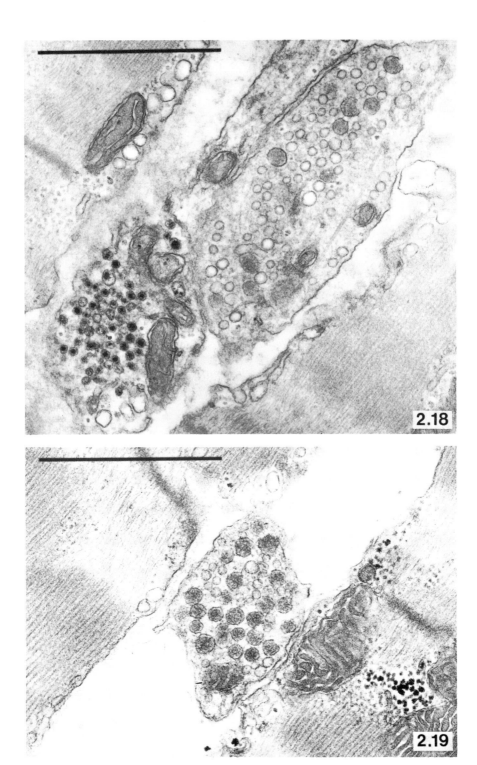

2.8.2 The Avian Cranial Autonomic System

Cranial autonomic fibres occur in the oculomotor (III) and vagus (X) nerves, and most likely also in the facial (VII) and glossopharyngeal nerves (IX). Autonomic fibres from the Edinger-Westphal nucleus of the avian midbrain run in the oculomotor (III) nerve to the ciliary ganglion, from which the long and short ciliary nerves run to the eye. The ciliary ganglion also receives contributions from the ophthalmic branch of the trigeminal (V) nerve, and from the superior branch of Vidian's nerve. Vidian's nerve is composed of branches from the facial (VII) nerve and the cephalic nerve from the superior cervical ganglion (Bolton 1971a, Bennett 1974).

Branches of Vidian's nerve and of the ophthalmic (V) nerve also enter the ethmoidal ganglion which sends fibres to the periorbita and lacrymal, Harderian and nasal glands. There is an anastomosis between the ethmoidal ganglion and the sphenopalatine ganglion, and the latter also receives preganglionic fibres from Vidian's nerve. Postganglionic fibres from the sphenopalatine ganglion innervate the lachrymal and nasal glands, and the palatine and maxillary salivary glands.

The several small submandibular ganglia lie in the chorda tympani branch of the facial (VII) nerve, and send fibres to the anterior submandibular salivary glands. Fibres from the glossopharyngeal nerve (IX) innervate the ventromedial and posterior submandibular, spheno-pterygoid, lingual and crico-arytenoid glands. As in reptiles, there is no description of on otic ganglion in birds (Bolton 1971a, Bennett 1974).

Vagal (X) fibres innervate the heart, trachea, lung and gut from the oesophagus down to the duodenum or small intestine. Vagal fibres mix with spinal autonomic fibres in the cardiac, coeliac and pulmonary plexuses, and enter Remak's nerve. As in the reptiles, the vagal trunk ganglion (nodose ganglion) is located in the lower neck region (Stiemens 1934, Bolton 1971a, Bennett 1974).

2.8.3 The Avian Enteric Nervous System

There are very well-developed ganglionated myenteric and submucous plexuses in the avian gut. A special arrangement, by definition not a part of the enteric system but probably of great importance in the control of the gut, is the ganglionated nerve of Remak. There is a possibility that adrenergic cell bodies occur within the enteric nervous system in birds, something not normally seen in mammals (Bennett 1974).

Fig. 2.20. Stretched preparation of the lizard (*Trachydosaurus rugosus*) lung wall. Note the blood vessel (*bv*) in the *left part* of the picture, which is made visible by its surrounding network of varicose nerve terminals. Varicose fluorescent terminals can also be seen running in the cross-work of the smooth muscle bands in the lung wall. Calibration bar = 100 µm. Falck-Hillarp fluorescence histochemistry. (Reproduced with permission from McLean and Burnstock 1967c)

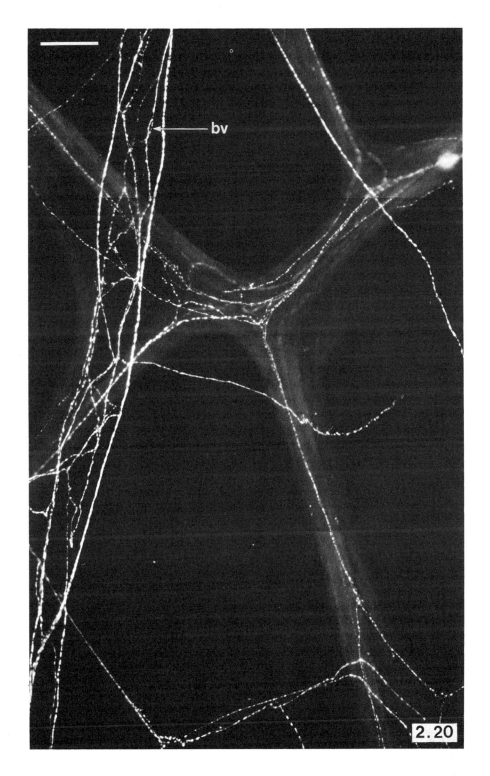

2.9 Conclusions

There is a striking similarity in the general anatomical plan of autonomic nervous systems in all vertebrate groups, with the exception of the cyclostomes. Paravertebral ganglia are closely associated with chromaffin tissue in some of the lower vertebrates (elasmobranchs, urodele amphibians), but later in the vertebrate evolution the two types of tissue become more separated.

All vertebrates, with the possible exception of the cyclostomes and elasmobranchs, have an anatomical arrangement that allows a contribution from the paravertebral ganglia or sympathetic chains to the cranial nerves. The contribution by spinal autonomic fibres is very prominent in the vagi of teleosts and amphibians, but also sufficiently in the other vertebrate groups to justify the designation of "vago-sympathetic trunks".

The spinal autonomic outflow occurs in both the ventral and dorsal spinal nerves of cyclostomes, but the arrangement in other vertebrates in still open to question. Anatomical claims of a dorsal root outflow of autonomic fibres in some amphibians are not supported by functional studies in others. Further experimen-

Fig. 2.21. Stretched preparation of lung wall of the lizard (*Trachydosaurus rugosus*). At the *top* a branching blood vessel, and in the *centre* a bundle of varicose nerve terminals in a smooth muscle septum. Calibration bar = 100 µm. Falck-Hillarp fluorescence histochemistry. (Reproduced with permission from McLean and Burnstock 1967c)

Fig. 2.22. Fluorescent neurons in the myenteric plexus of the stomach from the dogfish, *Squalus acanthias*. Fluorescent fibres form a dense network throughout the myenteric plexus. Calibration bar = 50 µm. Falck-Hillarp fluorescence histochemistry

Fig. 2.23. Fluorescent fibres in the wall of a small artery running in the stomach wall of the rainbow trout (*Salmo gairdneri*). Calibration bar = 50 µm. Falck-Hillarp fluorescence histochemistry. (Courtesy of Dr. Susanne Holmgren)

Fig. 2.24. Immunohistochemical visualization of VIP-immunoreactive nerve fibres in the myenteric plexus of the stomach of the dogfish (*Squalus acanthias*). Calibration bar = 50 µm. (Courtesy of Dr. Susanne Holmgren)

Fig. 2.25. Immunohistochemical visualization of neurotensin-immunoreactive nerve fibres in the myenteric plexus of the intestine of *Lepisosteus platyrhincus*. Whole mount. Calibration bar = 50 µm. (Courtesy of Dr. Susanne Holmgren)

Fig. 2.26. Immunohistochemical visualization of VIP-immunoreactive nerve fibres forming a network in the myenteric plexus of the intestine of the rainbow trout (*Salmo gairdneri*). Whole mount. Calibration bar = 50 µm. (Courtesy of Dr. Susanne Holmgren)

Fig. 2.27. Chromaffin tissue lining the anterior part of the left posterior cardinal vein (azygos vein) of the African lungfish (*Protopterus aethiopicus*). Calibration bar = 200 µm. Falck-Hillarp fluorescence histochemistry. (Reproduced with permission from Abrahamsson et al. 1979a)

Fig. 2.28. Axillary body from the dogfish (*Squalus acanthias*). In the *left part* of the preparation the intensely fluorescent catecholamine-storing chromaffin cells can be seen. The ganglion cells of the postganglionic adrenergic neurons present in the paravertebral ganglionic mass associated with the chromaffin tissue can be seen in the *middle* and *right* part of the preparation. Fluorescent nerve terminals surround some of the ganglion cells (*arrows*). Calibration bar = 200 µm. Falck-Hillarp fluorescence histochemistry. (Reproduced with permission from Nilsson et al. 1975)

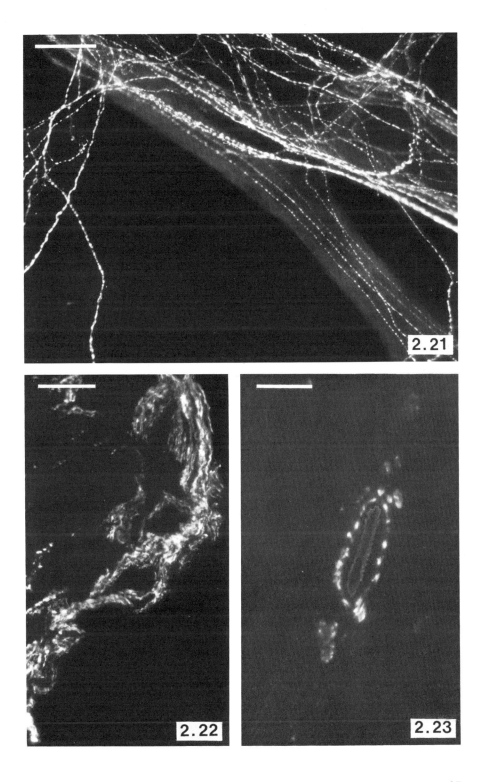

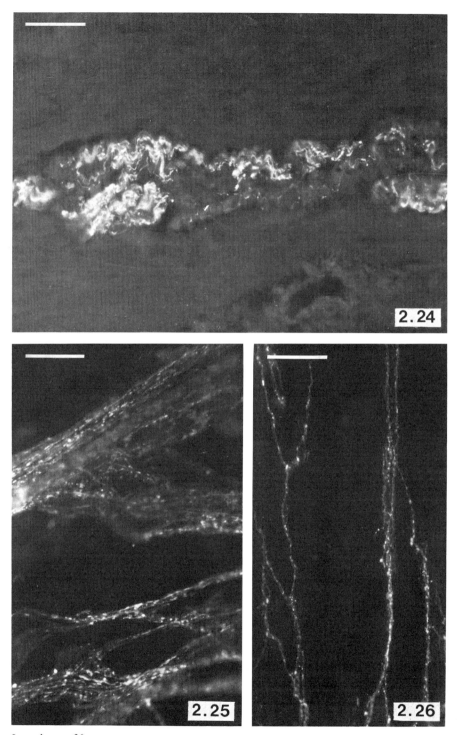

Legends see p. 36

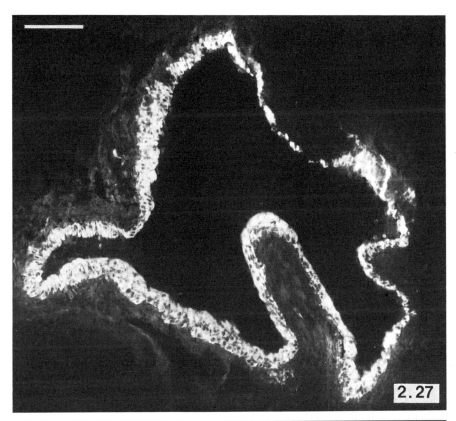

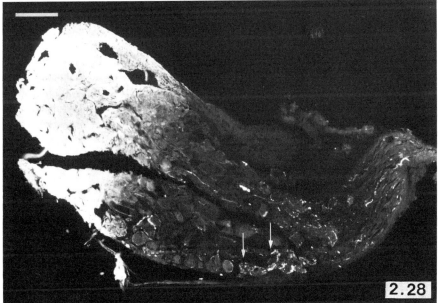

Legends see p. 36

tation, bearing in mind the possibilities of reflexogenic responses, is clearly indicated to settle this question.

The cranial autonomic outflow is very similar throughout the vertebrate series. In vertebrates possessing eyes, oculomotor fibres to the iris and ciliary muscles synapse in a ciliary ganglion of very similar arrangement. Some of the cranial nerve ganglia [e.g., in the facial (VII) and glossopharyngeal (IX) nerves] may be lacking or have other functions in fish than in mammals, since fish lack lachrymal and salivary glands.

The question of a separate "parasympathetic" system in the posterior end of the spinal cord ("sacral region") is avoided by the terminology used in the present descriptions. As a matter of academic interest it could be speculated that the vasicular nerve ganglion along the posterior splanchnic nerve in some teleosts represents the first sign of a "sacral" system, since this ganglion contains many non-adrenergic ganglion cells. It is essential, however, not to confuse functional terms (adrenergic, cholinergic etc) with anatomical terms (spinal autonomic, cranial autonomic etc), and it is hoped that the anatomical terminology will be successively supplemented by a functional description of the autonomic neurons in different parts of the vertebrate body.

Chapter 3

Neurotransmission

The idea of a chemical transmission at the autonomic nerve junctions emerged at the end of the 19th century (DuBois Reymond 1877), and more specifically around the turn of the century from the work of Elliott (1904), who suggested that the "sympathetic" nerve endings might release small amounts of an adrenaline-like substance in close contact with the effector cells. A similar argument was put forward by Dixon (1906, 1907), who first noted the similarities between the effects of vagus stimulation and the alkaloid muscarine. About the same time Langley (1905) advanced the theory of excitatory and inhibitory receptive substances (receptors) on the target cells, and some years later Dale (1914) demonstrated the great similarities between acetylcholine and "parasympathetic" nerve effects.

The first clear evidence in favour of a chemical mediation of the nerve impulses comes from the work of Cannon and Uridil (1921), and the brilliant work of Loewi (1921) on the amphibian heart. In his experiments Loewi used two frog hearts (from *Rana temporaria* or *R. esculenta*) or toad hearts (from *Bufo* spp.), which he suspended with Straub cannulas in such a way that the Ringer's solution filling the first heart could be removed at intervals and transferred to the second heart. Thus the Ringer's solution filling the first heart could be bioassayed for its content of cardio-active substances using the second heart. The first heart was suspended with its autonomic ("vago-sympathetic") nervous supply intact, and Loewi could demonstrate that prolonged electrical stimulation of the nerve produced a release of an inhibitory substance which he called "Vagusstoff". The "Vagusstoff" produced an atropine-sensitive inhibition of the second heart.

Stimulation of the "vago-sympathetic" nerve supply to the toad's heart, arranged in a similar manner, instead produced an increase in the force of beat of the heart. This difference between frogs and toads is due to seasonal variations which at any given time will cause a difference in the relative importance of the excitatory and inhibitory innervation. In the toad system, the release of a substance that enhanced the force of beat of the second heart could be demonstrated by the same bioassay method. This excitatory substance was tentatively labelled "Förderungssubstanz" or "Acceleransstoff". The "Vagusstoff" was later shown to be identical with acetylcholine, and the "Förderungssubstanz" is now known to be adrenaline (which is the dominant catecholamine in the adrenergic neurons of amphibians) (Loewi and Navratil 1926a, b). In mammals, adrenaline dominates over noradrenaline in the adrenal medulla, but in the adrenergic neurons von Euler (1946) was able to provide evidence that noradrenaline is the adrenergic transmitter substance.

The presence and functions of the two types of "classical" transmitters (acetylcholine and adrenaline/noradrenaline) has since been well established. The neurons responsible for the release of these substances have been called *cholinergic* and *ad-*

renergic respectively by Dale (1933), and the postsynaptic cholinoceptive and adrenoceptive sites (Dale 1954) are now called cholinoceptors and adrenoceptors respectively. Although the term *noradrenergic*, rather than adrenergic, would more accurately describe the neurons of the higher vertebrates, the term adrenergic is retained in this text. The reason for this is that in many vertebrate groups mixtures of adrenaline and noradrenaline are released from the adrenergic neurons, with the proportion of the two catecholamines varying among the species and even between different organs within the same species (e.g., teleosts: Abrahamsson and Nilsson 1975, 1976, birds: de Santis et al. 1975, Komori et al. 1979).

It is now quite clear that all responses to autonomic nerve stimulation cannot be explained within the framework of cholinergic and adrenergic neurotransmission (Campbell 1970a, Burnstock 1972, 1981, Campbell and Gibbins 1979), and the nature and properties of the transmitter substances responsible for the non-adrenergic, non-cholinergic (NANC) transmission are now subject to vigorous research activities. Several likely, but as yet not universally accepted candidates for the roles as NANC transmitters have emerged, the most well-supported being ATP (adenosine triphosphate) in the *purinergic* neurons (Burnstock 1972, 1981, Burnstock et al. 1970, 1972), 5-hydroxytryptamine (serotonin) in the *serotonergic* neurons (Gershon 1970, 1981) and various polypeptides in *peptidergic* neurons (Hökfelt et al. 1980a, b, Furness and Costa 1980, Sundler et al. 1980) (see also Chap. 9.1.4).

The available information is compatible with the view of not one but *several* NANC transmitters in the autonomic nervous system, and the number of putative transmitters in the enteric nervous system of mammals is at least a dozen. The current view of the different types of autonomic neurons is considered in more detail in Chaps. 3.2.6–3.2.10. Before doing so, it is practical to outline some general features of the autonomic neurons, and also the criteria used to establish a new substance as a transmitter in the autonomic nervous system.

3.1 Transmitter Substances

It is customary to consider five criteria that must be fulfilled to establish a new substance as a neurotransmitter. The following criteria are often used (Iversen 1979, Burnstock 1981).

1. Synthesis and Storage. The substance must be synthesized and stored within the neuron. A neuron may, however, be able to function well without being able to synthesize the transmitter, provided that the substance is manufactured elsewhere and actively taken up and stored in a releasable form by the neuron. Although there is no known system in which an adrenergic neuron is *entirely* dependent on catecholamines synthesized in and released from the adrenomedullary chromaffin cells, such a system could well work.

2. Release. The substance should be released from the neuron during nerve stimulation in quantities large enough to produce a *postsynaptic* response (response of the effector cell). In all known cases, the neuronal release of transmitter is dependent on Ca^{2+} ions.

3. Mimicry. Exogenous application of the substance should produce the same effect as nerve stimulation on the effector cell (organ).

4. Inactivation. A system for the inactivation of the transmitter should be present. Such systems include neuronal reuptake of the transmitter and degradation by enzymes.

5. Pharmacological Identity. Potentiating or antagonistic effects of drugs should be similar on the nerve response and the response to exogenous substance. The difficulty in fulfilling this criterion for the NANC-transmitter candidates is due to the lack of specific "chemical tools" that can be used to enhance or block the effect of these substances. The problem is further discussed in Chaps. 4 and 5.

If the quantity of a substance released from a nerve terminal is not high enough to affect the postsynaptic effector cell, the substance may still be of importance in the autoregulation of the nerve terminal. Such substances, which are released together with a "proper" transmitter substance, and modulate the release of this main transmitter by a presynaptic feed-back control, are called *neuromodulators* (see Sect. 3.2.4). In some cases the nerve terminal is able to release more than one transmitter substance with true postsynaptic effects. In this case the transmitters are referred to as *co-transmitters* (see Sect. 3.2.5).

3.2 The Autonomic Neuron

The aim of this section is to outline the general arrangement of the autonomic neurons of the five main types described below. It will be immediately obvious that the main part of the knowledge comes from studies of mammalian adrenergic neurons. The adrenergic, cholinergic, purinergic, serotonergic and peptidergic types of neuron are described separately in Sects. 3.2.6–3.2.10.

3.2.1 General Arrangement

Contrary to the arrangement at the neuromuscular junction in skeletal muscle, the autonomic neuron branches into a terminal apparatus of considerable extension, forming an "autonomic ground plexus" in the effector organ (Hillarp 1946). Of the five types of neurons described below, all terminate in a multitude of fine branches, each making "en passant" contacts with several effector cells. The total length of the terminals of a single adrenergic neuron may be several cm, although many are shorter (Olson and Malmfors 1970, Pick 1970, Benett 1972, Burnstock and Costa 1975). The terminals of the preganglionic neurons are generally shorter and make contact at several points on the postganglionic cell body and dendrites. At least in amphibians, preganglionic fibres spiralling around the axon-hillock region of the postganglionic neuron are also frequently seen (Fig. 2.11).

The terminal region of the adrenergic neuron has the appearance of a pearl necklace, with a pearl or *varicosity*, 0.5–2 µm in diameter, every few micrometers. Along the mammalian adrenergic nerve terminals there are a few hundred varicosities per mm of nerve terminal (see also Fig. 2.14) (Burnstock and Costa 1975).

The varicose nerve terminal is not unique to the adrenergic neuron, but is also present in cholinergic (Richardson 1964, Bell 1967, McLean et al. 1967), purinergic (Burnstock 1972, Burnstock et al. 1978 a, b, Campbell and Gibbins 1979), serotonergic (Gershon et al. 1979, Gershon 1981) and peptidergic neurons (Furness and Costa 1979 a, b, Hökfelt et al. 1980 a, b, Schultzberg et al. 1980, Sundler et al. 1980). In all cases studied, the appearance of the autonomic nerve terminals in non-mammalian vertebrates is similar to that in mammals.

The autonomic nerve terminals innervating smooth muscle fall into two main categories regarding the types of contact with the effector cells. Thus the transmitter substance may be released either from *small axon bundles* (Fig. 2.17), running between the smooth muscle cells enclosed in a common Schwann cell sheath, or from *close contact varicosities* in which the naked terminal fibre is indented into the effector cell membrane. The synaptic cleft in the former type of synapse is in the order of 100–200 nm in mammalian gastrointestinal tract, or even more in large blood vessels. The latter type of varicosity is separated from the effector cell membrane by a uniform gap of 20 nm. There are also several examples of intermediate types of contact between nerve fiber and smooth muscle (Bennett 1972, Burnstock 1979). It is of interest to note that some small axon bundles contain preterminal or terminal fibres of more than one type of neuron, e.g., both cholinergic and adrenergic fibres (see Sect. 3.2.1) (Yamauchi 1969, Burnstock and Costa 1975, Burnstock 1980 b, 1981).

The transmitter substance is stored within the varicosities in storage vesicles (or granules) with distinct features in the electron microscope depending on the type of neuron examined. The EM profiles of the "classical" autonomic neurons (cholinergic and adrenergic) show storage vesicles of two different size classes: one larger type with an electron-dense core and a diameter of 60–150 nm, which is usually referred to as large granular vesicles (LGV), and a smaller type, which may or may not show an electron-dense core, with a diameter of 30–60 nm. The small vesicles with electron-dense cores (small granular vesicles; SGV) are generally characteristic of the adrenergic nerve profiles ("a-type" nerve terminals; Fig. 2.18), while cholinergic nerve profiles ("c-type" nerve terminals; Fig. 2.18) typically contain small clear vesicles (SCV). The SGV and SCV are present mainly in the nerve terminals, while LGV are found also in the nerve cell body and along the axon (Geffen and Ostberg 1969, Geffen and Livett 1971, Bennett 1972, Burnstock and Costa 1975, Campbell and Gibbins 1979).

The electron microscope nerve profiles of the NANC neurons contain storage vesicles similar to the LGV of the cholinergic and adrenergic terminals. Their similarity to the peptide hormone-storing granules has led to the term "p-type" vesicles (Baumgarten et al. 1970). The diameters of the "p-type" vesicles range from 100–120 nm (small "p-type" vesicles) or 160–200 nm (large "p-type" vesicles) (Campbell and Gibbins 1979; Gibbins 1982). ATP (Burnstock 1972), 5-hydroxytryptamine (Gershon 1981) and peptides (Larsson 1977, Johansson and Lundberg 1981, Probert et al. 1981) appear to be stored mainly in "p-type" vesicles, and it is therefore practical to refer to the LGV-containing nerve profiles of the NANC neurons as "p-type" nerve terminals (Figs. 2.15 and 2.19). It should be noted, however, that storage of NANC transmitters in SCV is also possible (Gibbins 1982), and that the distinction of "a-type", "c-type" and "p-type" nerve terminals is a gross oversimplification. Some authors distinguish between several types of nerve profiles in the enteric nervous system of mammals (guinea-pig, 10 types: Cook and Burnstock 1976a; rabbit colon, 6 types: Komuro et al. 1982), while others find evidence for very few separate types (guinea-pig ileum, 2 main types: Wilson et al. 1981b). In a study of several organs from mammals and the toad (*Bufo marinus*) Gibbins (1982) concluded that the "c-type" and the "p-type" nerve terminals may represent extremes of the same class of fibres.

A possibility to relate the electron microscope nerve profile with the transmitter substance(s) stored within that particular nerve terminal is offered by ultrastructural immunohistochemical techniques (Larsson 1977). In these techniques, the antibodies for a certain substance (amine, peptide, vesicle-bound enzyme such as dopamine-β-hydroxylase) are "labelled" with electron-dense material (e.g., colloidal gold particles) and brought in contact with the tissue sections. The method has been used to distinguish two sub-populations of "p-type" neurons in the guinea-pig colon, one storing vasoactive intestinal polypeptide (VIP) and the other storing substance P (Probert et al. 1981) (see Sect. 3.2.10).

3.2.2 Release of Transmitter

In the mammalian adrenergic neurons and in the chromaffin cells of the adrenal medulla, catecholamines are stored in the storage vesicles together with ATP, dopamine-β-hydroxylase and chromogranin A (a protein). During nerve activity or secretion of catecholamines these substances are all released from the adrenergic cell together with the catecholamine(s). The failure of cytoplasmic enzymes, such as lactate dehydrogenase, to be released has, together with these observations, led to a theory of exocytosis as the mechanism of catecholamine release. This means that the storage vesicle membrane fuses with the neuronal membrane or the membrane of the chromaffin cell and empties part of its content into the extracellular space (Blaschko et al 1967, Geffen and Livett 1971, Smith and Winkler 1972, De Potter 1973, Burnstock and Costa 1975, Trifaró and Cubeddu 1979). The strength of the binding of catecholamines and the proteins to the matrix of the storage vesicle limits the release at each exocytotic fusion. In mammalian adrenergic neurons only 2%–3% of the total noradrenaline store, a fraction that consists mainly of newly synthesized noradrenaline, is available for release. It is possible that noradrenaline is stored within the nerve terminal in several compartments or pools, only some of which are available for release at any given time (Iversen 1967, Folkow et al. 1968, Kopin et al. 1968).

The exocytotic release of catecholamines from the adrenergic cells is dependent on the presence of Ca^{2+} in the external medium, and it is believed that Ca^{2+} entering during the action potential is responsible for the initiation of exocytosis (Douglas 1968, Bennett 1972, Axelrod 1972). In this context it is of interest to note that tyramine, which acts indirectly by releasing catecholamines from the nerve terminals, does so by a Ca^{2+} independent process that is different from the exocytotic release during nerve stimulation (see Chap. 5.2.4) (Burnstock and Costa 1975).

Little is known about the mechanism of transmitter release by autonomic nerve terminals other than the adrenergic type. By analogy with the knowledge of the mechanisms of acetylcholine release at the neuro-muscular junction in skeletal muscle, and some observations of autonomic cholinergic fibres, it seems likely that exocytosis is also the mechanism involved in acetylcholine release from the autonomic cholinergic neurons (Bennett 1972). In the serotonergic neurons, 5-hydroxytryptamine is released together with a specific protein (serotonin binding protein, SBP), which again suggests an exocytotic release (Jonakait et al. 1979, Gershon 1981).

3.2.3 Axonal Transport

In adrenergic, cholinergic, purinergic and serotonergic neurons, the transmitter substance can be synthesized in the nerve terminals, but enzymes, storage vesicles and other organelles are all produced in the nerve cell body and transported to the nerve terminals by a proximodistal (orthograde) axonal transport mechanism involving the axonal microtubular system (Fig. 2.17).

The first demonstration of an orthograde axonal transport of catecholamine-storing vesicles was made by Dahlström and co-workers using mammalian systems (Dahlström and Fuxe 1964, Dahlström 1965, 1971, Dahlström and Häggendal 1966, 1967). Axonal transport of storage vesicles, including the catecholamines and dopamine-β-hydoxylase and other enzymes involved in catecholamine synthesis have since been demonstrated in both mammalian and other vertebrate systems (Wooten and Coyle 1973) (see Sect. 3.2.6.1).

There are some indications of axonal transport of cholinergic storage vesicles (Dahlström et al. 1981) and the enzymes responsible for synthesis and degradation of acetylcholine (Häggendal et al. 1971, Lubinska and Niemierko 1971, Dahlström et al. 1974) (see Sect. 3.2.7).

A specific problem occurs in the peptidergic neurons, where the transmitter itself is synthesized in the cell body only and must be transported to the nerve terminals at a rate high enough to replenish the transmitter stores (Hökfelt et al. 1980 a, b). The rate of orthograde axonal transport of VIP was estimated to 9 mm/h in the sciatic nerve of the cat, which corresponds to a turn-over time of some 5 days (Lundberg 1981, Lundberg et al. 1981 a). For the neurons innervating the cat submandibular gland it has been calculated that the release of VIP at stimulation frequencies below 6 Hz can be compensated for by axonal transport of the peptide alone (Lundberg et al. 1981 b).

The rates of transport vary considerably between the different components. It has been suggested that the main cause of the differences in transport rates is the size of the transported material, where large particles and substances associated with these (e.g., storage vesicles with dopamine-β-hydroxylase and catecholamines) are transported faster than the soluble proteins (e.g., cytoplasmic enzymes such as phenylethanolamine-N-methyl transferase in the adrenergic neurons of the lower vertebrates) (Oesch et al. 1973). There is also a marked temperature dependence of the transport rates (Edström and Hanson 1973).

3.2.4 Presynaptic Receptor Systems (Auto-Receptors)

In addition to the more obvious receptors of the neuron which are directly involved in the transfer of nerve impulses from the preganglionic to the postganglionic neuron (e.g., nicotinic cholinoceptors, see Chap. 4.2.2), there are presynaptic receptors which are involved in modulation of the amount of transmitter released per impulse. The presynaptic modulation can be of four basic types:

1. Automodulation. This means that the transmitter substance itself combines with presynaptic receptors in the nerve terminal and thereby modulates its own release. A special case of automodulation is seen when a neuromodulator, which in itself has no postsynaptic effect, is co-released with the transmitter.

2. Heteromodulation. This means that the release of a transmitter substance is affected by presynaptic modulation by other substances released from surrounding neurons, e.g., interaction of cholinergic and adrenergic neurons within the same Schwann cell sheath.

3. Transsynaptic Modulation. The release of transmitter is affected by substances released into the synaptic cleft from the effector cell, e.g., prostaglandins.

4. Humoral Modulation. Modulation of transmitter release can be effected through the action of circulating substances, e.g., catecholamines, angiotensin, opiate peptides etc.

It was recognized early on that the overflow of noradrenaline from adrenergic neurons in the cat spleen during nerve stimulation could be enhanced by addition of α-adrenoceptor antagonists (Brown and Gillespie 1957). There is now excellent evidence that this effect is due to the presence of presynaptic α-adrenoceptors (Westfall 1977, Langer 1979). These adrenoceptors inhibit further release of the catecholamine from the nerve terminal (automodulation), thus forming a negative feed-back "loop". With the introduction of a subclassification of the adrenoceptors, the presynaptic α-adrenoceptors have been placed in the $α_2$-category (Berthelsen and Pettinger 1977, see also Chap. 4.2.1.1). Interestingly enough, there is now also evidence for the presence of presynaptic $β$-adrenoceptors, which instead mediate facilitation of the release and thus provide a positive feed-back "loop" (Adler-Graschinsky and Langer 1975, Dahlöf et al. 1975, 1978 a, b, Stjärne and Brundin 1976, Dahlöf 1981).

Presynaptic α-adrenoceptors that inhibit neurotransmitter release are not only present in the adrenergic nerve terminals, but are also involved in a heteromodulation of acetylcholine release (Kilbinger and Wessler 1979, Gustafsson 1980), and may indeed be a feature of all autonomic neurons.

In addition to the presynaptic adrenoceptors, there is now very good evidence for presynaptic muscarinic cholinoceptors, which inhibit the release of transmitter both in cholinergic (automodulation) and adrenergic (heteromodulation) neurons (Starke 1977, 1979, Westfall 1977, Muscholl 1979, Stjärne 1979, Kilbinger and Wessler 1980). Furthermore, evidence is rapidly accumulating for an ever-increasing number of other receptor systems, including such for dopamine, histamine, angiotensin, 5-hydroxytryptamine, prostaglandins, adenosine and peptides (Starke 1977, 1979, Westfall 1977, Stjärne 1979, Gustafsson 1980, Hedqvist et al. 1980).

The modification of the transmitter release via the presynaptic receptor systems probably involves interference with some step in the Ca^{2+}-dependent release mechanism. With the recognition of the co-existence of several transmitters and/or neuromodulators within the same neuron (Sect. 3.2.5), a potentially complex picture of synaptic regulation emerges. The presence of different types of nerve fibres within the same Schwann cell sheath further increases the possibilities of a varied peripheral heteromodulation of transmitter release (Richardson 1964, Yamauchi 1969, Westfall 1977, Burnstock 1980b, 1981, Fredholm and Hedqvist 1980). An important case of presynaptic heteromodulation is found in the enteric nervous system of mammals, where adrenergic nerve terminals at the synapses of cholinergic preganglionic neurons inhibit the release of acetylcholine in the synapse (Gershon et al. 1979).

An attempt to summarize some of the presynaptic receptor types known in the different types of autonomic nerve terminals has been made in Figs. 3.1, 3.4, 3.7, 3.9 and 3.10.

Very little is known about the possible functions or even the existence of presynaptic receptor systems in the lower vertebrates. A single observation on the effect of the α-adrenoceptor antagonist phentolamine on ^{3}H-noradrenaline overflow from a perfused teleost spleen indicates that presynaptic α-adrenoceptors play a modulating role similar to that seen in mammals (Nilsson and Holmgren 1976). It may therefore be speculated that the presynaptic receptors at the nerve terminals are a feature that is not unique to mammals, but rather a mechanism for the modulation of transmitter release generally present in vertebrate autonomic nerve terminals.

3.2.5 Neurons with More Than One Transmitter

The principle that each neuron is capable of synthesizing and releasing only one transmitter substance (somewhat improperly known as Dale's principle) was first placed in doubt by the "cholinergic link hypothesis" (see below), and has been challenged repeatedly in recent years (Burnstock 1976b, 1978b, Osborne 1979). The presence of acetylcholine and noradrenaline within the same neuron in tissue cultures of neurons from sympathetic chain ganglia of the rat has been elegantly demonstrated (Furshpan et al. 1976) and the requirements for the differentiation into an adrenergic or a cholinergic neuron are becoming clear (Hill and Hendry 1977, Patterson et al. 1978, Hill et al. 1980).

The "cholinergic link hypothesis" was advanced by Burn and Rand (1960) to explain the effects of certain drugs on the release of noradrenaline from adrenergic neurons. The hypothesis suggests an intraneuronal release of acetylcholine at the arrival of the action potential, and that this acetylcholine, in turn, releases Ca^{2+} from intracellular stores which causes the final release of noradrenaline from the nerve terminal (Burn and Rand 1960, 1965, Burn and Froede 1963, Ferry 1966, Burn 1977a, b). The "cholinergic link hypothesis" has had a great stimulating effect on the research concerning the transmitter functions in autonomic nerve terminals, but the ideas as such are not supported by more recent research (Fozard 1979).

The presence of both acetylcholine and catecholamines within the same nerve terminals in vivo has been suggested by Holmgren and Nilsson (1976), working with the splanchnic innervation of the cod (*Gadus morhua*) spleen. The results suggest that both transmitters are released to act as independent (in the sense that no "link" is assumed) transmitters on the splenic smooth muscle (Nilsson and Grove 1974, Holmgren and Nilsson 1976, Winberg et al. 1981). There is a possibility that the cod possesses a type of primitive, undifferentiated spinal autonomic neuron.

There are now several reports that show the presence of peptide transmitter candidates within the "classical" adrenergic and cholinergic neurons, and within other peptidergic neurons (Lundberg et al. 1979, Hökfelt et al. 1980 a, b, Schultzberg et al. 1980, Uddman et al. 1980, Campbell et al. 1982). A very appealing model of co-transmission is that proposed by Lundberg (1981) for the "classically" cholinergic neurons to the salivary glands of the cat. In this model, both acetylcholine and VIP, are stored in and released from the same neurons, and act synergistically to increase the salivary secretion. The main effect of acetylcholine is at the gland cells, and the role of VIP is chiefly to dilate the vasculature of the gland (Fig. 3.11, Lundberg et al. 1979, 1981 a, b, c, Lundberg 1981).

A second elegant demonstration of the co-existence of acetylcholine and a peptide, in this case somatostatin, in the postganglionic vagal fibres in the heart of the toad (*Bufo marinus*) has been made by Campbell et al. (1982) (see Chap. 7.6.1).

3.2.6 Adrenergic Neurons

The cell bodies of the adrenergic neurons are located in the sympathetic chain ganglia or in the prevertebral (collateral) ganglia such as the coeliac and mesenteric ganglia. The axons of the postganglionic adrenergic neurons leave the ganglion either as a direct nerve to the effector organ (e.g., cardiac and pulmonary nerves) or, in the case of the sympathetic chain ganglia, may rejoin the spinal nerves via the grey *rami communicantes* to be distributed to the skin and somatic structures (blood vessels, glands, hair or feather muscles, melanophores, photophores etc).

Most adrenergic fibres are non-myelinated, but there are exceptions to this rule, e.g., the adrenergic fibres to the nictitating membrane of the cat (Thompson 1961, Kosterlitz et al. 1964) and most postganglionic adrenergic fibres in birds (Langley 1904). The adrenergic fibres leaving the ganglia are often enclosed in small bundles within the same Schwann cell sheath (see Sect. 3.2.1).

The adrenergic neurons in the pelvic ganglia, which innervate the urogenital organs (e.g., the vas deferens) have comparatively short processes and are consequently called "short adrenergic neurons", as opposed to the "long adrenergic neurons" of the prevertebral and sympathetic chain ganglia (Sjöstrand 1965). An attempt to summarize the features of the adrenergic neuron is shown in Fig. 3.1.

3.2.6.1 Synthesis of Catecholamines

Adrenaline and noradrenaline are synthesized in the chromaffin cells and in the adrenergic neurons by an identical process. The precursor of the catecholamines is L-tyrosine, which is taken up by the neuron from the extracellular fluid. Adrenaline

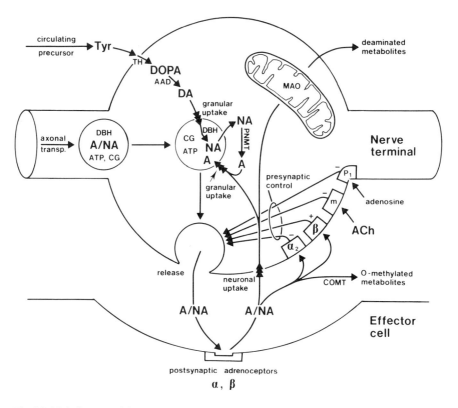

Fig. 3.1. Main features of the adrenergic nerve terminal. Abbreviations: *A*, adrenaline; *AAD*, aromatic-L-aminoacid decarboxylase; *ACh*, acetylcholine; *ATP*, adenosine triphosphate; *CG*, chromogranins; *COMT*, catechol-O-methyl transferase; *DA*, dopamine; *DBH*, dopamin-β-hydroxylase; *DOPA*, 3,4-dihydroxyphenylalanine; *MAO*, monoaminoxidase; *NA*, noradrenaline; *TH*, tyrosine hydroxylase; *Tyr*, tyrosine; α,β, adrenoceptors; *m*, cholinoceptor; P_1, purinoceptor

is the dominant catecholamine in some adrenergic neurons in holosteans, teleosts, amphibians and birds, while noradrenaline is dominant in other vertebrate groups and also in some neurons in teleosts and birds. However, both are present in the chromaffin tissue of all vertebrate groups (von Euler and Fänge 1961, Falck et al. 1963, Angelakos et al. 1965, Burnstock 1969, Holtzbauer and Sharman 1972, de Santis et al. 1975, Abrahamsson and Nilsson 1976, Saurbier 1977, Komori et al. 1979).

The pathway for the synthesis of the catecholamines (the so-called "Blaschko pathway") was originally proposed for adrenaline in the mammalian adrenal medulla, but has since been shown to be identical for the neuronal synthesis of catecholamines in the adrenergic neurons (Holtz 1939, Blaschko 1939, von Euler 1972).

The rate of catecholamine synthesis is determined by the activity of the enzyme responsible for the first step in the "Blaschko pathway": tyrosine hydroxylase. The various steps in the synthesis of adrenaline are summarized in Fig. 3.2.

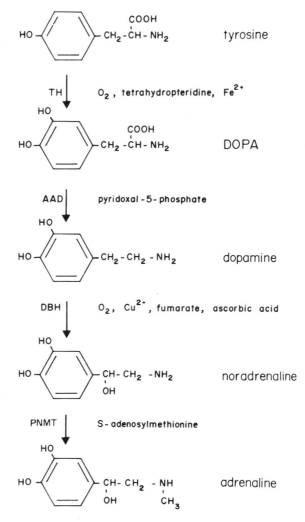

Fig. 3.2. Catecholamine biosynthesis along the "Blaschko pathway". The co-factors required for enzymatic activity are shown. Enzymes: *TH*, tyrosine hydroxylase; *AAD*, aromatic L-amino acid decarboxylase (DOPA decarboxylase); *DBH*, dopamine-β-hydroxylase; *PNMT*, phenylethanolamine-N-methyl transferase

Tyrosine hydroxylase (TH; E.C. 1.14.16.2; tyrosine 3-monooxygenase). This enzyme catalyses the hydroxylation of L-tyrosine to L-DOPA (3,4-dihydroxyphenylalanine). The reaction requires oxygen and a tetrahydropteridine co-factor, and is stimulated by Fe^{2+} (Nagatsu et al. 1964, Udenfriend 1966). Measurements of axonal transport rates for mammalian TH shows stationary, slowly transported (soluble) and rapidly transported (organelle-associated) fractions (Brimijoin 1975, Brimijoin and Wiermaa 1977).

The hydroxylation of tyrosine is the rate-limiting step in catecholamine biosynthesis, and the enzymatic activity of TH is rapidly regulated by a negative feedback from synthesized noradrenaline (Nagatsu et al. 1964, Spector et al. 1967, Weiner 1970, Cotten 1972). Thus an increased catecholamine release (reduction in the intraneuronal catecholamine levels) will rapidly lead to an increased rate of formation of DOPA. There is also a more slowly acting regulation via induction of the enzyme during increased nerve activity (Thoenen et al. 1969 a, b, Thoenen 1970, Burnstock and Costa 1975).

Tyrosine hydroxylase has not been studied extensively in non-mammalian vertebrates, but activity of the enzyme in vitro has been described in homogenates from the chromaffin tissue of the cod, *Gadus morhua* (Jönsson and Nilsson 1982).

Aromatic-L-amino-acid decarboxylase (AAD; E.C. 4.1.1.28). The second step in catecholamine synthesis is the formation of dopamine by decarboxylation of DOPA. Since DOPA decarboxylase, originally described by Holtz (Holtz et al. 1938, Holtz 1939) and Blaschko (1939), decarboxylates a wide variety of aromatic L-amino acids, the name aromatic L-amino acid decarboxylase is more appropriate (Lovenberg et al. 1962). The enzyme decarboxylates DOPA, 5-hydroxytryptophan, histidine, tryptophan and tyrosine (o-, m-, and possibly p-) (Hagen 1962, Lovenberg et al. 1962). AAD requires pyridoxal phosphate as a co-factor. Axonal transport rates for the enzyme are lower than for the fast fraction of tyrosine hydroxylase or dopamine-β-hydroxylase, suggesting a mainly cytoplasmic localization of AAD (Brimijoin and Wiermaa 1977, Starkey and Brimijoin 1979).

Dopamine-β-hydroxylase (DBH; E.C. 1.14.17.1; dopamine β-monooxygenase). The conversion of dopamine to noradrenaline is catalyzed by dopamine-β-hydroxylase which is present within the storage vesicles, and thus transported rapidly in the adrenergic axons (400 mm/d; 37 °C, mammals) (Brimijoin and Wiermaa 1977). Dopamine is taken up by the transport system of the storage vesicles (granular uptake mechanism) in which formation of noradrenaline occurs. Part of the DBH is firmly bound to the vesicular matrix, while some is released together with the catecholamine(s), ATP and chromogranin A during exocytosis and can be demonstrated in the blood.

DBH is a copper-containing molecule that requires ascorbic acid, Cu^{2+}, oxygen and a dicarboxylic acid, such as fumarate, for activity. Apart from dopamine, a number of other amines, including tyramine, are hydroxylated by DBH (Levin et al. 1960, Pisano et al. 1960, Kaufman and Friedman 1965, Goldstein 1966).

In comparison with mammalian DBH, the properties of this enzyme from elasmobranch (*Squalus acanthias*), teleost (*Gadus morhua*) and amphibian sources (*Bufo marinus*) are very similar, except that the temperature optimum for the fish enzymes is lower (Wooten and Saavedra 1974, Jönsson and Nilsson 1978, Jönsson 1982). Immunological cross-reactions of bullfrog (*Rana catesbeiana*) DBH with antiserum against bovine DBH has been demonstrated (Nagatsu et al. 1979).

DBH has been demonstrated also in chromaffin tissue from a cyclostome (*Myxine glutinosa*), a holocephalan elasmobranch (*Chimaera monstrosa*), a holostean ganoid (*Lepisosteus platyrhincus*) and a lungfish (*Protopterus aethiopicus*) (Abrahamsson et al. 1979a, 1981, Jönsson, unpublished).

The rate of axonal transport of DBH in adrenergic neurons of the cod (*Gadus morhua*) is very similar to the transport rate for the catecholamines in the same system (Abrahamsson 1979a, Jönsson and Nilsson 1981), suggesting a vesicular localization of the enzyme. The enzyme has been demonstrated in low quantities in blood plasma from an elasmobranch (*Squalus:* Jönsson 1982), which indicates an exocytotic release. In the cod, *Gadus morhua*, there are no measurable amounts of

DBH in the plasma, suggesting that the enzyme is very firmly bound to the storage vesicles (Jönsson and Nilsson 1978).

Phenylethanolamine-N-methyl transferase (PNMT; E.C. 2.1.1.28). The final step in the biosynthesis of adrenaline is the N-methylation of noradrenaline catalyzed by phenylethanolamine-N-methyl transferase (PNMT). The methylation of noradrenaline by adrenal homogenates was first described by Bülbring (1949), and it was later shown that S-adenosylmethionine serves as the methyl donor for the reaction (Kirshner and Goodall 1957). The enzyme is specific for phenylethanolamines, and is thus different from the non-specific N-methyl transferase of some tissues (Axelrod 1962, 1966, Märki et al. 1962, Saavedra et al. 1973).

PNMT is a cytoplasmic enzyme, although some activity remains in the particulate fraction of a homogenate even after thorough washing (Joh and Goldstein 1973). In the adrenal medulla, PNMT is activated by two systems: one neuronal and one hormonal (adrenocortical steroids). The latter system, which acts by preventing degradation of the enzyme, is absent in amphibians and fish, and it appears that the PNMT in these groups is independent of cortical steroids (Wurtman and Axelrod 1965, Wurtman 1966, Wurtman et al. 1968, Mazeaud 1972, Ciaranello 1978).

PNMT has been demonstrated in chromaffin tissue from representatives of all the major vertebrate groups (Märki et al. 1962, Wurtman et al. 1967, 1968, Peyrin et al. 1969, 1971, 1972, Mazeaud 1971, 1972, Wooten and Saavedra 1974, Zachariasen and Newcomer 1974, Lindmar and Wolf 1975, Abrahamsson and Nilsson 1976, Abrahamsson 1979b, Abrahamsson et al. 1979c, 1981, Jönsson, unpublished).

In some vertebrates, especially holosteans, teleosts and amphibians, adrenaline dominates over noradrenaline in the adrenergic neurons (Burnstock 1969, Holtzbauer and Sharman 1972). In these groups, PNMT is also present in the adrenergic neurons, which means that adrenaline can be synthesized intraneuronally and so act as an adrenergic neurotransmitter together with noradrenaline (Wurtman et al. 1968, Wooten and Saavedra 1974, Abrahamsson and Nilsson 1976, Abrahamsson et al. 1981). The adrenaline is synthesized in the cytoplasm from noradrenaline leaking from the storage vesicles, and then taken up again and stored in vesicles before release. This is the same order of events as described for the mammalian adrenomedullary chromaffin cells (Douglas 1975, Viveros 1975, Abrahamsson 1979a).

Axonal transport of PNMT in amphibian and teleost adrenergic neurons is much slower than the transport of DBH and catecholamines in the same systems, which is compatible with the view of a cytoplasmic localization of PNMT also in the adrenergic neurons (Wooten and Saavedra 1974, Abrahamsson 1979a).

3.2.6.2 Neuronal Uptake of Catecholamines

Catecholamines released from the adrenergic nerve terminal, or reaching the neuron from the circulation, are rapidly and specifically taken up into the terminal by a neuronal uptake mechanism. One estimation for the nictitating membrane of the cat shows that as much as 70% of the noradrenaline released from the adrenergic nerve terminals is inactivated by re-uptake into the nerve terminal, while 23% is taken up by the muscle cells (extra-neuronal uptake) (Langer 1970, Bennett 1972). In the terminology of Iversen (1967, 1974) the neuronal uptake has become known as "uptake$_1$," while the second uptake process, "uptake$_2$," originally described by Iversen (1967) is now known to be identical with the extraneuronal uptake (Iversen 1974, Westfall 1977).

Degeneration of the nerve terminals induced by surgical or chemical "sympathectomy" (Chap. 5.2.3), or inhibition of the neuronal uptake mechanism by drugs such as cocaine (Chap. 5.2.5) produces a dramatic increase in the apparent sensitivity of an effector organ to exogenous catecholamines. This phenomenon, called presynaptic supersensitivity, is due to the lack of efficient removal of the amine from the extraneuronal space, which allows a higher concentration of the applied amine to reach the adrenoceptors of the effector cells (Macmillan 1959, Trendelenburg 1963, Iversen 1967).

The neuronal uptake mechanism in mammalian adrenergic neurons is most effective for noradrenaline, less so for adrenaline and entirely ineffective for the structurally related substances isoprenaline and methoxamine (Chap. 5.2.7, Iversen 1967, Trendelenburg et al. 1970). In amphibians, in which adrenaline dominates in the adrenergic neurons, the neuronal uptake mechanism seems instead to be most effective for adrenaline, while the removal of noradrenaline is claimed to take place chiefly by an extraneuronal uptake mechanism (Stene-Larsen and Helle 1978a, Ask et al. 1980, Stene-Larsen 1981). It is clear, however, that a substantial neuronal uptake of noradrenaline and α-methylnoradrenaline exists in amphibians, judged from the enhancement of the fluorescence of the nerve fibres after pretreatment with these substances (see Sect. 3.2.6.4).

The lower vertebrates also exhibit a slight difference in the mechanism of catecholamine release by "indirectly acting amines" such as tyramine (Chap. 5.2.4). In mammals, tyramine is thought to be accumulated in the nerve terminal by the neuronal uptake mechanism, so that the release of the catecholamines is due to a "displacement" of the transmitter from the storage sites (Iversen 1967, Burnstock and Costa 1975). In adrenergic neurons of the cod spleen, tyramine is also effective in releasing stored catecholamines. The uptake of tyramine into the nerve terminals is not related to the neuronal uptake mechanism for noradrenaline, since the effect of noradrenaline is not disturbed by addition of tyramine and the effect of tyramine is not blocked by addition of cocaine (Nilsson and Holmgren 1976, Holmgren and Nilsson 1982).

3.2.6.3 Metabolic Degradation of Catecholamines

The major mechanism for the termination of the action of catecholamines released during nerve activity is the neuronal re-uptake described above, which removes as much as 70% of the transmitter released (Langer 1970, Bennett 1972). The fraction of the neuronally released catecholamines which is not taken up again and stored in the storage vesicles of the nerve terminal, as well as catecholamines released into the blood from the chromaffin tissue eventually undergoes metabolic degradation before final excretion.

The two enzymes that are mainly responsible for the catabolism of adrenaline and noradrenaline are the mitochondrial monoaminoxidase (MAO; E.C. 1.4.3.4) which attacks the amino group on the side chain, and catechol-O-methyl transferase (COMT; E.C. 2.1.1.6), which methylates the ring hydroxyl group in the 3-position (Fig. 3.3). Both enzymes are present in a wide variety of tissues, and MAO is of particular importance in the degradation of neuronal catecholamines, while COMT (which is largely extraneuronal) is responsible for the degradation of circulating catecholamines (Kopin 1964, Fänge and Hanson 1973). The pathways of

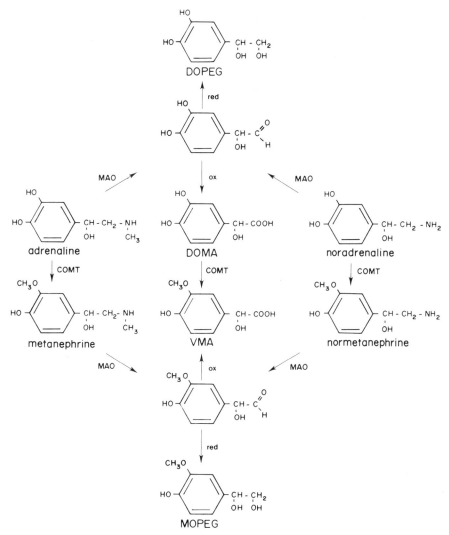

Fig. 3.3. Catabolism of adrenaline and noradrenaline by deamination (*MAO*, monoaminoxidase) and O-methylation (*COMT*, catechol-O-methyl transferase). Oxidases (*ox*) and reductases (*red*) responsible for the oxidation and reduction respectively of the aldehyde intermediates are shown. Abbreviations: *DOMA*, 3,4-dihydroxymandelic acid; *DOPEG*, 3,4-dihydroxyphenyl glycol; *MOPEG*, 3-methoxy-4-hydroxyphenyl glycol; *VMA*, vanillylmandelic acid

Table 3.1. Major urinary catabolites of adrenaline in some vertebrate species. A large, often major, portion of the catabolites is excreted in conjugated form

Elasmobranchs		
Scyliorhinus canicula	VMA, MN	Mazeaud and Mazeaud (1973)
Teleosts		
Salmo gairdneri	MOPEG, VMA	Mazeaud and Mazeaud (1973)
Gadus morhua	MN	Ungell and Nilsson (1979)
Amphibians		
Bufo marinus	MOPEG, VMA, MN	Ungell and Nilsson (1982b)
Reptiles		
Anolis carolinensis	MOPEG, VMA, MN	Scott and Neudeck (1972)
Birds		
Gallus domesticus	MOPEG, VMA	Rennick et al. (1965)
Mammals		
Rattus norvegicus	MOPEG, MN	Kopin et al. (1961)
Homo sapiens	VMA, MN	LaBrosse et al. (1961)

Abbreviations: *MN*, metanephrine; *MOPEG*, 3-methoxy-4-hydroxyphenyl glycol; *VMA*, vanillylmandelic acid

deamination and O-methylation of the catecholamines, and the catabolites formed are summarized in Fig. 3.3.

The relative proportions of the two enzymes in the tissues, the availability of the enzymes for the catecholamines and also the efficiency of renal excretion of the individual catabolites produces a urinary catabolite pattern that is constant in any one species. The variability among some of the vertebrate groups in the pattern of urinary catabolites of adrenaline is summarized in Table 3.1. It should be noticed that the main portion of each catabolite is usually excreted into the urine as a conjugate (glucuronide or sulphate).

3.2.6.4 Histochemical Demonstration of Adrenergic Neurons

The first demonstration by fluorescence histochemistry of the noradrenaline-storing varicose nerve terminals took place in August 1961 in the laboratory of Nils-Åke Hillarp and Bengt Falck in Lund (Owman and Björklund 1978). Following the very first successful preparation of Hillarp and Falck on a whole-mount of the rat iris, the Falck-Hillarp fluorescence histochemical technique has been used to demonstrate the neuronally stored monoamines (adrenaline, noradrenaline, dopamine and 5-hydroxytryptamine) in an impressive variety of tissues from both vertebrates and invertebrates (see Figs. 2.13, 2.14, 2.20–2.23, 2.27, 2.28). At the time of the discovery by Falck and Hillarp, it had been known that a formaldehyde solution produces a fluorescent compound from stored catecholamines (Eränkö 1952, 1955), and when the formaldehyde reaction was performed on tissues that had been rapidly air-dried or freeze-dried to avoid diffusion of the amines, the method became a useful tool for the demonstration of amines stored in the fine terminals of the adrenergic neurons (Carlsson et al. 1962, Falck 1962, Falck et al. 1962, Falck and Owman 1965, Björklund et al. 1972). The chemical background to the reaction was subsequently worked out, showing that during the exposure of the catecholamines to formaldehyde vapour at controlled temperature and humidity, the neuronal catecholamines condense to form fluorescent 3,4-dihydroisoquinolines (Corrodi and Jonsson 1967, Björklund et al. 1972).

Later a slight modification of the original Falck-Hillarp technique was introduced. In this "ALFA"-method the preparations are brought into contact with a solution of an aluminium salt (e.g., aluminium sulphate) prior to freeze-drying, which enhances the fluorescence yield significantly (Ajelis et al. 1979, Björklund et al. 1980, Lorén et al. 1980).

Another method based on the same general principle uses glyoxylic acid instead of formaldehyde vapour (Axelsson et al. 1973, Furness and Costa 1975). The main advantages are an increased fluorescence intensity and its applicability to non-freeze-dried tissues. However, the reagent penetrates tissue blocks poorly, and is best used with stretched preparations (whole-mounts) or cryostat sections.

Immunohistochemical demonstration of adrenergic fibres makes use of the specific localization in the adrenergic neurons of certain components, such as the enzymes involved in catecholamine synthesis (e.g., Nagatsu and Kondo 1975, Nagatsu et al. 1979; Schultzberg et al. 1980).

3.2.7 Cholinergic Neurons

The knowledge of autonomic cholinergic nerve functions is sparse compared to the impressive knowledge of adrenergic functions. It is reasonable to assume similarities between cholinergic transmission at the neuromuscular junction in skeletal muscle and the autonomic cholinergic nerve terminals, and by the use of analogy a picture of the functions of cholinergic autonomic neurons can be pieced together. The main anatomical difference between the motor end-plate and the autonomic nerve terminal is the extension of the terminal region, and the "en passant" contacts of the autonomic cholinergic terminals with a large number of effector cells (Burnstock 1970, 1979, Bennett 1972). A synaptic specialization showing some similarities with the motor end-plate is present at cholinergic junctions in ganglia (Fig. 2.16). A summary of the functions in the cholinergic nerve terminal is attempted in Fig. 3.4.

Acetylcholine is synthesized intraneuronally from choline and acetyl-coenzyme A by the action of choline acetyltransferase (choline acetylase) (ChAT; E.C. 2.3.1.6; Nachmansohn and Machado 1943) (Fig. 3.5). This enzyme is present in the placenta of primates, but not of other mammals (Comline 1946, Hebb and Ratkovič 1962), and in all tissues other than the placenta the enzyme is unique to the cholinergic neurons and thus a good marker for cholinergic fibres (Ehinger et al. 1966, Ekström and Elmer 1977, Roskoski et al. 1977, Lund et al. 1979). Choline is actively taken up from the extracellular space by a Na^+-dependent mechanism and the acetylation takes place in the cytoplasm of the neuron (Birks and MacIntosh 1961, Marchbanks 1968). The acetylcholine formed is actively taken up and stored in small clear storage vesicles (SCV, see Sect. 3.2.1), and released by a Ca^{2+}-dependent exocytosis during nerve activity (Potter 1970, Bennett 1972, Burnstock 1979, 1981). A release of ^3H-acetylcholine synthesized from ^3H-choline has been demonstrated (Szerb 1975, 1976, Wikberg 1977).

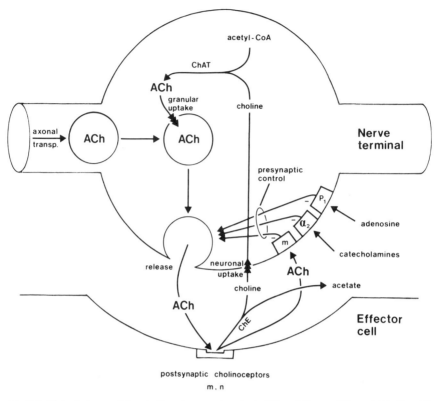

Fig. 3.4. Main features of the cholinergic nerve terminal. Abbreviations: *ACh*, acetylcholine; *ChAT*, choline acetyltransferase; *ChE*, cholinesterase; α, adrenoceptor; *m*, *n*, cholinoceptors; P_1, purinoceptor

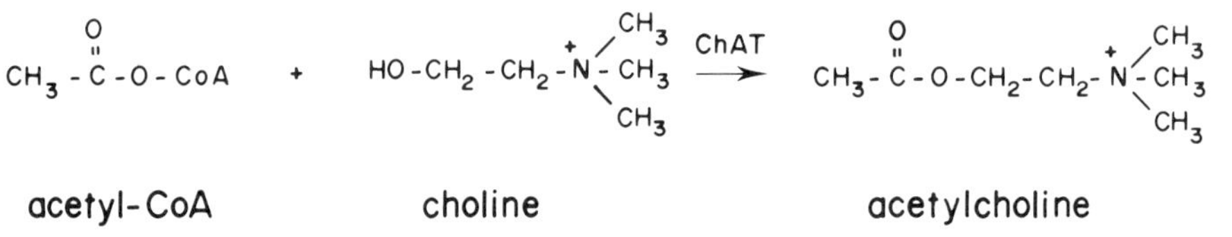

$$CH_3-\overset{O}{\underset{\|}{C}}-O-CoA \quad + \quad HO-CH_2-CH_2-\overset{+}{N}\underset{CH_3}{\overset{CH_3}{<}}CH_3 \quad \xrightarrow{ChAT} \quad CH_3-\overset{O}{\underset{\|}{C}}-O-CH_2-CH_2-\overset{+}{N}\underset{CH_3}{\overset{CH_3}{<}}CH_3$$

acetyl-CoA   choline   acetylcholine

Fig. 3.5. Synthesis of acetylcholine. *ChAT*, choline acetyltransferase

ChAT is transported into the nerve terminals by the slow axonal transport typical of cytoplasmic enzymes (Oesch et al. 1973, Tuček 1975).

The effect of the transmitter released during nerve activity is terminated by the action of acetylcholinesterase (AChE; specific cholinesterase E.C. 3.1.1.7), which is present both pre- and postsynaptically in the synaptic cleft. The choline formed by the degradation of acetylcholine is taken up again by the nerve terminal and "recycled". An unspecific cholinesterase (pseudo-cholinesterase; E.C. 3.1.1.8), which hydrolyses all choline esters, occurs in blood plasma.

Acetylcholinesterase is a glycoprotein which is present in several forms. The axonal transport rate in mammals varies among the different forms, with the fastest transport reaching 400 mm/d. The rapidly transported fraction is associated with the smooth endoplasmic reticulum, and thus organelle-bound. A large fraction of the acetylcholinesterase is stationary, bound to the outer surface of the neuron (Lubinska and Niemierko 1971, Tuček 1975, Brimijoin and Wiermaa 1978, Brimijoin et al. 1978, Brimijoin 1979, Couraud and Di Giamberardino 1980).

3.2.7.1 Histochemical Demonstration of Cholinergic Neurons

The presence of acetylcholinesterase in the cholinergic neurons can be demonstrated histochemically by using acetyl*thio*choline as a substrate for the enzyme (Koelle 1950, Coupland and Holmes 1957, Holmstedt 1957a, b, Karnovsky and Roots 1965). The activity of acetylcholinesterase liberates the SH-group in the thiocholine, which reacts in later steps of the reaction procedure to form a brown or black precipitation of copper salts.

Although the cholinergic neurons are particularly rich in AChE, the method is not entirely specific, since neurons other than the cholinergic variety also contain cholinesterase. The possible presence of both cholinergic and non-cholinergic neurons within the same Schwann cell sheath may also complicate the interpretation (Eränkö et al. 1970, Barajas and Wang 1975, Burnstock 1979).

A more specific method for the demonstration of cholinergic neurons was developed by Kasa et al. (1970), in which the "marker" enzyme ChAT is localized histochemically. The method is, unfortunately, not sensitive enough to be used for demonstration of autonomic neurons (Burnstock 1979).

A third method involves the localization of immunochemically reactive components unique to the cholinergic neurons, e.g., ChAT (KarenKan and Chao 1981, Peng et al. 1981) or cholinergic storage vesicles (Dahlström et al. 1981). Development of these methods will produce highly specific techniques for the localization of cholinergic neurons in tissues.

3.2.8 Purinergic Neurons

Although responses to electrical stimulation that were mediated by neurons but resistant to cholinergic and adrenergic antagonists have been recognized for a long time, it was not until the early sixties that the presence of non-adrenergic, non-cholinergic (NANC) nervous mechanisms was more firmly established. The initial evidence came from work on the mammalian gut, especially the stomach and taenia coli of the guinea-pig (Greeff et al. 1962, Burnstock et al. 1963a, 1964, Bennett 1966, Campbell 1966b, see also Chap. 9.1.3).

Attempts to identify the transmitter substance(s) in the NANC neurons were made, and led to the presentation of the "purinergic nerve hypothesis", which suggests that adenosine 5' triphosphate or a related substance is the NANC transmitter (Burnstock 1972). Later it was found that several of the gastro-intestinal peptides are stored within neurons, and the possibility of peptidergic nerves was suggested (Hökfelt et al. 1980a, b, Schultzberg et al. 1980; see Chaps. 3.2.9 and 9.1.4). Although there are now a number of convincing pieces of evidence for several of

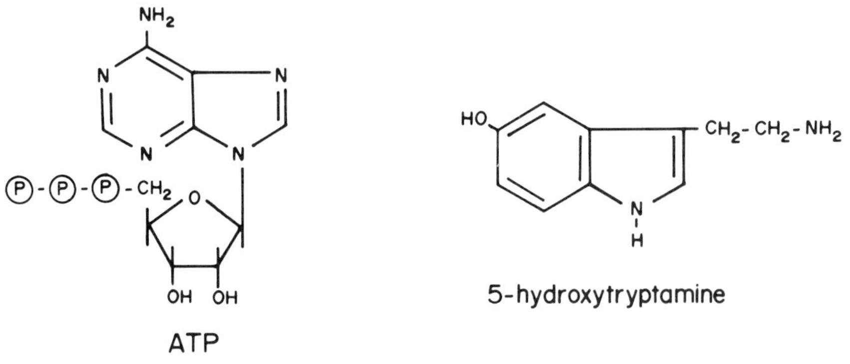

His - Ser - Asp - Ala - Val - Phe - Thr - Asp - Asn
                                                  |
Ala - Met - Gln - Lys - Arg - Phe - Arg - Ser - Tyr
 |
Val - Lys - Lys - Tyr - Leu - Asn - Ser - Val - Leu - ThrNH$_2$

VIP (chicken)

Fig. 3.6. Three examples of putative NANC-transmitters: ATP, adenosine 5′-triphosphate; 5-hydroxytryptamine and VIP, vasoactive intestinal polypeptide. The amino acid sequence in VIP is from the domestic fowl (chicken) (*Gallus domesticus*) (Nilsson 1975)

the putative NANC transmitters, it must again be emphasized that none of these has yet reached the acceptance of the "classical" transmitters (acetylcholine and catecholamines). Three of the NANC transmitter candidates, ATP, 5-hydroxytryptamine and VIP, are shown in Fig. 3.6.

ATP and the enzymatic equipment for its synthesis and degradation can be found in all living cells. It has been possible to show that ^{3}H-adenosine is taken up and stored mainly as ^{3}H-ATP in neurons (Su et al. 1971). Electrical stimulation of the guinea-pig taenia coli after pretreatment with ^{3}H-adenosine produced a release of radioactivity that could be blocked by tetrodotoxin, showing that the release was nervous (Su et al. 1971, Burnstock 1981). An attempt to summarize the features of the purinergic nerve terminal is made in Fig. 3.7.

Responses of the isolated taenia coli of the guinea-pig to exogenous ATP and nerve stimulation are very similar (Fig. 3.8), thus providing an elegant fulfillment of the mimicry criterion stated above (Chap. 3.1) (Cocks and Burnstock 1979, Jager and Schevers 1980, MacKenzie and Burnstock 1980). There are also several pieces of evidence that point to similarites in the effects of "chemical tools" on nerve stimulation on one hand and exogenous ATP on the other (see Chaps. 4.2.3 and 5.4).

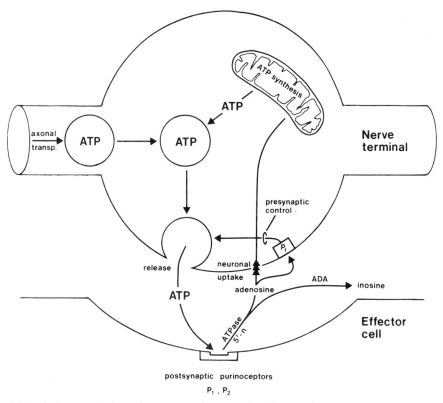

Fig. 3.7. Main features of the purinergic nerve terminal. 5'-n, 5'-nucleotidase; *ADA*, adenosine deaminase; *ATP*, adenosine triphosphate

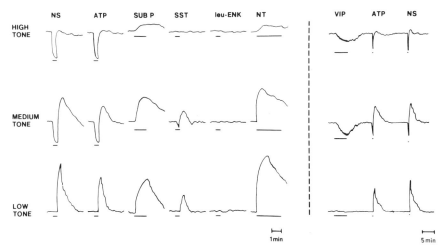

Fig. 3.8. A comparison of the effect of transmural nerve stimulation (*NS*) and the effects of a series of putative NANC transmitter substances on the isolated taenia coli from the guinea-pig at three different base-line tensions. *ATP*, adenosine triphosphate; *SUB P*, substance P; *SST*, somatostatin; *leu-ENK*, leu-enkephalin; *NT*, neurotensin; *VIP*, vasoactive intestinal polypeptide. (Reproduced with permission from Cocks and Burnstock 1979)

An important role of adenosine released together with other transmitters (or formed from ATP released with other transmitters) could be that of a presynaptic modulator (Chap. 3.2.4) (Lokhandwala 1979, Burnstock 1980a, 1981, Fredholm and Hedqvist 1980). According to Burnstock (1980a), it is thus possible: "(1) that ATP, which is a molecule that appeared early in biochemical evolution and was involved in intracellular communication, was used as a primitive neurotransmitter, (2) that it has been retained as the principal transmitter in some nerves in mammals and (3) that is has been retained as a co-transmitter in other nerves (including some cholinergic nerves) where, upon release, it is utilized largely as a modulator of release of the principal transmitter via presynaptic purinergic receptors".

3.2.8.1 Histochemical Demonstration of Purinergic Neurons

A fluorescence histochemical method that may be of help in demonstrating ATP-rich nerve fibres has been developed by Olson et al. (1976). In this method, quinacrine is taken up by the neurons and binds to ATP, and there appears to be a certain specificity for the quinacrine binding in neurons which, from other types of studies, are thought to be purinergic (Olson et al. 1976, Burnstock et al. 1978a, b, Burnstock 1981). However, the specificity of the method is low, since there may also be an unspecific quinacrine binding (Burnstock 1981). Furthermore Ekelund et al. (1980) have concluded that quinacrine binds also to peptidergic cells.

3.2.9 Serotonergic Neurons

Serotonin was originally discovered as a vasoactive principle present in blood serum (Rapport et al. 1948) and also in the gastric mucosa ("enteramine") (Erspamer 1954). It has now been shown that both these principles are identical to 5-hydroxytryptamine (5-HT; Fig. 3.6). The major source for 5-HT present in the blood, and indeed the major site of 5-HT storage in the mammalian body, is the enterochromaffin cells (EC) in the mucosa of the gut (Erspamer 1966). The non-EC stores of 5-HT in the gut are comparatively small (Feldberg and Toh 1953, Gershon 1981), and although the presence of serotonergic neurons in the central nervous system of mammals is well documented, the idea of serotonergic neurons in the mammalian enteric nervous system has not so far won general acceptance. As with the purinergic neurons, the lack of a specific histochemical demonstration and the low selectivity of the chemical tools used in the study of serotonergic transmission have prevented the establishment of a "serotonergic nerve" concept (Gershon 1981).

The documented ability of a group of gut neurons to take up and metabolize amine precursors, and ultrastructural and autoradiographic studies of 5-HT uptake and storage, indicate that 5-HT is taken up and stored in large granular vesicles ("p-type" vesicles; diameter 65–100 nm) of the mammalian enteric neurons (Rothman et al. 1976, Dreyfus et al. 1977). In analogy with the APUD concept of Pearse (1968) (Chap. 9.1.4.4), these neurons have been called "intrinsic amine-handling neurons" (Costa et al. 1976, Furness and Costa 1978, 1980). Although def-

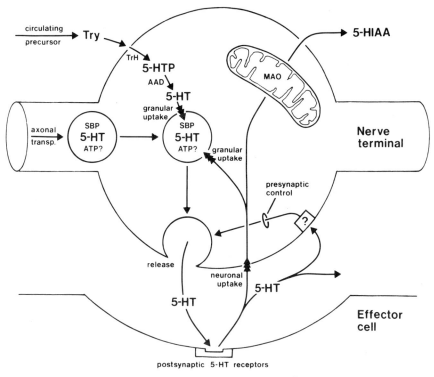

Fig. 3.9. Main features of the serotonergic nerve terminal. Abbreviations: *AAD*, aromatic-L-aminoacid decarboxylase; *ATP*, adenosine triphosphate; *5-HIAA*, 5-hydroxyindole acetic acid; *5-HT*, 5-hydroxytryptamine; *MAO*, monoaminoxidase; *SBP*, "serotonin binding protein"; *TrH*, tryptophan hydroxylase; *Try*, tryptophan

inite evidence is wanting, there is a possibility that the "amine-handling neurons" are related to the serotonergic neurons postulated by Gershon (1981).

5-HT is synthesized from L-tryptophan, which is taken up into the nerve terminal and initially hydroxylated to 5-hydroxytryptophan (5-HTP) by the enzyme tryptophan hydroxylase (tryptophan 5-monooxygenase; E.C. 1.14.16.4). This enzyme requires a tetrahydropteridine co-factor and oxygen for activity (cf. tyrosine hydroxylase, Sect. 3.2.6.1). 5-HTP is subsequently decarboxylated by aromatic-L-amino-acid decarboxylase (AAD, which is identical to the enzyme responsible for the decarboxylation of DOPA in the adrenergic nerve terminal (Chap. 3.2.6.1). Although tryptamine is present in tissues, there is no conversion of this substance to 5-HT (Udenfriend 1959).

5-HT is taken up and stored in storage vesicles by a reserpine-sensitive granular uptake similar to that in the adrenergic neurons, and forms a complex with a specific serotonin binding protein (SBP; see below) and possibly also ATP. The presence of ATP in neuronal storage vesicles has not been confirmed, but the substance is present in the 5-HT storing vesicles of the enterochromaffin cells (Udenfriend 1959, Douglas 1970, Gershon 1981). The possible features of a serotonergic nerve terminal are summarized in Fig. 3.9.

Termination of the 5-HT effect is mainly due to re-uptake into the nerve terminal and degradation of the 5-HT in the mitochondria, where 5-hydroxyindole acetic acid (5-HIAA) is formed by the action of MAO (cf. Chap. 3.2.6.3).

The Falck-Hillarp fluorescence histochemical technique mentioned above (Chap. 3.2.6.4) can also be used for the cellular localization of 5-HT in EC and neurons. Although the presence of 5-HT-containing neurons in the mammalian enteric nervous system can be confirmed with this method only after pretreatment with a precursor (L-tryptophan) and/or blockade of MAO (Dreyfus et al. 1977, Gershon 1981), there is good histochemical evidence for the presence of 5-HT-containing neurons in the gut of both cyclostomes (Baumgarten et al. 1973, Goodrich et al. 1980) and teleosts (Watson 1979). In mammals, the presence of the enzyme tryptophan hydroxylase, which may be specific for serotonergic neurons, has been demonstrated in enteric neurons by an immunohistochemical technique (Gershon et al. 1977). Specific antisera against 5-HT itself have also been produced, and show great promise for the future immunohistochemical detection of 5-HT-storing neurons (Consolazione et al. 1981).

Further evidence for the presence of 5-HT in the mammalian gut comes from studies on the uptake of L-tryptophan and 5-HT uptake (Gershon and Altman 1971, Rothman et al. 1976, Dreyfus et al. 1977, Gershon 1981), demonstration of a specific "serotonin-binding protein" (SBP) similar to that in neurons of the central nervous system (Jonakait et al. 1977) and a Ca^{2+}-dependent release of 5-HT possibly by exocytosis (Jonakait et al. 1979, Gershon and Tamir 1981).

The demonstration of release of 5-HT from *neuronal* stores in the gut is difficult due to the large stores of releasable 5-HT in the enterochromaffin cells. One way around this problem has been to use a low concentration of radiolabelled 5-HT, which is taken up and stored by the serotonergic neurons but not by the EC which lack a specific uptake mechanism (Rubin et al. 1971, Jonakait et al. 1979). By turning the piece of intestine inside-out and perfusing through the "serosal lumen". the 5-HT diffusion to the EC (which are now on the outside) can be further reduced. The release of label from this type of preparation is then likely to be chiefly neuronal (Jonakait et al. 1979, Gershon 1981). The introduction of a very sensitive and selective radio-immunochemical assay technique for 5-HT has greatly increased the possibility of demonstrating the release of minute amounts of 5-HT from neurons (Delaage and Puizillout 1981).

3.2.10 Peptidergic Neurons

Immunohistochemical techniques have proved very elegant and specific for the demonstration of various compounds in tissues. These methods have now been used to demonstrate the presence and localization of a large number of polypeptides. It has now become clear that several of the gastro-intestinal peptide hormones, as well as other peptides, are stored in neurons of the gut, brain and other organs. The specificity of the methods is entirely dependent on the immunological identity of the antisera used, and the cautious terms "VIP-like immunoreactivity" ("VIP-like IR") etc. are used as a reminder that the selectivity of the method is only relative.

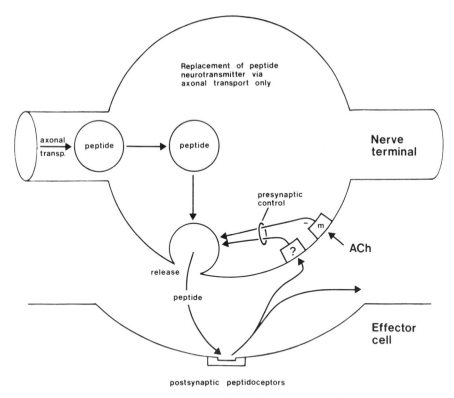

Fig. 3.10. Main features of the peptidergic nerve terminal. Abbreviation: *m*, cholinoceptor

The neuronal peptides include vasoactive intestinal polypeptide (VIP), substance P (which was the first "brain-gut" peptide, described by Euler and Gaddum 1931), somatostatin, enkephalins, gastrin/cholecystokinin, angiotensin, neurotensin, bombesin, ACTH-like peptide and others (Dockray et al. 1979, Fahrenkrug 1979, Furness and Costa 1980, Hökfelt et al. 1980a, b, Larsson 1980, Schultzberg et al. 1980, Said et al. 1980, Sundler et al. 1980, Leander et al. 1981a, Malmfors et al. 1981) (see also Chap. 9.1.4).

Although the neuronal localization of the peptides is well established, there are many other of the criteria outlined in Chap. 3.1 that are not as yet fulfilled. As with the purinergic and serotonergic types of transmission, there is a particular lack of drugs that can interfere specifically with the peptidergic transmission, and which can be used as "chemical tools" to demonstrate the pharmacological identity of the nerve response. One possible technique would be to use peptides that retain the affinity for the receptor, but with impaired stimulating activity caused by exchange of one or more amino acids in the sequence (see Chap. 5.6).

There are now several demonstrations of the release of peptides during nerve stimulation. The earliest of these studies include the demonstration by radioimmunoassay of VIP release into the blood during stimulation of the vagal nervous supply to the gut (Fahrenkrug et al. 1978a, b). Release of VIP from neurons innervating the mammalian salivary glands has also been concluded (Lundberg et al.

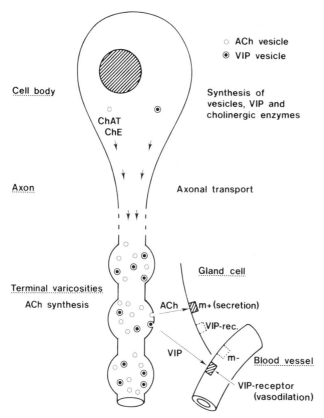

Fig. 3.11. A mixed function neuron according to the model of Lundberg (1981). This type of cholinergic neuron innervates the mammalian salivary glands, and is able to release both acetylcholine and VIP, which act synergistically to increase the secretion of saliva. The effect of acetylcholine may be chiefly to stimulate the gland cell (but also cause vasodilation), and the effect of VIP to cause vasodilation (but also affect the gland cells). (Redrawn with permission from Lundberg 1981)

1980a, Uddman et al. 1980, Lundberg 1981). There are several cases of neurons with more than one peptide and of peptides acting as co-transmitters in "classically" cholinergic and adrenergic neurons (Fig. 3.11) (Hökfelt et al. 1977b, 1980 a, b, Lundberg et al. 1979, 1981 a–c, Schultzberg et al. 1980, Lundberg 1981, Campbell et al. 1982). Some peptides (e.g., substance P) have also been demonstrated in primary sensory neurons, where an involvement in axonal reflexes has been suggested (Hökfelt et al. 1975, 1980, Delbro 1981). An attempt to summarize the features of the peptidergic nerve terminal is made in Fig. 3.10.

3.3 Conclusions

The classical view of the autonomic nervous system comprising only cholinergic and adrenergic neurons had to be abandoned after the first clear evidence of non-adrenergic, non-cholinergic (NANC) neurons in the mammalian gut which came forward in the 1960's. Among other important later findings that must be considered to understand fully the mechanisms of autonomic nerve function are the concept of presynaptic receptor systems (autoreceptors), and the recognition of neurons with more than one transmitter substance. Introduction of electron microscope techniques have also been of importance for the understanding of structural backgrounds to observed phenomena, but there is still a long way to go before the ultrastructural nerve profiles can be combined with functional observations of the transmitter substance of a neuron.

There is now sound and generally accepted evidence for catecholamines and acetylcholine as transmitter substances in the adrenergic and cholinergic neurons, respectively. How, then, do the putative NANC transmitters fulfill the criteria outlined in Chap. 3.1?

The criteria of synthesis, storage, release, mimicry and inactivation are fulfilled for ATP as a transmitter in autonomic neurons, but selective chemical tools (Chap. 5) are still wanting to test decisively the criterion of "pharmacological identity". A specific histochemical method to localize the presumed purinergic neurons would also be welcome.

The evidence for storage and release of 5-HT in the mammalian gut is still open to challenge due to the possible interference by 5-HT from enterochromaffin cells. The presence of neurons in fish gut that exhibit the yellow fluorescence characteristic of 5-HT studied by Falck-Hillarp fluorescence histochemistry, speaks in favour of serotonergic neurons in fish. The pharmacological identity of the mammalian "amine-handling neurons" has been difficult to establish due to the lack of selective chemical tools.

The evidence in favour of a peptidergic neurotransmission is based on demonstrations of synthesis, storage, release and mimicry. The immunochemical methods used to demonstrate the peptides histochemically or in tissue perfusates and extracts depend entirely on the specificity of the antisera used, which should be remembered when evaluating the evidence. There is so far no detailed description of a system for inactivation of any of the neuropeptides, but the small amounts released may be removed by extracellular peptidases or simply by diffusion. As with the other NANC-transmitter systems, there is a lack of selective chemical tools useful for the study of peptidergic neurotransmission (Chap. 5.6).

Chapter 4

Receptors for Transmitter Substances

4.1 Drug–Receptor Interaction

The concept of a receptive substance, later simply called *receptor*, was developed by Langley (1905) from his experiments with nicotine and curare. Dale (1906) also referred to the neuromuscular junction as "the receptive mechanism for adrenaline." Among the substances that exert their action by interacting with tissue receptors are the neurotransmitters, which diffuse the distance to the effector cell after release from the nerve terminal. At the effector cell the neurotransmitter interacts with its appropriate receptor and induces a stimulus in the effector tissue. The stimulus in turn triggers a series of events that eventually causes an effect (or response) in the effector cell (Chap. 4.1.10). The effect may be *excitatory* (contraction of smooth muscle, increased glandular activity, increased lipolysis or glycolysis, increased heart rate and force, pigment aggregation in melanophores etc.) or *inhibitory* (relaxation of smooth muscle, decreased glandular activtiy, decreased heart rate and force, dispersion of melanophore pigment etc.). Regardless of the type of effect, or whether it is excitatory or inhibitory, the magnitude of the effect is in one way or another related to the number of receptors being stimulated by the substance.

A very large part of the knowledge of autonomic nerve function comes from experiments in which pharmacologically active substances have been used as "chemical tools" to elucidate the pattern and nature of the innervation of individual cells, tissues, organs and organ systems (see also Chap. 5). In order to appreciate the great advantages of the use of chemical tools in the study of autonomic nerve function, but also the grave dangers associated with the uncritical application of drugs, a brief account of the mechanisms of drug receptor interaction is present in this chapter. Also presented is a summary of the criteria used for the classification and sub-classification of the receptors for the different neurotransmitters.

4.1.1 Specificity and Selectivity

As will be shown in Chap. 5, chemical tools are available that can be used to interfere with neurotransmission at many different levels (synthesis, storage, release, uptake, degradation, interaction with pre- and postsynaptic receptors etc.). Many of the pharmacologically active substances (*drugs*) used as chemical tools in the study of autonomic nerve function are chosen because of their more or less well-defined and documented (usually in mammals) *specificity* as stimulators (*agonists*) or blockers (*antagonists*) of certain receptors. It is very true, however, that no substance has a single effect: administration of a high enough concentration (or dose) of any substance may well produce unwanted, and often unknown, side effects that can invalidate its use as a specific tool. Examples of such side effects are plentiful and include, for instance, the anti-cholinergic action of some adrenoceptor antagonists and adrenergic neuron blocking agents, or local anaesthetic effects of the widely used adrenoceptor antagonist propranolol. Furthermore, clonidine, which is an adrenergic agonist in most mammalian systems is an adrenoceptor antagonist in fish (Benfey and Grillo 1963, Boyd et al. 1963, Åberg and Welin 1967, Åberg et al. 1969, Gulati et al. 1973, Johansson 1979) (see also Chap. 5).

To avoid these traps, it is necessary to establish concentrations (or doses) of the chemical tools that are *selective*. This means that the unwanted effects are minimized and *known* to such an extent that the specific action of the chemical tools can be utilized in the experiment.

4.1.2 The "Dose-Response" Concept

The relationship between the dose of a drug administered and the effects caused can be conveniently described by a *dose–response curve* or, if the effects of various concentrations of the drug are measured, by a *concentration–response curve*. Practically, such curves can be constructed from measurements of the effects of exposure to single concentrations of the drug with a wash-out and a resting period between each addition of drug, or by a step-wise increase in the drug concentration without resting periods. The latter type is known as the cumulative concentration-response curve (van Rossum and van den Brink 1963, van Rossum 1963, Fig. 4.1). A diagrammatic representation of the concentration-response curve is usually made by plotting the effect as a function of the logarithm for the concentration of the agonist (Fig. 4.2).

The theoretical model for drug-receptor interaction implies a steady-state situation, where the receptors are equilibrated with a constant concentration of the drug(s). Therefore, the term "concentration-response" is more appropriate here than the commonly used term "dose–response". Several of the concepts presented

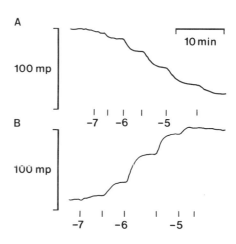

Fig. 4.1. Cumulative concentration–response curves for adrenaline producing relaxation of an isolated strip preparation from the posterior part of the swimbladder mucosa of *Ctenolabrus rupestris* (*A*), and contraction of the coeliac artery of *Pollachius pollachius* (*B*). Adrenaline was added to give final bath concentrations of 10^{-7}, 3×10^{-7}, 10^{-6}, 3×10^{-6}, 10^{-5} and 3×10^{-5} M as indicated. Vertical tension calibration = 100 mp (1 mp = 9.81×10^{-6} N)

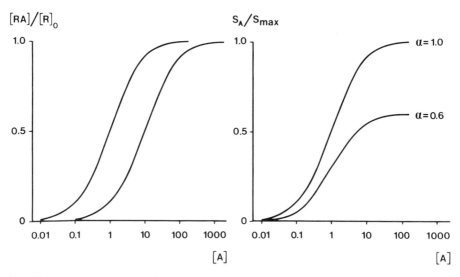

Fig. 4.2. *Left diagram:* Theoretical "concentration–response curves" expressed as fraction of the receptors occupied by the agonist [RA] of the total "concentration" of tissue receptors [R]$_0$ at different concentrations of the agonist A. The curves, which show the effects of two agonists with different affinities for the receptors, have been calculated according to Eq. (4). *Right diagram:* Theoretical "concentration–response curves" this time expressed as fraction of the maximal stimulus possible in the receptor system (S$_{max}$) which can be achieved with the agonist A, S$_A$, at different concentrations of A. The two curves show the effects of two agonists with different intrinsic activities: one full agonist ($\alpha = 1.0$) and one partial agonist ($\alpha = 0.6$)

below will also be very useful in *dose*–response evaluations (injections of single doses in vivo, exposure to brief "pulses" of drugs in vitro, etc.), provided due attention is paid to the non-equilibrium status of the system.

It was already suspected by Langley (1905) that the drug-receptor interaction could be interpreted in the terms of the law of mass action. The concept was elaborated in some detail by Clark (1926, 1937). The presentation herein is based on the later developements of these early ideas. It is an introduction to the field of quantitative pharmacology applied to the study of neurotransmission. For a more extensive treatment of this subject, see, e.g., Ariëns (1954, 1966), Ariëns and van Rossum (1957), Ariëns and Simonis (1961, 1964a, b), van den Brink (1969), Michelson and Zeimal (1973).

4.1.3 The Theoretical Model for Drug–Receptor Interaction

The receptors of an effector cell are similar to the active site of an enzyme molecule in being able to recognize and combine with structurally distinct drug molecules. In the case of enzyme–substrate interactions, the substrate will be metabolized, while in the case of drug–receptor interaction the drug (if it is an agonist) will induce an alteration in the receptor molecule (R). This will lead to a stimulus (S) that eventually produces an effect (E). A schematic representation of drug–receptor interactions is given in Fig. 4.3.

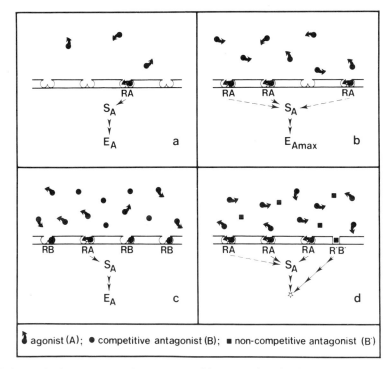

Fig. 4.3 a-d. The interaction between an agonist (A), a competitive antagonist (B) and a non-competitive antagonist (B') at the tissue receptors (R).

In **a**, the agonist is present in low concentration, which means that only a few receptors will have formed a complex with the agonist at any given time. The complex (RA) will induce a stimulus (S_A) proportional to the number of RA formed. The stimulus triggers off a chain of events which will lead to an effect (E_A).

In **b** the concentration of agonist is higher, and therefore the number of RA formed is higher than in A [Eq. (1)]. This means that the stimulus S_A will be greater, which may lead to a stronger effect. If a sufficient number of agonist-receptor complexes are formed, the effect becomes maximal for the agonist A (E_{Amax}). This maximal effect for A may occur even if not all receptors are occupied (receptor reserve).

In **c** the concentration of A is the same as in **b**, but the stimulus (and hence the effect) is weaker because of the presence of a competitive antagonist (B) which competes with A for the receptor sites. The antagonist-receptor complex (RB) is incapable of producing a stimulus (i.e., intrinsic activity of $B=0$). By further increasing the concentration of A, the competitive blockade will be overcome, and a maximal effect can be obtained. The effect of B is manifested as a parallel shift to the right of the concentration-response curve for A (Figs. 4.4 and 4.6).

In **d**, finally, a non-competitive (metactoid) antagonist (B') forms a complex with a receptor (R') which is different from the agonist receptors (R). The formation of the $R'B'$-complexes blocks the chain of events that allows a stimulus to manifest itself as an effect. Despite increasing the concentration of agonist, the maximal effect produced by A (E_{Amax}) will be reduced or, in the presence of a high enough concentration of B', abolished. The effect of B' is thus manifested as a depression of the maximal effect seen in the concentration-response curve for A (Figs. 4.5 and 4.6)

The interaction between receptors (R) and agonist (A) may be simply written as

$$R + A \underset{k_2}{\overset{k_1}{\rightleftarrows}} RA. \tag{1}$$

The rate of formation of the receptor–agonist complex (RA) is dependent on the rate constant k_1, and the rate of disintegration of RA is dependent on k_2. The ratio between these (k_2/k_1) is the *dissociation constant* (K_A) for the receptor-agonist complex RA. Assuming a homogenous distribution of agonist *and receptor* in the biophase, which in itself is very doubtful, the amounts of agonist and receptor molecules can be expressed as concentrations, and assuming further a great excess of A ($[A] \gg [R]$), the law of mass action will give

$$K_A = \frac{[A] \times [R]}{[RA]}. \tag{2}$$

The concentration of free receptors ($[R]$) can be substituted by $[R] = [R]_0 - [RA]$ where $[R]_0$ is the total concentration of receptors, and Eq. (2) becomes

$$K_A = \frac{[A] \times ([R]_0 - [RA])}{[RA]}. \tag{3}$$

The fraction of receptors occupied by A at equilibrium will then be

$$\frac{[RA]}{[R]_0} = \frac{[A]}{K_A + [A]}. \tag{4}$$

This equation is comparable to the Michaelis-Menten equation for enzyme-substrate binding (Michaelis and Menten 1913).

Theoretical "concentration-response" curves based on this equation are presented in Fig. 4.2.

The stimulus (S) produced by the agonist is proportional to the concentration of RA formed

$$S_A = i_A [RA]$$

and the maximal stimulus possible in the system (S_{max}) is likewise proportional to the total concentration of receptors (Ariëns 1966, see, however, Chap. 4.1.4)

$$S_{max} = i_{max}[R]_0.$$

The fraction of the maximal stimulus produced by the agonist A is then

$$\frac{S_A}{S_{max}} = \frac{[RA]}{[R]_0} \times \alpha \tag{5}$$

and combining Eqs. (4) and (5) (Ariëns 1966)

$$\frac{S_A}{S_{max}} = \frac{\alpha}{\frac{K_A}{[A]} + 1}. \tag{6}$$

The proportionality constant (α) determines the maximal response possible with agonist A in the receptor system, and is called the *intrinsic activity* or *efficacy*. Not all agonists have the ability to produce a maximal stimulus (i.e., $\alpha < 1$) despite having high affinity for the receptors. Such agonists are called *partial agonists* (Ariëns 1954, 1966, Ariëns et al. 1955, Stephenson 1956) (Fig. 4.2).

Equation (6) has been reached by assuming that the magnitude of the stimulus is determined by the *number of receptors occupied* (occupation theory). It has also been proposed that the *rate of drug-receptor combination* determines the stimulus (rate theory) (Paton 1961, 1964). Equation (6) also covers this possibility (Ariëns and Simonis 1961, Ariëns 1966).

Assuming that the stimulus (S_A/S_{max}) is proportional to the effect (E_A/E_{max}) (Ariëns 1966), Eq. (6) can be transformed to

$$\frac{E_A}{E_{max}} = \frac{\alpha}{\frac{K_A}{[A]} + 1} \tag{7}$$

and at high concentrations of A, the intrinsic activity (α) is determined by E_{Amax}/E_{max} where E_{Amax} is the maximal effect of the agonist A, and E_{max} is the maximal effect that can be obtained in the receptor system.

When $E_A/E_{Amax} = \frac{1}{2}$ (50% of maximal effect; EC_{50}), then $K_A = [A]$, if it is assumed that *all* receptors must be occupied to produce maximal effect. It was shown by Stephenson (1956) and subsequently by several other workers that this is very often not the case, requiring introduction of the concept of a receptor reserve (spare receptors).

4.1.4 Receptor Reserve

The concept of a receptor reserve (spare receptors) implies that the maximal effect can be produced by an agonist occupying only a fraction of the receptors (Stephenson 1956). Thus E_{Amax} is reached before all the receptors are occupied (see Fig. 4.3 b), and in this case $K_A \neq [A]$ at the EC_{50} (see above). The presence of a receptor reserve has been firmly established by the use of modern radio-ligand binding techniques, which allow a determination of the actual number (concentration) of receptors in a tissue (see Chap. 4.1.9).

While the effects of a receptor reserve will affect the upper part of the concentration–response curve, there may be other manifestations of a lack of a direct relationship between receptor occupation and effect in the lower part of the concentration-response curve. Such *threshold phenomena* imply that the number of receptors occupied by the agonist must reach a certain threshold before the effect is expressed (Ariëns 1966). Both receptor reserve and threshold phenomena will increase the steepness of the concentration-response curve, and affect the displacement of the curve in the presence of non-competitive (but not of competitive) antagonists (Ariëns 1966).

4.1.5 Affinity and Relative Intrinsic Activity

The affinity between agonist and receptor is determined by the reciprocal of the dissociation constant ($1/K_A$). Considering the existence of a receptor reserve and threshold phenomena, the direct determination of K_A is difficult, and the affinity of the agonist may instead be represented by pD_2, a value derived from the concentration of A producing 50% of the maximal response (EC_{50}). The negative logarithm of the EC_{50} is used in a way analogous with the pH concept (Ariëns and van Rossum 1957):

$$pD_2 = -\log(EC_{50}).$$

In the experimental situation, determination of affinity (as pD_2-value) and relative intrinsic activity (α-value of the agonist compared to a reference agonist) sometimes involves difficulties. For example, the presence of effective uptake mechanisms for catecholamines in the adrenergic nerve terminals will reduce the concentration of adrenergic agonist near the receptors. "Sympathectomy" (surgical or chemical) in such preparations will produce an increase in the apparent affinity (presynaptic supersensitivity; Chap. 3.2.6.2) that is not related to the adrenoceptors of the effector cells (Trendelenburg 1963). Likewise, the presence of cholinesterases in a tissue will impair the effect of acetylcholine, and thus alter the apparent affinity of this drug for its receptors.

In spite of the obvious difficulties in determining the absolute affinity for an agonist, concentration–response data are of vital importance in the assessment of the specificity and selectivity of "chemical tools."

4.1.6 Competitive Antagonism

In this type of antagonism the antagonist (B) competes with the agonist (A) for the same receptor sites (Fig. 4.3). The equilibrium reaction can thus be written:

$$R + A + B \rightleftarrows RA + RB.$$

The agonist A will induce a stimulus in the effector cell, while B is inactive (intrinsic activity of B = 0). It can be shown the concentration of A necessary to produce a certain effect (normally 50% of maximal for experimental purposes) is altered by addition of B according to the equation (Ariëns and van Rossum 1957)

$$\frac{[A]_B}{[A]_0} = \frac{[B]}{K_B} + 1 \tag{8}$$

where $[A]_0$ is the concentration of A necessary to produce a certain effect (e.g., EC_{50}) before addition of B, and $[A]_B$ is the concentration of A necessary to produce the same effect in the presence of B in the concentration [B]. K_B is the dissociation constant for B in the receptor system. The relationship (8) means that the concentration–response curve for A will be displaced to the right in a parallel fashion by

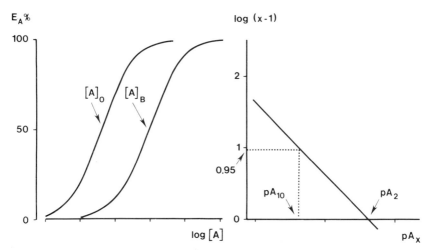

Fig. 4.4. Competitive antagonism. *Left diagram:* theoretical concentration–reponse curves for the agonist A before ($[A]_0$) and after ($[A]_B$) addition of the competitive antagonist B. The effect of B is manifested as a parallel shift of the concentration-response curve to the right. *Right diagram:* the graphical method of estimating values of pA_2 and pA_{10} with the so called "Schild plot" [Eq. (9)]. pA_x ($-\log [B]$) is plotted against $log\ (x-1)$, where x is obtained from the left diagram as the concentration ratio $[A]_B/[A]_0$. Note that the letter A in the term pA_x really refers to the *antagonist* B

addition of B (Figs. 4.4 and 4.6), and the concentration ratio $[A]_B/[A]_0$ is dependent on [B] only.

A clear and simple expression of the competitive antagonistic properties of a substance can be made by the use of pA_x-values introduced by Schild (1947). pA_x is the negative logarithm for the concentration of B that will produce a concentration ratio $[A]_B/[A]_0 = x$. Equation (8) can be transformed to a linear line equation (Ariëns and van Rossum 1957), which is useful in the experimental determination of antagonist properties

$$pA_x = -\log K_B - \log (x-1) \qquad (9)$$

where K_B is the dissociation constant for the antagonist B. By plotting $\log (x-1)$ as a function of pA_x it is possible to determine graphically the values of pA_2 and pA_{10} for the antagonist (Fig. 4.4). For a pure competitive antagonist, $pA_2 - pA_{10} = 0.95$. For higher or lower values, higher order reactions or non-competitive antagonism may be involved (Marshall 1955, Ariëns and van Rossum 1957, Benfey and Grillo 1963, van Rossum 1963).

4.1.7 Non-Competitive Antagonism

An antagonist that blocks the action of an agonist drug by interfering with receptors other than the agonist receptors is a non-competitive antagonist (Fig. 4.3). The "classical" type of non-competitive antagonism (metactoid antagonism) is due to a blockade, by the antagonist of the chain of events induced by formation of the agonist-receptor complex before an effect is reached (Fig. 4.3). There is also a sec-

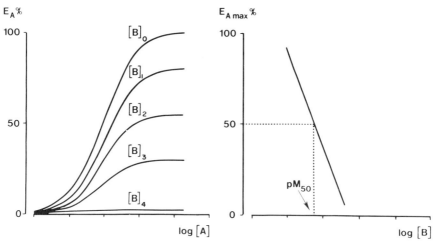

Fig. 4.5. Non-competitive antagonism. *Left diagram:* theoretical concentration-response curves for the agonist A in the absence ($[B]_0$) and presence ($[B]_{1-4}$) of a non-competitive antagonist B. The effect of B is manifested as a reduction in the maximal effect that can be obtained with A. *Right diagram:* graphical method of estimating the values of pD_2'- or pM_{50} for the antagonist B, by plotting the maximal effect of A obtained at different concentrations of B against log[B]

ond type of non-competitive antagonism (metaffinoid antagonism) in which the antagonist instead alters the affinity between the agonist and its receptors. The theoretical models for the two types of non-competitive antagonism have been discussed by Ariëns et al. (1956), van den Brink (1969) and Brandt and Offermeier (1977) and will not be dealt with here.

The metactoid type of non-competitive antagonism is manifested as a decrease in the maximal effect which can be obtained with the agonist (Figs. 4.5 and 4.6). If there is no receptor reserve, the depression of the maximal effect will be immediately evident. In the case of a receptor reserve the maximal effect can be reached by addition of higher concentrations of the agonist as long as the number of "receptor-effect" pathways is higher than the number required for maximal effect. When this limit is reached by addition of still higher concentrations of the antagonist, a decrease in the maximal response starts. In this case there is also a non-parallel shift to the right of the concentration-response curve with increasing concentrations of the antagonist before the depression in the maximal effect is manifested (Ariëns 1966).

The non-competitive antagonistic properties of a substance is described by the pD_2'-value (Ariëns and van Rossum 1957), or its equivalent pM_{50} (Guimarães 1969) by plotting the percentual $E_{A\max}$ against the antagonist concentration (Fig. 4.5).

4.1.8 Functional Interaction

Very often a tissue is equipped with more than one receptor system: for example acetylcholine and adrenaline (or noradrenaline) may exert their effects by combin-

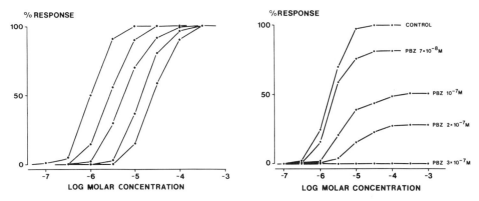

Fig. 4.6. Experimental concentration–response curves from cod (*Gadus morhua*) coeliac artery showing the effects of increasing concentrations of the competitive antagonist yohimbine (*left diagram*) and the non-competitive antagonist phenoxybenzamine (*right diagram; PBZ*) on the concentration–response curve for noradrenaline. (Reproduced with permission from Holmgren and Nilsson 1974)

ing with cholinoceptors and adrenoceptors respectively in the same organ. The interaction of two substances with different receptor systems may either be synergistic (*functional synergism*) or antagonistic (*functional antagonism*). Theoretical models for the functional interaction of drugs has been presented by van den Brink (1973 a, b).

4.1.9 Radio-Ligand Binding Studies

This modern method to establish the number (or concentration) of receptors in a tissue utilizes the specific binding of radio-labelled agonists and antagonists to the receptor sites in a tissue. The method can be used to provide accurate data for receptor concentrations and affinities between receptors and drugs, but does not give any measurement of the intrinsic activity of the agonists.

Radio-labelled agonists or antagonists with very high specific activity are equilibrated with tissue homogenates, and the binding of the drug to the tissue can be estimated after washout of free drug. By "protecting" the receptors with well-known and highly specific agonists or antagonists, the magnitude of specific and non-specific binding can be estimated. The methods have been used extensively for the study of mammalian adrenoceptors (Lefkowitz 1978, Triggle and Moran 1979, Hedberg 1980), but have not so far been used for the characterization of non-mammalian receptor systems.

4.1.10 Stimulus–Effect Coupling

The effect caused by a drug is the result of a series of events triggered by the drug–receptor interaction. These events involve systems of secondary messengers (within the effector cell) and effects on the permeability of the cell membrane to certain ions (see below). Of the secondary messengers, cyclic 3′,5′-adenosine monophosphate (c-AMP) has received particular attention. It is now generally accepted that adenyl cyclase, which is an enzyme responsible for the formation of c-AMP from ATP, is closely associated with the β-adrenoceptor (see Chap. 4.2) and that stimulation of β-adrenoceptors causes an increase in the intracellular levels of c-AMP. This in turn produces an altered Ca^{2+}-availability within the effector cell, leading to changes in the cell function (e.g., relaxation of smooth muscle) (Robison et al. 1971, Burnstock and Costa 1975). In the heart, however, a positive inotropic effect can also be seen without a change in the total c-AMP level of the tissue, showing that c-AMP may not be a necessary component in all β-adrenoceptor systems (McNeil 1979).

Other substances that could possibly be involved as secondary messengers are the prostaglandins, a group of compounds derived from a fatty acid (arachidonic acid). Changes in the synthesis of prostaglandins can be induced by several receptor systems, and a relationship with the c-AMP system has been described (e.g., Hedqvist 1977, Stjärne 1979). The prostaglandins are potent inhibitors of transmitter release from adrenergic nerve endings, and a major role of these compounds may be that of trans-synaptic modulators of transmitter release (Chap. 3.2.4). The importance of this mechanism in vivo is, however, unclear (Stjärne 1979).

The effects of drug–receptor interaction, direct or via secondary messengers, may lead to changes in the electrical potential across the membrane of the effector cell (membrane potential). Detailed accounts of the electrical events in excitable cells are given by Katz (1966), Miles (1969), Bennett (1972) and Schmidt (1978).

During rest, the inside of a neuron or smooth muscle cell is in the order of 30–60 mV negative compared to the outside. The cause of this membrane potential is the uneven distribution of certain ions, particularly Na^+, K^+, Cl^- and Ca^{2+} between the inside and the outside of the cell. The difference in ion concentration is due to the presence of large undiffusible ions (proteins) within the cell, and an active Na^+/K^+ exchange (in smooth muscle also Ca^{2+}/K^+) which creates a high K^+- and a low Na^+-concentration within the cell compared to the extracellular fluid. The membrane potential at rest is mainly determined by the K^+-gradient, since the permeability of this ion is higher than that of Na^+. During an excitatory stimulus of the cell membrane, the permeability of Na^+ increases, causing a reduction in the absolute value of the membrane potential (depolarization). Such transient potential changes are known as excitatory postsynaptic potentials (EPSPs), and depending of the time course of these potentials it is possible to distinguish between fast and slow EPSPs (Bennett 1972). The fast EPSP appears in synapses with close contact varicosities, while the slow EPSP is characteristic of the small axon bundle synapses (Chap. 3.2.1) (Bennett 1972).

If the EPSP reaches a certain threshold value, an action potential ensues. The first phase of the action potential is a rapid depolarization of the membrane caused by a general increase in the Na^+-permeability, followed by a second phase when the membrane potential is restored by an increase in the K^+-permeability (Miles 1969). During the action potential, the level of intracellular Ca^{2+} increases causing an alteration in cell functions (e.g., contraction of smooth muscle), which can also be caused by a passive depolarization of the cell membrane (e.g., by K^+-rich solutions) (Bennett 1972).

When the transmitter released from the presynaptic nerve terminals is inhibitory, there is instead a hyperpolarization of the effector cell membrane (an increase in the absolute value of the membrane potential). Such inhibitory postsynaptic potentials (IPSPs) are caused by an increase in the K^+- and Cl^--permeability of the cell membrane, and produce a decreased excitability of the postsynaptic cell. The synaptic transmission is thus dependent on the sum of the EPSPs and the IPSPs (Miles 1969). An impaired or enhanced release of transmitter from the presynaptic nerve terminals due to presynaptic modulation (Chap. 3.2.4) will also affect the transmission in the synapses.

4.2 Classification of Receptors for Neurotransmitters

After the idea of a receptive substance or receptor became established (Langley 1905, Dale 1906), Dale (1914) showed that the cholinergic receptors (cholinoceptors) could be subdivided into two distinct groups: "nicotinic" and "muscarinic" cholinoceptors. The multiplicity of receptors for the different neurotransmitters is now increasingly evident so that a classification of adrenergic receptors (adrenoceptors), cholinergic receptors (cholinoceptors), purinergic receptors (purinoceptors) and receptors for 5-hydroxytryptamine (serotonergic receptors) can be recognized. The receptors for peptides are (so far) referred to only as VIP-receptors, somatostatin receptors etc.

The classification and subclassification of receptors for neurotransmitters does not in itself say anything about the type of effect caused by receptor stimulation, or the physiological significance of the receptor system in question. Nevertheless the pharmacological classification of receptors for neurotransmitters is a very convenient way of describing the pharmacological properties of a population of receptors in a tissue. The various drugs used in the classification of the receptors are further described in Chap. 5.

4.2.1 Adrenoceptors

In his study of the actions of ergot-alkaloids, Dale (1906) recognized the difference in blocking effects of these compounds on the excitatory compared to the inhibitory responses to adrenaline. For a long time the adrenoceptors were classified simply as excitatory and inhibitory.

The fundamental work of Ahlquist (1948) demonstrated two types of adrenoceptors, without necessarily implying a connection with exclusively inhibitory or excitatory effects. Ahlquist studied the effects of a series of sympathomimetic amines (adrenoceptor agonists) for which he found a consistent order of potency for vasoconstriction, contraction of the uterus and ureters, contraction of the nictitating membrane, dilation of the pupil and inhibition of the gut. Ahlquist also found an entirely different order of potency for the agonistic drugs producing vasodilation, relaxation of the uterus and stimulation of the heart. He called the adrenoceptors α- and β-adrenoceptors respectively, a classification that has withstood well the trial of time (Table 4.1).

Table 4.1. Some examples of the distribution of postsynaptic α- and β-adrenoceptors in different vertebrate organs. + indicates excitatory and − indicates inhibitory effects as defined in Sect. 4.1. More detailed accounts of the distribution and physiological role of the adrenoceptors are given in Chapters 6–13. "Generally" below means that the adrenoceptor has been demonstrated in several, but not necessarily all, vertebrate groups

Tissue	Type of adrenoceptor and effect
Heart (inotropic effects)	β+ (generally) In addition: α+ (mammals)
Heart (chronotropic effects)	β+ (generally) In addition: α− (elasmobranchs)
Blood vessels	α+ (generally) β− (generally)
Uterus (mammals only)	α+ (non-pregnant) β− (pregnant)
Ureters	α+ (mammals) β− (teleosts)
Vas deferens (mammals)	α+
Sphincter of bladder (mammals)	α+
Detrusor of bladder (mammals)	β−
Gut	α−, β− (generally) α+ (sphincters) α+ (some teleosts)
Swimbladder (teleosts)	α+ (secretory mucosa) β− (resorbent mucosa)
Lung wall, bronchi, trachea	β− (generally) α− (toad lung wall)
Iris sphincter	β− (generally) α+ (teleosts)
Iris dilator	α+ (generally)
Spleen	α+ (generally)
Lipolysis, glycolysis	β+ (generally)
Melanophores (amphibians, teleosts)	α+ (mainly) β−
Pilomotor muscles (mammals)	α+
Pennamotor muscles (birds)	α+ (?)

Of the substances used by Ahlquist in his original study, adrenaline, noradrenaline and isoprenaline are currently used for adrenoceptor identification, and the specific α-adrenoceptor agonist phenylephrine has been added to the list. By definition, α-adrenoceptors are stimulated by the amines with the following order of potency (Furchgott 1967):

adrenaline ≥ noradrenaline > phenylephrine > isoprenaline.

The β-adrenoceptors are instead stimulated by the amines with the following order of potency:

isoprenaline > adrenaline > noradrenaline > phenylephrine.

In addition to this basic criterion for identification of adrenoceptors, specific α- and β-adrenoceptor antagonists are now available. The first α-adrenoceptor antagonists used (although not called so at the time), were the ergot alkaloids studied by Dale (1906). Since the synthesis of the first β-adrenoceptor antagonist dichloroisoprenaline (DCI) (Moran and Perkins 1958, Powell and Slater 1958), a great number of specific adrenoceptor antagonists have been synthesized (see also Chap. 5).

A third type of aberrant adrenoceptors, which are probably closely associated with the adrenergic nerve terminals ("junctional adrenoceptors") has been demonstrated in smooth muscle of blood vessels and in the heart of the toad (*Bufo marinus*) (Hirst and Neild 1980, Morris et al. 1982).

A separate receptor system for dopamine has been demonstrated in vascular smooth muscle (Goldberg 1972, Setler et al. 1975, Bell 1982).

4.2.1.1 Subdivision of α- and β-Adrenoceptors

By the same kind of reasoning as that originally used by Ahlqvist (1948), Lands and co-workers were able to show a difference in the potency ratios for a series of β-adrenoceptor agonists when studied in different tissues (Lands et al. 1967a, b). It could be shown that, according to the responsiveness of the tissues to the agonists, the β-adrenoceptors could be put into two separate groups named β_1- and β_2-adrenoceptors (see Table 4.2). As with the α- and β-adrenoceptors, both types may be present within the same organ, and can be distinguished by the use of specific agonists and antagonists (Daly and Levy 1979). In the mammalian heart, a functional difference in the β-adrenoceptor populations has been postulated. According to this hypothesis, the β_1-adrenoceptors are "neural receptors" mainly involved in neuronal cardio-regulation, while the β_2-adrenoceptors, which are relatively less sensitive to noradrenaline, are "humoral receptors" mainly affected by circulating adrenaline (Carlsson et al. 1972, Carlsson and Hedberg 1976). The same hypothesis has been applied to the teleost and amphibian cardiac adrenoceptors. In these animals, adrenaline dominates in the adrenergic neurons while, at least in some amphibians, noradrenaline is present in high concentrations in blood plasma. The demonstration of β_2-adrenoceptors in the heart of trout (*Salmo*) and frogs (*Rana* spp.) is compatible with the β_2-adrenoceptors as the "innervated receptors" in these species, and it was therefore proposed that the β_2-adrenoceptors should be called "adrenaline receptors", as opposed to the β_1-adrenoceptors which are "noradrenaline receptors" (Stene-Larsen and Helle 1978a, Ask et al. 1980, Stene-Larsen 1981).

Table 4.2. Subclassification of mammalian α- and β-adrenoceptors according to Lands et al. (1967a, b), Berthelsen and Pettinger (1977) and Wikberg (1979)

Type of adrenoceptor	Effector tissue
α_1	Heart (positive inotropic effects) Intestinal smooth muscle (contraction and relaxation) Vas deferens (contraction)
α_2	Presynaptic α-adrenoceptors Amphibian melanophores (aggregation)
β_1	Heart (positive chronotropic and inotropic effects) Lipolysis Intestinal smooth muscle (relaxation)
β_2	Heart (positive chronotropic and inotropic effects) Vascular smooth muscle (relaxation) Uterus smooth muscle (relaxation) Bronchial smooth muscle (relaxation)

The clinical value of being able to affect selectively only one or the other type of β-adrenoceptor in cardiac and pulmonary disorders has led to the development of a great variety of specific β_1- and β_2-adrenoceptor agonists and antagonists. The drugs used for the identification of the different sub-types of β-adrenoceptors are β_1-specific agonists, such as prenalterol [(−)-H 80/62] and antagonists such as practolol, and the β_2-specific agonists salbutamol and terbutaline and antagonists butoxamine and H 35/25 (see Stene-Larsen 1981).

Differences in the ratios of effects of agonists at pre- and postsynaptic α-adrenoceptors have also been observed (Starke et al. 1974, 1975a, b, Westfall 1977). Similar to the subdivision of the β-adrenoceptors, the mammalian α-adrenoceptors have now been subdivided into α_1- and α_2-adrenoceptors (Berthelsen and Pettinger 1977, Wikberg 1979). The α_1-adrenoceptors are associated with vascular and intestinal smooth muscle, the vas deferens and the anococcygeus muscle, while the α_2-adrenoceptors are present at the presynaptic nerve terminals of at least adrenergic and cholinergic neurons, postsynaptically in amphibian melanophores and in the central nervous system (Table 4.2, Berthelsen and Pettinger 1977, Wikberg 1979). Specific α_1-adrenoceptor agonists include phenylephrine, methoxamine and amidephrine, with prazosin as a specific antagonist. α_2-Adrenoceptor agonists include methyl-noradrenaline, clonidine and tramazoline, with yohimbine and rauwolscine as specific antagonists (Berthelsen and Pettinger 1977, Wikberg 1979, Davey 1980, Glossman et al. 1980, Starke and Docherty 1980, Flavahan and McGrath 1981, Kobinger and Pichler 1981) (see also Chap. 5).

The original classification of the adrenoceptors into α- and β-categories made by Ahlquist (1948) appears to be applicable to all vertebrate groups studied. There are, however, certain differences in the actions of some of the agonists and antagonists in non-mammalian species. Again this emphazises the necessity of a careful pharmacological elucidation of putative chemical tools when applied in new species (or even organs). Examples of deviations from the "standard mammalian picture" include the high affinity and intrinsic activity of isoprenaline, a β-adrenoceptors agonist, on teleost vasculature and toad lung and spleen α-adrenoceptors (Holmgren and Campbell 1978, Nilsson 1978, Wahlqvist and Nilsson 1981), the antagonistic effect of clonidine on α-adrenoceptors of fish (Johansson 1979) and α-adrenoceptor antagonistic properties in fish arteries of the β-adrenoceptor antagonist propranolol (Holmgren and Nilsson 1974).

There is some evidence that the β-adrenoceptor of the hearts of trout (*Salmo*) and frog (*Rana*) belong to the β_2-category (Stene-Larsen and Helle 1978a, Ask et al. 1980). With these few exceptions, attempts to subclassify the α- and β-adrenoceptors in lower vertebrates have not been made with sufficient pharmacological insight to conclude whether a subdivision of the adrenoceptors into α_1, α_2, β_1 and β_2-adrenoceptors is justified in non-mammalian vertebrates.

4.2.1.2 The Adrenoceptor Interconversion Hypothesis

The concentration ratio of functioning α- to β-adrenoceptors in a tissue varies according to the state of the tissue; the influence of endocrine factors may for instance produce changes in the receptor populations (Dale 1906, Miller 1967, Triggle and

Moran 1979). The possibility that this variation is due to the presence of *one single type* of adrenoceptor which changes its configuration between an α- and a β-state according to the external influence was proposed by Kunos and Szentivanyi (1968). Their experiments showed a decrease in the antagonistic potency of β-adrenoceptor antagonists in the frog heart when the temperature was lowered from 22°–24 °C to 5°–10 °C, and in the rat heart below 24°–27 °C. The hypothesis that a single type of adrenoceptor in the heart of frogs and mammals is altered from a β-type at higher temperatures to an α-type at lower temperatures was later formulated by Kunos et al. (1973) and Kunos and Nickerson (1976, 1977).

The experiments have been criticized for poor experimental design, lack of pharmacological precision and badly controlled conditions of radio-ligand studies (Stene-Larsen and Helle 1978 b, Benfey 1979 a, b). At the moment, therefore, it appears that the evidence in favour of an inter-convertible one-receptor system is scanty, and in need of new careful experimentation (Benfey 1979 a, b).

4.2.2 Cholinoceptors

From the work of Dale (1914) stems the concept of "muscarinic" and "nicotinic" receptors for acetylcholine in the autonomic neurotransmission. The nicotinic cholinoceptors participate in the cholinergic neurotransmission at the nerve-nerve synapses in all autonomic ganglia, and at the nerve-chromaffin cell synapses in chromaffin tissue. Nicotinic cholinoceptors are also present along the axon and in the nerve terminals of autonomic neurons (Westfall 1977).

The nicotinic cholinoceptors are characterized by the agonistic action of nicotine and DMPP (1,1-dimethyl-4-piperazinium) and antagonistic effects of excessive nicotine and ganglionic blocking agents such as hexamethonium (Chap. 5.3).

The muscarinic cholinoceptors are stimulated by muscarine and pilocarpine, and very specifically and competitively blocked by atropine and hyoscine. Muscarinic cholinoceptors are found in the effector organs (smooth muscle, heart, glands), and at the presynaptic nerve terminals of at least cholinergic, adrenergic and VIP neurons. The effect of muscarinic stimulation is often excitatory (contraction of smooth muscle, increased glandular activity etc.) but may also be inhibitory (inhibition of heart rate and force, dilation of certain mammalian blood vessels and gastrointestinal sphincters, presynaptic inhibition of transmitter release) (Koelle 1970a, Carrier 1972, Muscholl 1979). A careful pharmacological analysis, including receptor-ligand studies, has demonstrated two sub-populations of muscarinic cholinoceptors in some mammalian tissues (Vanhoutte 1977, Hulme et al. 1981).

4.2.3 Purinoceptors

A classification of the receptors for purine derivatives was proposed by Burnstock (1976a, 1978a, 1979). According to this classification P_1-purinoceptors are stimulated with the following order of potency of the agonists:

adenosine $\geq$ AMP $>$ ADP $\geq$ ATP.

The P_1-purinoceptors are blocked by the methylxanthines (such as caffeine and theophylline). They affect the production of cyclic, 3',5'-adenosine monophosphate (c-AMP), but do not induce synthesis of prostaglandins.

The P_2-purinoceptors, on the other hand, are stimulated by the agonists with the following order of potency:

$$ATP \geq ADP > AMP \geq adenosine.$$

The P_2-purinoceptors are blocked by high concentrations of quinidine, by imidazolines (such as phentolamine, which is also a widely used α-adrenoceptor antagonist) and 2,2'-pyridylisatogen. They do not affect the production of c-AMP, but induce prostaglandin synthesis (Burnstock 1979, Coleman 1980). The two types of purinoceptors are found at different locations: P_1-purinoceptors are thought to mediate the presynaptic effects of adenosine, while the postsynaptic effects of ATP on smooth muscle (e.g., inhibition of the guinea-pig taenia coli) are mediated by P_2-purinoceptors (Brown and Burnstock 1981, Burnstock 1981). Burnstock and Meghji (1981) concluded that P_1-purinoceptors are present in the atria of the guinea-pig heart, but that purinoceptors are absent in the ventricle. In the heart of the frog (*Rana pipiens*), both P_1- and P_2-purinoceptors could be demonstrated.

4.2.4 Receptors for 5-Hydroxytryptamine (Serotonergic Receptors)

A differentiation of receptors for 5-hydroxytryptamine (5-HT) into "M-receptors" and "D-receptors" was originally made to explain the effects of 5-HT on mammalian gut preparations (Gaddum and Picarelli 1957, Born 1970). Recognition of the sites of action of 5-HT has now led to the distinction of "neural 5-HT receptors" and "muscular 5-HT receptors" (Gyermek 1966, Gershon 1981). The specificity of the 5-HT antagonists is low, and classification is therefore difficult. It appears that the neural 5-HT receptors are insensitive to tryptamine, but affected by phenylbiguanide (at least in some mammalian species), while the muscular excitatory 5-HT receptors are stimulated by tryptamine, blocked by methysergide in selective concentrations, but insensitive to phenylbiguanide (Born 1970, Costa and Furness 1979a, b, Gershon 1981).

Chapter 5

Chemical Tools

The experimental approach to the study of autonomic nerve function in mammals and other vertebrates involves the use of biologically active substances, "chemical tools", which exert a specific action to interfere with the neurotransmission at some level. Thus, there are compounds available that affect adrenergic transmission by interfering with synthesis, storage, release, uptake, degradation or at pre- or postsynaptic receptor levels (adrenoceptor agonists and antagonists). There is also a wide variety of compounds useful for the study of cholinergic neurotransmission, but there are few drugs indeed available with selective effects on the different types of non-adrenergic, non-cholinergic (NANC) transmission.

Many of the chemical tools in use for elucidation of autonomic nerve functions are plant and animal toxins (e.g., ergotamine, atropine, cocaine, tetrodotoxin). Modern pharmacology owes much to the elaborate and ingenious unravelling of the effects of these compounds by physiologists and phamacologists around the turn of this century. Despite these early contributions to our understanding of drug actions, it is of crucial importance that the selectivity of a chemical tool is confirmed by careful pharmacological analysis (see Chap. 4). As with other types of tools, uncritical use may create severe side effects that lead to erroneous conclusions about the mechanisms studied. Particularly important is the critical evaluation of drugs when applied to new animal groups or species.

The use of pharmacologically active substances in vivo creates additional problems, since the concentration at the site of action cannot be estimated and the duration of the drug action may be unknown. Other possible complications in vivo are direct effects on the central nervous system by drugs that penetrate the blood-brain barrier. Triggering of reflexes that affect the mechanism studied primarily may also interfere with the results.

The following account of chemical tools relevant to the study of autonomic nerve function is by no means a complete list of the drugs available. The drugs chosen for this survey have been and/or are currently used as aids in the elucidation of both mammalian and non-mammalian nerve function. Some new substances which may prove valuable for future studies on non-mammalian autonomic nerve functions have also been included. Knowledge of the drug effects stems mainly from work on mammals, and it is only in very few cases in which selectivity in non-mammalian systems has been established. Some examples of *known* differences in effect between mammalian and non-mammalian application have been included.

For extensive treatment of mammalian autonomic pharmacology see Goodman and Gilman (1970 and later editions), Carrier (1972) and Day (1979).

5.1 Drugs That Affect Ion Permeability of Cell Membranes

A specific inhibition of neuronal conduction can be achieved by the puffer fish poison, tetrodotoxin (TTX). TTX acts by specifically blocking the Na^+-channels of the nerve fibres, and can thus be used to create "nerve-free" preparations (Narahashi et al. 1964, Gershon 1967). The effector organ (smooth muscle) retains its functions due to an exchange of Ca/K ions in the membranes. A similar ion exchange system possibly occurs also in some neurons, e.g., the cell bodies of the so called type 2 (AH) neurons in the enteric nervous system which are resistant to TTX (North 1973, Gershon 1981). A high resistance to TTX is also shown by neurons of the puffer fish itself (*Tetraodon* and *Sphaeroides* spp.), and the Californian newt (*Taricha torosa*) in which TTX also occurs (Evans 1972).

Apamin, a polypeptide of 18 amino acids from bee venom, has been shown to block the inhibitory effects of certain substances on the smooth muscles of the gut (Vladimirova and Shuba 1978, Jenkinson 1981). This effect of apamin is due to a specific blockade of the K^+-channels in the cell membranes (cf. Chap. 4.1.10). Apamin thus non-competitively blocks the effect of substances that act by increasing the K^+-conductance of the smooth muscle cell membrane, e.g., the inhibitory response of the guinea-pig taenia coli to α-(but not β-)adrenoceptor stimulation (Jenkinson 1981), P_2-(but not P_1-)purinoceptor stimulation (Brown and Burnstock 1981) and neurotensin (Kitabgi and Vincent 1981).

5.2 Drugs That Affect Adrenergic Transmission

There are chemical tools available that interfere with many steps in the adrenergic transmission. In older literature the term *sympathicomimetic* was used to describe a substance with effects mimicking those of the adrenergic neurons, whereas the terms *sympathicolytic* was used for an agent which blocked the adrenergic transmission. Agents that interfere with catecholamine synthesis, storage and degradation and with adrenoceptors are also useful tools in the study of the functions of humoral (circulating) catecholamines released from the chromaffin tissue.

5.2.1 Synthesis

All steps in the biosynthesis of adrenaline can be specifically inhibited to manipulate the rate of synthesis (Chap. 3.2.6.1). The first enzymatic step in the "Blaschko pathway" catalyzed by tyrosine hydroxylase can be inhibited by a number of compounds similar to L-tyrosine in structure. The most efficient and widely used of these compounds are 3-monoiodo-L-tyrosine (Fig. 5.1), 3,5-diiodo-L-tyrosine and 3-iodo-α-methyl-L-tyrosine (Spector et al. 1965, Udenfriend et al. 1965). Administration of these drugs in vivo is a useful way to empty the stores of catecholamines in different tissues.

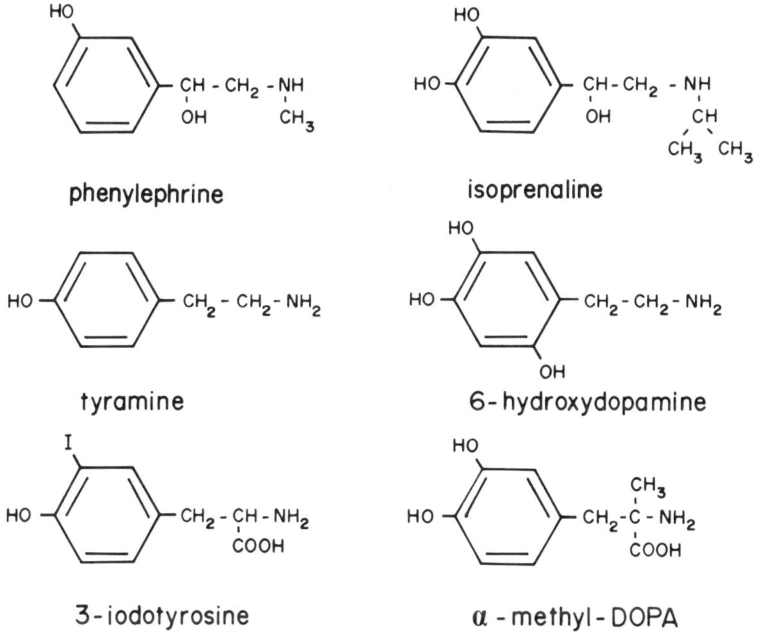

Fig. 5.1 phenylephrine, isoprenaline, tyramine, 6-hydroxydopamine, 3-iodotyrosine, α-methyl-DOPA

The second step in the catecholamine biosynthesis is the decarboxylation of DOPA to form dopamine. The enzyme responsible for this step, aromatic L-amino-acid decarboxylase (AAD), can be blocked both in vivo and in vitro by two compounds, carbidopa (MK486) and benserazid (Ro4-4602). Carbidopa can be used in vivo specifically to inhibit AAD of peripheral tissues of mammals and amphibians, while in fish there is also a slight inhibition of AAD in the central nervous system. Benserazide inhibits both peripheral and central AAD in all three groups (Henning 1969, Wurtman and Watkins 1977, Henning and Johansson 1981, Johansson and Henning 1981). A new compound that may prove to be a valuable tool as an AAD inhibitor is α-monofluoromethyldopa, which is a potent in vivo inhibitor of both peripheral and central AAD, and efficiently reduces the peripheral catecholamine stores (Fozard et al. 1980, Jung et al. 1980).

A marked, if not total (in vivo), suppression of the enzymatic activity of dopamine-β-hydroxylase (DBH) can be achieved by disulfiram (tetraethylthiuram disulphide). As a tool to inhibit catecholamine synthesis in vivo, this substance is however of limited value compared with the AAD and, especially, with the TH inhibitors (van der Schoot and Creveling 1964).

The final step in the synthesis of adrenaline, the methylation of noradrenaline by PNMT (phenylethanolamine-N-methyl transferase), can be inhibited by two specific compounds: SK & F 64139 and SK & F 29661 (Pendleton et al. 1976, 1980). The first of these compounds also produces a blockade of PNMT in a teleost (Fig. 5.2, Abrahamsson 1980).

In addition to the drugs that inhibit the individual steps in catecholamine formation, α-methylated precursors for the catecholamines (α-methyltyrosine; α-

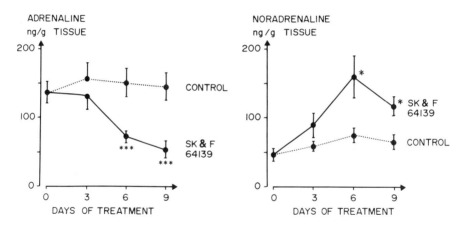

Fig. 5.2. Effect of daily intramuscular injection of the PNMT-inhibitor SK & F 64139 on the adrenaline and noradrenaline content in the spleen of the cod, *Gadus morhua*. Each value is the mean for 10–16 animals. Significance levels (SK & F 64139-treated animals compared to saline treated controls): *$p<0.05$, ***$p<0.002$. (Reproduced with permission from Abrahamsson 1980)

methylDOPA, Fig. 5.1; α-methyldopamine) have been used to impair formation of noradrenaline and adrenaline. The drugs act as competitive substrates for the enzymes, and are metabolized to "false transmitters" with lower agonistic activity than the naturally occurring catecholamines (Spector et al. 1965, Kopin 1968, Muscholl 1972). The accumulation of α-methylnoradrenaline by the neuronal uptake mechanism has been used as a specific marker for adrenergic neurons studied by fluorescence histochemistry (Read and Burnstock 1969b).

5.2.2 Storage

Reserpine (Fig. 5.3), which is obtained from the roots of the Indian plant *Rauwolfia serpentina*, has been shown to block specifically the uptake mechanism of the adrenergic storage granules (von Euler and Lishajko 1963, Carlsson 1965, 1966, Iversen 1967, Nickerson 1970, Burnstock and Costa 1975). The drug has been widely used to deplete adrenergic neuronal catecholamine stores and is also effective in reducing the catecholamine content of chromaffin tissue. Higher doses and prolonged treatment are necessary to produce depletion of the chromaffin stores both in mammals and in teleosts (Lee 1967, Grove et al. 1972). It should be noticed that adrenergic nerve function may persist until very low levels of catecholamines in the neurons have been reached by the reserpine treatment (Carlsson 1965, 1966, Andén and Henning 1966, Antonaccio and Smith 1974, Burnstock and Costa 1975).

5.2.3 Chemical Sympathectomy

Destruction of the adrenergic nerve terminals ("sympathectomy") can be achieved in vivo by injections of 6-hydroxydopamine (6-OH-DA; Fig. 5.1). The drug is ac-

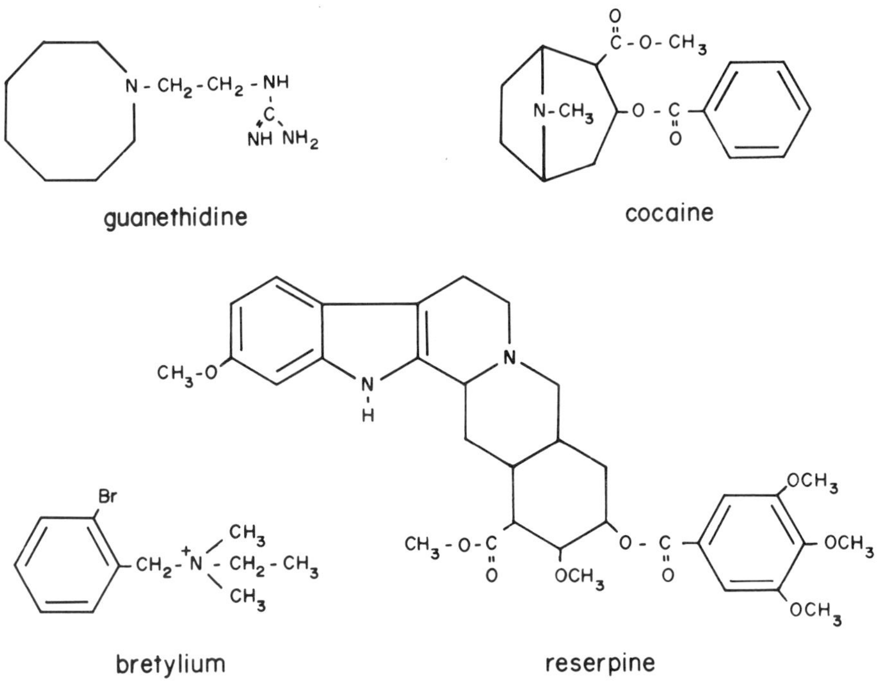

Fig. 5.3

guanethidine   cocaine   bretylium   reserpine

cumulated specifically in the adrenergic nerve terminals by the neuronal uptake mechanism ("uptake$_1$") after which the non-specific cytotoxic effect of the oxidation products of 6-OH-DA destroys the nerve terminals. The accumulation of 6-OH-DA in the axon and cell body is not sufficient to produce destruction of these parts of the neuron, and regeneration of the nerve terminals takes place after cessation of the 6-OH-DA treatment (Malmfors and Sachs 1968, Thoenen and Tranzer 1968, Tranzer and Thoenen 1968, Thoenen 1972, Finch et al. 1973). 6-OH-DA does not affect chromaffin tissue directly but a compensatory increase in the adrenomedullary release of catecholamines has been demonstrated in dogs (Porlier et al. 1977).

The storage of catecholamines is also impaired by prolonged administration of guanethidine (see Sect. 5.2.4), which exerts a neurotoxic action specifically in adrenergic neurons (Burnstock et al. 1971, Heath and Burnstock 1977).

5.2.4 Release

There are two groups of drugs that specifically affect release of catecholamine from the adrenergic nerve terminals. The first group, often referred to as "indirectly acting amines", produce release of the stored catecholamines; the second group, the adrenergic neuron blocking agents, prevent release of transmitter from the nerve terminal. In addition to these two groups of specific drugs, nicotinic cholinoceptor agonists and antagonists (see Sect. 5.3.3 and 5.3.4) will induce or prevent release of catecholamines in vivo, by affecting ganglionic transmission. The regulatory action of a wide variety of substances at the presynaptic level should also be considered (see Chap. 3.2.4).

The indirectly acting amines include tyramine (Fig. 5.1), amphetamine and ephedrine. The effect of these substances is exerted by displacement of the stored catecholamines from the adrenergic nerve terminals by a process different from the exocytotic release induced by nerve stimulation or nicotinic cholinoceptor agonists (see Chap. 3.2.2). It should be pointed out that the term "indirectly acting amine" is not absolute. Ideally, this group of substances have no direct agonistic effect on the adrenoceptors, but there is a gradual transition from a purely indirect effect (tyramine) via mixed effects (ephedrine, phenylephrine) to purely direct adrenoceptor agonistic effects (adrenaline, noradrenaline) (Trendelenburg 1972). Tyramine and related compounds are very useful chemical tools in the demonstration of the presence of a functional adrenergic innervation of an organ. After "sympathectomy" (surgical or chemical) or reserpine treatment no effect of these drugs persists.

The adrenergic neuron blocking agents bretylium and guanethidine (Fig. 5.3) prevent release of catecholamines from the adrenergic nerve terminals during nerve activity. The exact details of their mode of action are not known, but their specific action at adrenergic neurons may at least partly depend on a specific accumulation in these neurons (Boura et al. 1961, Boura and Greene 1965, Nickerson 1970). Prolonged treatment with guanethidine will also produce a reserpine-like depletion of catecholamines and retraction of the neurons from the effector cells (Burnstock et al. 1971, Evans et al. 1972, Health et al. 1973).

5.2.5 Neuronal Uptake

The classical drug used to inhibit the neuronal uptake of catecholamines ("uptake$_1$" is cocaine (Fig. 5.3), which is obtained from the leaves of South American bushes (*Erythroxylon coca, Erythroxylon* spp.). Cocaine shows a variety of pharmacological actions, the most significant being a local anaesthetic effect and the inhibition of the neuronal uptake mechanism for catecholamines (Trendelenburg 1959, 1966, Whitby et al. 1960, Muscholl 1961).

An example of a newer specific drug that blocks the neuronal uptake of catecholamines is desmethylimipramine (DMI; desipramine) which is effective in a lower concentration than cocaine. Possible side effects of this drug include α-adrenoceptor antagonistic properties (Ursillo and Jacobsen 1965, Iversen 1967).

In addition to the blockade of neuronal uptake ("uptake$_1$"), the extraneuronal uptake ("uptake$_2$") is also susceptible to blockade by compounds such as metanephrine and corticosterone (Iversen 1967, 1974, Stene-Larsen 1981).

5.2.6 Degradation

The oxidation of catecholamines by monoaminoxidase can be inhibited by several compounds including pargyline, nialamide and tranylcypromine. An important use of these compounds is in the histochemical demonstration of catecholamines, where they can be used to increase the fluorescence in the adrenergic neurons by inhibiting the degradation of intraneuronal catecholamines (Iversen 1967, Jarvik 1970).

5.2.7 Adrenoceptor Agonists

The naturally occurring adrenoceptor agonists are the catecholamines adrenaline and noradrenaline. Both catecholamines are present in the chromaffin tissue of all vertebrates so far studied and adrenaline is also stored as a transmitter substance together with noradrenaline in some groups (teleosts, amphibians). The name of these substances is derived from their presence in the adrenals (Latin: *ad*, near; *renes*, kidney) (Takamine 1901), in which adrenaline activity was demonstrated by Oliver and Schäfer (1895). The prefix *nor-* in noradrenaline stands for the German *N*itrogen *O*hne *R*adikal (nitrogen without radical, cf. Fig. 3.2). In American terminology the terms epinephrine (Abel 1899, 1901, Aldrich 1901) and norepinephrine (arterenol) have been retained (Greek: *epi*, on; *nephros*, kidney). Both adrenaline and noradrenaline have α- and β-adrenoceptor agonist properties.

The subclassification of adrenoceptors makes use of specific α- or β-adrenoceptor agonists, such as phenylephrine and methoxamine (α-adrenoceptor agonists) and isoprenaline (isoproterenol; Fig. 5.1) (β-adrenoceptor agonist; see also Chap. 4.2.1). The further classification of adrenoceptors is based on the comparison of agonist and antagonist properties of several drugs (Chap. 4.2.1.1).

5.2.8 Adrenoceptor Antagonists

The first experiments with a specific blockade of the excitatory effects of adrenaline were performed by Dale (1906), using ergotoxin as the antagonist. Ergotoxin is obtained from a fungus, *Claviceps purpurea*, which grows on rye and other grains in damp conditions. The toxin was later shown to be a mixture of several alkaloids, including mainly ergokryptine, ergocristine and ergocornine. Stoll (1950) was the first to isolate ergot alkaloids in pure form (ergotamine). Dihydrogenated derivatives of the naturally occurring ergot alkaloids (dihydroergotamine etc.) show greater α-adrenoceptor blocking activity and less direct smooth muscle stimulating effects (Nickerson 1970).

Other competitive α-adrenoceptor antagonists are the 2-substituted imidazoline derivatives phentolamine and tolazoline, of which the first has been widely used experimentally as a specific α-adrenoceptor antagonist (Fig. 5.4). Another naturally occurring alkaloid that has been used experimentally is yohimbine (Fig. 5.4), an antagonist, which in the new sub-classification of α-adrenoceptors, shows some preference for the α_2-type adrenoceptor (Chap. 4.2.1.1). Yohimbine is a compound obtained from the bark of a West African tree, *Corynanthe johimbe*.

The most widely used non-competitive α-adrenoceptor antagonist is phenoxybenzamine (Dibenzyline). As with the related compound, dibenamine, the action is due to the formation of a strong "non-equilibrium" bond with the adrenoceptor

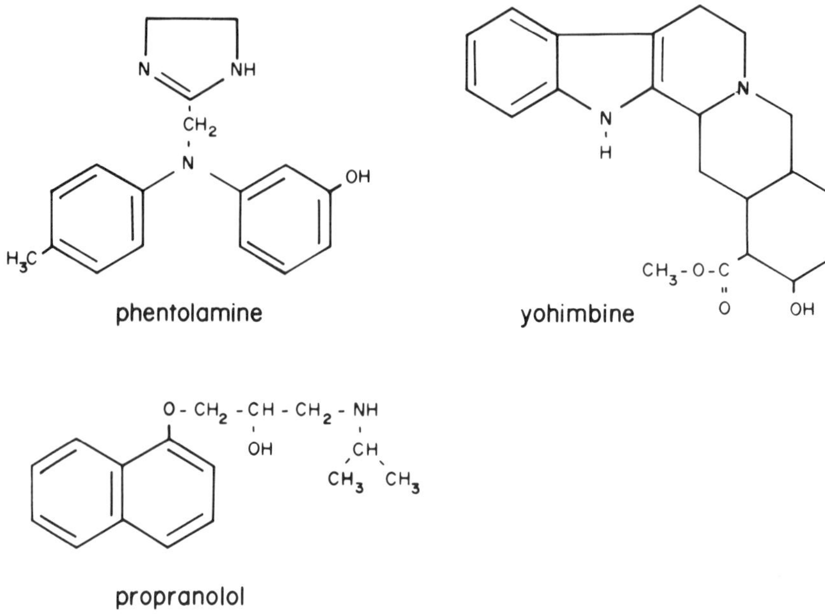

Fig. 5.4

(Nickerson 1970). The formation of the "non-equilibrium" bond is a relatively slow process, and therefore not only the concentration of phenoxybenzamine, but also the time allowed for "incubation" of the tissue must be considered when comparing the antagonistic properties of this drug. Several potent side effects of phenoxybenzamine, including cholinergic, serotonergic and histaminergic blockade, limits its use as a selective tool (Boyd et al. 1963, Nickerson 1970, Carrier 1972, Day 1979).

To achieve specific blockade of β-adrenoceptors, propranolol (Fig. 5.4) has been used in numerous studies of both mammals and non-mammalian vertebrates. The time required to obtain full antagonistic activity of propranolol is in the order of one hour or more (Potter 1967, Furchgott et al. 1973) and the possibility of unwanted side effects in the form of local anaesthetic actions of the drug should be kept in mind. The related β-adrenoceptor antagonist sotalol lacks the local anaesthetic effects (Åberg et al. 1969, Nickerson 1970).

Some of the compounds which are used in the subclassification of β-adrenoceptors into β_1- and β_2-adrenoceptors are summarized in Chap. 4.2.1.1.

5.3 Drugs That Affect Cholinergic Transmission

In analogy with the adrenergic system, the drugs affecting the cholinergic system have in the older literature been referred to as *parasympathicomimetics* (for the ag-

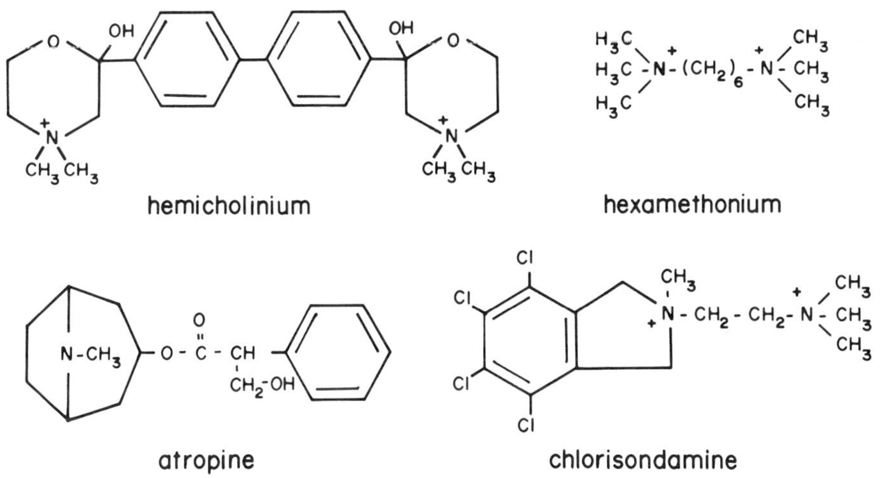

Fig. 5.5

onist drugs) and *parasympathicolytics* (for the antagonist drugs). In the elucidation of cholinergic pathways in vertebrates, the main part of the experimental evidence comes from the use of the specific cholinoceptor antagonists and cholinesterase blocking agents. The chemical tools available to manipulate the cholinergic transmission at other levels is therefore only briefly summarized below.

5.3.1 Release and Neuronal Uptake

The release of acetylcholine from the cholinergic nerve terminals in both the nerve-muscle synapses in skeletal muscle and in autonomic cholinergic synapses can be inhibited by botulinum toxin. This toxin, formed by the anaerobic bacterium *Clostridium botulinum*, exists in several forms (A–D). Treatment with the toxin inhibits the release of acetylcholine from cholinergic nerve endings (Ambache 1951, Ambache and Lessin 1955) and in some cases the release of noradrenaline from adrenergic fibres is also inhibited. Since the effect of botulinum toxin has been considered specifically anticholinergic, the effects on adrenergic transmission have been interpreted in favour of the "cholinergic link hypothesis" (Burn and Rand 1965, Ferry 1966, Burn 1977a, b) (see Chap. 3.2.5).

Another drug that impairs cholinergic transmission, at least in some preparations, is hemicholinium (HC-3; Fig. 5.5). This substance is believed to compete with the neuronal uptake of choline into the cholinergic nerve terminal and thus prevent the synthesis of acetylcholine (MacIntosh et al. 1956, Scheuler 1960, Birks and MacIntosh 1961, Carrier 1972). Hemicholinium has also been shown to impair adrenergic nerve function in some preparations, which again has been interpreted in favour of the "cholinergic link hypothesis" (Burn and Rand 1965, Ferry 1966, Burn 1977a, b).

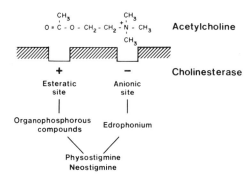

Fig. 5.6. Sites of interaction between the acetylcholine molecule and cholinesterase. The site of blockade by organophosphorous compounds and edrophonium respectively are shown. Physostigmine and neostigmine bind to both the esteratic and the anionic site

5.3.2 Degradation

Acetylcholine released from the autonomic nerve terminals is not taken up again as such into the nerve terminal. Instead it is rapidly hydrolysed by cholinesterase present at the synaptic cleft. The enzyme occurs in two forms: acetylcholinesterase (specific cholinesterase), which is mainly associated with the neuron, and pseudocholinesterase, which is found in non-neuronal tissue and blood plasma (see Chap. 3.2.7). The two forms of cholinesterase can be distinguished by the aid of specific inhibitors: BW 284 c51 [15-bis-(4-allyldimethylammoniumphenyl)-pentan-3-one dibromide] inhibits (true) acetylcholinesterase, while pseudocholinesterase can be blocked for example by Mipafox (Holmstedt 1957a, b).

It has been argued that the presence of pseudocholinesterase is a late development phylogenetically, since many organs of the lower vertebrates only show the presence of true cholinesterase (Sippel 1955, Bell and Burnstock 1965, Bell 1967, Burnstock 1969). A histochemical differentiation of the two types could, however, be made in a teleost (*Gadus morhua*) (Nilsson 1976, McLean and Nilsson 1981).

The interaction between the acetylcholine molecule and the active site of the cholinesterase molecule takes place at two points: the esteratic site and the anionic site (Fig. 5.6). The anionic site "fixes" the acetylcholine molecule by electrostatic binding, and the hydrolysis takes place at the esteratic site. The enzyme can be inhibited by the binding of inhibitors to either or both of these sites (Fig. 5.6).

Both forms of cholinesterase are susceptible to blockade by the "reversible" inhibitors physostigmine (eserine) (Fig. 5.7) which is derived from the Calabar bean (*Physostigma venenosum*), and the slightly more efficient synthetic compound neostigmine (Prostigmine) (Koelle 1970b, Carrier 1972, Day 1979). Both compounds block the effect of the cholinesterase by competing with acetylcholine for the active site of the enzyme (both esteratic and anionic sites). Another "reversible" anticholinesterase is edrophonium, which acts by binding only to the anionic site of the cholinesterase molecule (Fig. 5.6, Koelle 1970b, Carrier 1972).

5.3.3 Cholinoceptor Agonists

The naturally occurring transmitter substance in cholinergic synapses is, of course, acetylcholine (Fig. 3.5). The main disadvantage with this compound, when applied

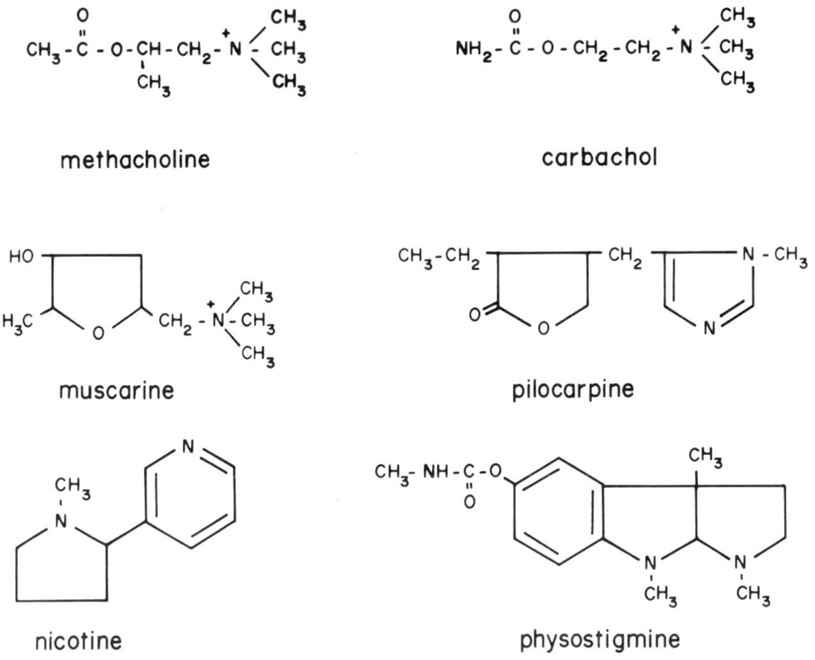

Fig. 5.7 methacholine, carbachol, muscarine, pilocarpine, nicotine, physostigmine

exogenously in the experimental situation, is its rapid degradation by cholinesterases. The concentration of acetylcholine actually reaching the receptors is therefore unknown, and the formation of choline, which is a partial agonist, introduces a further complication.

Cholinoceptor agonists with much lower susceptibility to the action of the cholinesterases are carbachol, which affects both muscarinic and nicotinic cholinoceptors, and methacholine and bethanecol with chiefly muscarinic activity (Fig. 5.7) (Koelle 1970c, Carrier 1972, Day 1979).

The muscarinic agonists also include some naturally occurring plant alkaloids, such as muscarine itself (Fig. 5.7). This drug is derived from the toadstool, *Amanita muscaria*, which was once added to the mead to produce the "berserk-inducing" drink of the Vikings. It is possible, however, that the hallucinogenic properties of the fungus are chiefly due to the related compound, ibotinic acid (Koelle 1970c). Other muscarinic agonists are pilocarpine (Fig. 5.7), which is a constituent of the leaves of South American species of *Pilocarpus*, and arecoline which is present in areca (betel) nuts (*Areca catechu*) (Koelle 1970c).

Although the muscarinic agonists are used experimentally to demonstrate muscarinic cholinoceptors, the chief argument for the present of this type of receptor usually comes from the use of antagonists like atropine and related drugs (see below).

Stimulation of nicotinic cholinoceptors can be made by nicotine (Fig. 5.7), but during prolonged or repeated treatment nicotine acts as an antagonist by producing sustained depolarization of the postsynaptic cell membrane. This drug is derived from the tobacco plant, *Nicotiana tabacum*, and the pure alkaloid is a liquid at room temperature. Other nicotinic cholinoceptor agonists are lobeline, an alkaloid from *Lobelia* species, tetramethylammonium and DMPP (1,1-dimethyl-4-phenylpiperazinium) (Volle and Koelle 1970).

5.3.4 Cholinoceptor Antagonists

The demonstration of muscarinic cholinoceptors is usually based on the antagonistic properties of alkaloids derived from members of the potato family (Solanaceae). The first of these "Belladonna alkaloids" is atropine, derived from the deadly nightshade, *Atropa belladonna*. The Latin name of the plant comes from *Atropos*, one of the Fates in Greek mythology who cut the thread of life, and the Italian *belladonna* (beautiful lady), which refers to the mydriatic effect of the drug on the iris (see Chap. 12). Atropine is the racemic mixture of d- and l-hyoscyamine, the antimuscarinic properties of the drug being due chiefly to the l-isomer (Fig. 5.5).

A second muscarinic cholinoceptor antagonist of great specificity is hyoscine, named after the henbane, *Hyoscyamus niger*. The name "henbane", death of hens, is supposed to originate from the use by thieves of the plant seeds to kill hens and chickens. By burning the seeds a thick poisonous smoke is produced (the Swedish name for the plant, "bolmört", means "smoke herb") which, when introduced by chicken-thieves into poultry houses secured the silent removal of the inhabitants. Hyoscine, like atropine, is a racemic mixture, in which l-hyoscine (also known as scopolamine) is the active principle.

Both atropine and hyoscine are present in *Atropa* and *Hyoscyamus* species, and also in the thorn apple, *Datura stramonium*. The drugs can pass the blood-brain barrier and induce several central nervous effects. The hallucinogenic properties of the drugs were once used in "witch's salve", and are reputed to produce a sense of weightlessness ("flying on a broom-stick").

In investigations of functions of the autonomic nervous system, atropine and hyoscine have been used in a wide variety of vertebrates in which they show a remarkably high specificity for the muscarinic cholinoceptors (Innes and Nickerson 1970, Carrier 1972, Day 1979).

Nicotinic antagonists, often referred to as ganglionic blocking agents, are of two types: depolarizing and non-depolarizing compounds. The first group consists of nicotine itself and also lobeline, but these drugs now have limited use as antinicotinic agents.

The non-depolarizing drugs can again be subdivided into two groups with reference to the structure of the molecules. The first "sub-group" consists of quaternary ammonium derivatives, which include hexamethonium, tetraethylammonium (TEA) and chlorisondamine (Fig. 5.5). Hexamethonium shows the highest ganglionic blocking capacity of a series of bisonium compounds with a polymethylene chain of varying length between the two quaternary ammonium groups (Paton and Zaimis 1949). The cholinoceptors of the neuromuscular junction in skeletal muscle, on the other hand, are most effectively blocked by decamethonium, which has ten methylene groups in the chain (Carrier 1972).

The second "subgroup" of the non-depolarizing ganglionic blocking agents, the "non-quaternary" compounds, include mecamylamine and pempidine (Stone et al. 1956, Carrier 1972). Both have been valuable tools in the study of ganglionic transmission in the vertebrates. In several cases curare (d-tubocurarine), which is normally regarded as a neuromuscular blocking agent, has been used successfully as a ganglionic blocking agent. It should be remembered that these ganglionic blocking agents inhibit the cholinergic transmission in all autonomic ganglia, i.e., also when the postganglionic neuron is non-cholinergic.

The selectivity of these agents in their capacity to block nicotinic cholinoceptors has not been extensively studied in vertebrates other than mammals. In one case, the stomach of *Pleuronectes*, a potent antimuscarinic action of hexamethonium has been described (Edwards 1972b, Stevenson and Grove 1977).

5.4 Drugs That Affect Purinergic Transmission

Compared to the great number of specific drugs that can be used in the study of adrenergic and cholinergic neurotransmission, there are relatively few drugs available that can be used to manipulate non-adrenergic, non-cholinergic transmission. Some substances have been tested and shown to have some degree of selectivity for the purinergic type of transmission but, as pointed out by Burnstock (1981): "Development of a specific competitive antagonist for ATP and purinergic transmission is of high priority and would allow a decisive test of this hypothesis".

5.4.1 Uptake and Degradation

The neuronal uptake of adenosine is inhibited by dipyridamole, a substance that has been shown to potentiate the effect of both ATP and NANC nerve stimulation in the taenia coli of the guinea-pig, while the response to adrenergic nerve stimulation is instead reduced (Satchell et al. 1972, Burnstock 1979).

An enhanced degradation of ATP by incubation with the enzyme nucleotide phosphatase reduces the effect of exogenous ATP and stimulation of the NANC inhibitory fibres of the taenia coli. The treatment does not affect the response to noradrenaline or stimulation of the adrenergic nerves (Satchell 1981).

5.4.2 Purinoceptor Agonists and Antagonists

The purinoceptors are stimulated by ATP (Fig. 3.6) and analogues ADP, AMP and adenosine, with the order of potency depending on the type of purinoceptor (P_1 or P_2; see Chap. 4.2.3).

ATP analogues have been synthesized in which one of the ester oxygens in the triphosphate chain has been replaced by a methylene group. Of these compounds, homo-ATP (6'-homo-adenosine 6'-pyrophosphorylphosphonate; ACPOPOP) was shown to have stronger agonistic effect on the guinea-pig taenia coli than ATP itself (Maguire and Satchell 1979). The effect of homo-ATP or the α,β-methylene derivative (adenosine 5'-α,β-methylenetriphosphate; AOPCPOP) is not enhanced by dipyridamole, which was taken as evidence that these compounds are not rapidly degraded to adenosine (Maguire and Satchell 1979).

Selective purinoceptor antagonists are still wanting but some drugs can be used to block the purinoceptors, provided attention is paid to the possible side effects. The antagonists include imidazoline derivatives such as antazoline and phentolamine, which are also a histaminic antagonist and an α-adrenoceptor antagonist, respectively, and 2,2'-pyridylisatogen (Satchell et al. 1973, Hooper et al. 1974, Burnstock 1979, 1981, Coleman 1980). In addition, the P_1-purinoceptors are blocked by methylxanthines, which antagonize the effects of adenosine (Burnstock 1979, Coleman 1980) (Chap. 4.2.3). A promising non-competitive antagonist for the P_2-purinoceptor-mediated inhibition of the guinea-pig taenia coli is apamin (Sect. 5.1), which acts by a general blockade of K^+-conductance (Banks et al. 1979, MacKenzie and Burnstock 1980, Brown and Burnstock 1981, Burnstock 1981, Jenkinson 1981).

5.5 Drugs That Affect Serotonergic Transmission

Again there are very few selective drugs that can be used to manipulate the serotonergic transmission. Early attempts to block M-receptors (with morphine) and D-receptors (with dibenamine and phenoxybenzamine) are now known to be very nonselective. Although the most convincing demonstration of 5-hydroxytryptamine containing autonomic neurons has been made in fish (Chap. 9.2 and 9.4), the few drugs mentioned below have mainly been tried in mammalian systems.

5.5.1 Uptake and Storage

The uptake of 5-HT into the nerve terminals can be inhibited by tricyclic antidepressant drugs such as chlorimipramine. This substance shows a slightly higher affinity for 5-HT uptake than for noradrenaline uptake, but is not selective in its action. Fluoxetine, at concentrations that do not affect noradrenaline uptake, is another compound that can be used to abolish the neuronal uptake of 5-HT (Gershon and Jonakait 1979, Gershon 1981).

Accumulation of 5,6- or 5,7-dihydroxytryptamine in intestinal serotonergic neurons produces cytotoxic effects which can be demonstrated by electron microscopy (Gershon et al. 1980, Gershon 1981).

As in the adrenergic neurons, uptake of 5-HT into the storage vesicles can be inhibited by reserpine.

5.5.2 5-HT Receptor Agonists and Antagonists

Apart from 5-HT itself (Fig. 3.6), tryptophan also acts as an agonist at the muscular 5-HT receptors. Antagonists for the muscular 5-HT receptors include methysergide (1-methyl-d-lysergic acid butanolamide) and 2-bromo-lysergic acid diethylamide, as well as other ergot derivatives. Neural 5-HT receptors are also affected by phenylbiguanide, at least in some mammalian species (Douglas 1970, Gershon 1981).

Another frequent approach to the problem of finding a selective antagonist to 5-HT action is to desensitize the receptors by keeping the tissue in contact with 5-HT for prolonged periods. This desensitization to 5-HT does not affect the response of the tissue to acetylcholine or other agonists (Bülbring and Gershon 1967, Furness and Costa 1973, Costa and Furness 1976, 1979 a, b, Gershon 1981).

5.6 Drugs That Affect Peptidergic Transmission

The pharmacology of peptidergic transmission, which is still very much in its infancy, suffers from a lack of drugs with a specific action on the peptidergic transmission. One approach to the problem is the design and synthesis of peptidoceptor antagonists by exchange of one or more amino acids in the natural peptide sequence for other amino acids or the stereoisomers of the same amino acid.

Competitive antagonists for substance P have been synthesized according to this philosophy, and show antagonistic effects towards substance P responses such as salivary secretion in rats, excitation of the guinea-pig ileum and excitation of neurons in the central nervous system (Leban et al. 1979, Rockur et al. 1979, Engberg et al. 1981, Folkers et al. 1981). A desensitization to substance P, obtained by repeated addition of the peptide, impairs the effect of stimulation of allegedly substance-P-releasing neurons in the cat stomach, and in the same preparation a marked reduction in the nerve response has been demonstrated after addition of the competitive substance P antagonist (D-Pro^2, D-Trp^{7,9})-substance P (Delbro 1981). The same substance is a partial agonist on guinea-pig taenia coli and rabbit iris sphincter. It also shows a marked inhibitory effect on the "rebound excitation" of the taenia coli (Chap. 9.1.3) following the stimulation of NANC inhibitory fibres, and on the excitatory NANC innervation of the rabbit iris sphincter (Leander et al. 1981 b).

The synthesis of peptide analogues with antagonistic properties thus gives some hope for future chemical tools that can be used specifically to manipulate the peptidergic transmission.

Chapter 6

Chromaffin Tissue

The term "chromaffin" was introduced by Kohn (1902) to describe cells that are stained brown by dichromate solutions, a phenomenon first described by Henle (1865). It was later shown that the chromaffin reaction is due to the oxidation of intracellular catecholamines (or 5-hydroxytryptamine) to the corresponding adrenochromes (see Coupland 1965, 1972).

According to Coupland (1965, 1972) a chromaffin cell develops from neuroectoderm, is innervated by preganglionic ("sympathetic") nerve fibres, synthesizes and releases catecholamines, and stores sufficient quantities of catecholamines to give a chromaffin reaction. The mammalian amine-storing cells that fulfill these criteria include the catecholamine-storing cells of the adrenal medulla and extra-adrenal catecholamine-storing cells in the autonomic ganglia and in the para-aortic, abdominal and carotid bodies (see below). Other monoamine-containing cells that do not fulfill all of these criteria are excluded, e.g., the monoamine-containing mast cells and enterochromaffin cells which may give a positive chromaffin reaction due to their content of 5-hydroxytryptamine or dopamine (Coupland 1972). Adrenergic neurons are also excluded from the category of chromaffin cells, since the intracellular levels of catecholamines are not sufficient to produce a chromaffin reaction. It should be pointed out, however, that there is in many respects a very fluid demarcation line between the different types of catecholamine-storing cells. These may be *endocrine*, releasing the catecholamines into the blood (adrenomedullary cells and some types of extra-adrenal chromaffin cells); *paracrine* with short processes releasing the catecholamines in the vicinity of neighbouring cells (some types of extra-adrenal chromaffin cells in ganglia, see Sect. 6.2.2) and *neurocrine* (some types of extra-adrenal chromaffin cells, "interneurons", see Sect. 6.2.2 and the long and short adrenergic neurons). Related to the chromaffin cell concept is that of "APUD" cells introduced by Pearse (1969, 1976). The APUD cell series comprises peptide hormone producing cells of neuroectodermal origin which share certain cytochemical characteristics, particularly the ability of taking up and decarboxylating amine precursors, hence the acronym *A*mine *P*recursor *U*ptake and *D*ecarboxylation (Pearse 1969, 1976, Pearse and Polak 1971). Both chromaffin and enterochromaffin cells (and many other types) belong to the APUD series (Pearse 1976), but it should be noted that the neuroectodermal origin of the enterochromaffin cells has been denied by some authors (Andrew 1974).

In the lower vertebrates there are many cases in which neither the embryological origin nor the presence of a positive chromaffin reaction has been established, but the term "chromaffin" is retained for convenience to describe the catecholamine-storing non-neuronal cells in these groups. True adrenal glands are absent in fish, in which the chromaffin tissue corresponding to the mammalian adrenal medulla can be located in many organs – heart, veins, arteries – or associated with

Table 6.1. Adrenaline and noradrenaline in the chromaffin tissue of some vertebrate species. The catecholamine levels are expressed in µg/g tissue. It should be remembered that the proportion of chromaffin tissue varies considerably among the different organs studied

	Adrenaline	Noradrenaline	Reference
Cyclostomes			
Myxine glutinosa			von Euler and Fänge (1961)
Atrium	8.1	18.0	
Ventricle	59.0	6.5	
Portal heart	3.1	58.0	
Kidney	–	16.0	
Lampetra fluviatilis			Stabrovskii (1967)
Atrium	127.1	16.0	
Ventricle	81.0	11.6	
Blood vessels	1.2	5.0	
Elasmobranchs			
Squalus acanthias			Abrahamsson (1979b)
Axillary body	445	2,139	
Scyliorhinus canicula			Mazeaud (1971)
Axillary body	14,670	20,410	
Chimaera monstrosa			Pettersson and Nilsson (1979b)
Axillary body	3,780	9,390	
Teleosts			
Salmo gairdneri			Nakano and Tomlinson (1967)
Head kidney	4.7	4.5	
Cyprinus carpio			Stabrovskii (1968)
Head kidney	0.05	0.84	
Gadus morhua			Abrahamsson and Nilsson (1976)
Posterior cardinal vein	38.2	14.3	
Ganoids			
Huso huso			Balashov et al. (1981)
Posterior cardinal vein	19.8	4.8	
Lepisosteus platyrhincus			Nilsson (1981)
Posterior cardinal vein	47.5	21.5	
Dipnoans			
Protopterus aethiopicus			Abrahamsson et al. (1979a)
Heart	4.2	70.8	
Proximal part of intercostal arteries	216	94	
Cardinal vein	0.55	0.03	
Amphibians			
Rana catesbeiana			Azuma et al. (1965)
Adrenal	1,100	1,830	
Bufo marinus			Ungell and Nilsson (1982a)
Adrenal	1,675	1,260	
Reptiles			
Chrysemys d'orbignyi			Marques and Serrano (1960)
(turtle) Adrenal	3,370	700	
Alligator sp.			Anton and Sayre (1962)
Adrenal	373	321	
Xenodon merremii			Houssay et al. (1962)
(snake) Adrenal	2,130	1,360	

Table 6.1 (continued)

	Adrenaline	Noradrenaline	Reference
Birds			
Columba sp.			Ljunggren (1969) cited in Fänge and Hanson (1973)
Adrenal (adult)	1,865	2,447	
Gallus domesticus			Lin and Sturkie (1968)
Adrenal	3,610	1,390	
Mammals			
Bos taurus			Shepherd and West (1953)
Adrenal medulla	4,000	1,500	
Homo sapiens			Shepherd and West (1953)
Adrenal medulla	1,260	314	

autonomic ganglia. Thus the chromaffin tissue of fish cannot be conveniently classified as "adrenal" and "extra-adrenal".

A summary of the catecholamine content (adrenaline and noradrenaline) in the chromaffin tissue of some vertebrates representing the major vertebrate groups is presented in Table 6.1. Further detailed information on the catecholamine content in chromaffin tissue from various vertebrates can be found in Goodall (1951), von Euler (1956), Anton and Sayre (1962), Burnstock (1969), Grove et al. (1972) and Holtzbauer and Sharman (1972). Reviews of chromaffin cell systems in vertebrates are given by Coupland (1965, 1972), Fänge and Hanson (1973) and in Coupland and Fujita (1976).

6.1 Histochemical Demonstration of Chromaffin Cells

Chromaffin cells can be readily visualized by fluorescence histochemistry, either by treatment with formaldehyde solution or formaldehyde vapour (Falck-Hillarp technique, Chap. 3.2.6.4) (Eränkö 1952, 1955, 1976).

For studies with light or electron microscopy the precipitation of monoamines with chromium salts offers a possibility to differentiate between primary (dopamine, noradrenaline and 5-hydroxytryptamine) and secondary (adrenaline) monoamines. During fixation with glutaraldehyde the primary amines from stable polymers (so called Schiff mono-bases), while adrenaline can be eluted (Coupland et al. 1964, Coupland and Hopwood 1966, Richards et al. 1973). The polymer formed reacts with chrome salts to give coloured products, which are thus absent in the adrenaline-storing cells. When the method is adopted for electron microscopy, the adrenaline-storing vesicles (granules) become characteristically less electron-dense than the storage vesicles for the primary monoamines (e.g., Coupland and Hopwood 1966, Tranzer and Richards 1976). The use of electron probe X-ray microanalysis of chromium has been used to quantify the noradrenaline content in chromaffin cells (Lever et al. 1977).

6.2 Chromaffin Tissue in Mammals

6.2.1 Adrenal Medulla

Adrenaline was first demonstrated in extracts from the adrenal glands by Oliver and Schäfer (1895), who also described the effects of the extracts on the blood pressure. Langley (1901) noted the similarity between the effects of the adrenal extracts and stimulation of "sympathetic" nerves, and suggested that the effect of the extracts was due to stimulation of "sympathetic" nerve endings. The active principle of the extracts was isolated and studied by Abel (1899) and Takamine (1901) who named it "epinephrine" and "adrenaline", respectively.

By the use of specific histochemical methods, including the aldehyde-chromaffin method mentioned in Section 6.1, two different cell types were described in the mammalian adrenal medulla: one noradrenaline storing and one adrenaline storing (Bänder 1954, Eränkö 1955, Coupland 1965, 1972). A third type of "small granule chromaffin cell" has been described in the mouse adrenal by Kobayashi et al. (1978). Studies on non-mammalian species have revealed a differential storage of adrenaline and noradrenaline in other vertebrates as well (except possibly cyclostomes) (Coupland 1965, 1972).

It has been suggested that the formation of adrenaline-storing chromaffin cells is dependent on the proximity of the corticosteroid-secreting adrenocortical cells, and there is good evidence that the enzyme responsible for the methylation of noradrenaline to form adrenaline (PNMT; see Chap. 3.2.6.1) is activated by corticosteroids (Shepherd and West 1951, Coupland 1965, 1972, Wurtman and Axelrod 1965). The effect of the steroids appears to be a prevention of enzyme degradation, rather than an induction of enzyme synthesis (Ciaranello 1978). It is notable that PNMT of fish and amphibians, in which the chromaffin tissue is often separated from the cortical cells and in which adrenaline dominates in the adrenergic nerve endings, seems to be insensitive to the steroids (Mazeaud 1971, 1972, Wurtman et al. 1968).

The synthesis of adrenaline and noradrenaline in the adrenomedullary cells follows the same pathways as in the adrenergic neurons (Chap. 3.2.6). Also the release of catecholamines takes place by exocytosis as in the neurons (Chap. 3.2.2, Stjärne 1972, Douglas 1975, Viveros 1975). The release is induced by the action of preganglionic cholinergic fibres which run in the splanchnic nerves (Elliott 1913a, Young 1939, Coupland and Holmes 1958, Coupland 1963, 1965).

VIP-immunoreactive nerve fibres and nerve cell bodies have been demonstrated in rat adrenal glands (Hökfelt et al. 1981) and VIP-immunoreactive nerve fibres are also present in the human adrenal gland (Linnoila et al. 1980). An involvement of the VIP-immunoreactive fibres in the control of blood flow in the adrenal gland has been proposed (Hökfelt et al. 1981). In addition to the demonstration of VIP-immunoreactivity in neurons within the adrenal gland, Linnoila et al. (1980) demonstrated enkephalin-like immunoreactivity in about one-third of the adrenomedullary cells in the human adrenal gland. The function of this peptide remains unknown.

6.2.2 Extra-Adrenal Chromaffin Cells

In adult mammals chromaffin cells are found mainly in the adrenal medulla, but during foetal life a mass of chromaffin tissue, the organ of Zuckerkandl (Zuckerkandl 1901), is found in the para-aortic region (Coupland 1965, Holtzbauer and Sharman 1972, Fänge and Hanson 1973). The total content of "adrenaline-like activity" in this organ exceeds that of the adrenals (Elliott 1913 b).

The organ of Zuckerkandl disappears more or less completely in the adult animal, leaving small abdominal clusters of chromaffin cells often known as paraganglia. The extra-adrenal chromaffin tissue of the adult mammal is found mainly in these abdominal paraganglia: the para-aortic bodies and the carotid bodies where the chromaffin cells may be involved in the sensory functions of these organs (Coupland 1965, 1972).

Extra-adrenal chromaffin cells have also been demonstrated within autonomic ganglia, particularly in prevertebral and sympathetic chain ganglia, of a number

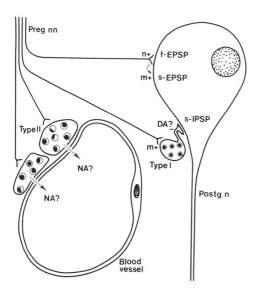

Fig. 6.1. Speculative figure outlining the possible arrangement of the SG cells in the autonomic ganglia of mammals (and possibly other vertebrates). Preganglionic cholinergic fibres (*Preg nn*) innervate the SG cells, and such fibres also synapse on postganglionic neurons (*Postg n*) of the ganglion. The cholinergic innervation of the postganglionic neuron can produce both fast excitatory postsynaptic potentials (*f-EPSP*) via nicotinic cholinoceptors, and slow excitatory postsynaptic potentials (*s-EPSP*) via muscarinic cholinoceptors (Chap. 4.1.10). Muscarinic cholinoceptors are also thought to be involved in the cholinergic control of the Type I SG cells, which may act as interneurons releasing a catecholamine (dopamine?). The catecholamine produces hyperpolarization (slow inhibitory postsynaptic potential, *s-IPSP*) of the postganglionic neuron, and the possibility of a preganglionic modulation of acetylcholine release from the preganglionic nerve terminals must also be considered. A facilitation of the s-EPSP by the release of a catecholamine from the SG cell has also been postulated (Tosaka and Kobayashi 1977). Preganglionic fibres also innervate the Type II SG cells, which may act as endocrine or paracrine cells releasing a catecholamine (noradrenaline?) in the vicinity of small blood vessels. Type I and Type II cells as described by Lu et al. (1976). [Based primarily on Libet (1970, 1976), Lu et al. (1976), Tosaka and Kobayashi (1977)]

of vertebrates (Figs. 2.12 and 2.13). Fluorescent histochemistry reveals these cells as small brightly fluorescent yellow and green cells (diameter 10–25 µm) which lie singly or in small groups within the ganglion. The cells have been called "small intensely fluorescent cells" ("SIF cells") (Eränkö and Härkönen 1963, 1965) and subsequent studies with the help of the electron microscope revealed these cells as small granulated cells ("SG-cells") (Grillo 1966, Williams 1967, Elfvin 1968, Grillo et al. 1974). Attempts to relate the small fluorescent cells of the prevertebral and sympathetic chain ganglia with chromaffin-positive cells have led to the conclusion that the numerical distribution of the *yellow* SIF cells follows closely that of the chromaffin-positive cells (Santer et al. 1975, Kemp et al. 1977).

Several types of SG cells have been identified by their histochemical reactions and ultrastructural appearance (Watanabe 1971, Elfvin et al. 1975, Lever et al. 1976, Lu et al. 1976). In the terminology of Lu et al. (1976), based on studies of autonomic ganglia from the rat (superior cervical and coeliac/mesenteric ganglia), three types are recognized. Type I cells are characterized by their content of round granules (vesicles) with a diameter of 50–150 nm. The core of these granules is surrounded by an electron-lucent halo of 15–20 nm. Type I cells were observed giving off processes to neighbouring nerve cell bodies. It is speculated that the Type I cells store dopamine (Lu et al. 1976) and act as interneurons (Fig. 6.1, Libet 1970, 1976).

Type II cells are characterized by irregular granules, often with an eccentric core of electron-dense material. The diameter of the granules range from 100–300 nm. Their similarity with the adrenomedullary noradrenaline storing cells led Lu et al. (1976) to speculate that they store noradrenaline, a suggestion supported by X-ray probe analysis (Lever et al. 1977). Type II cells were observed in the vicinity of blood vessels, suggesting an endocrine or sensory (?) function (Lu et al. 1976).

Type III cells, finally, were characterized by granules of variable shape and size, often elongated (170 × 50 nm) with a central electron-dense core surrounded by an electron-lucent halo (15–20 nm) (Lu et al. 1976). Nothing has been said about the possible functions of these cells.

All three types of SG cell receive a preganglionic innervation by "c-type" terminals but only Type I cells were observed to give off processes that make synaptic contact with the ganglion cells (Lu et al. 1976) (Fig. 6.1).

6.3 Chromaffin Tissue in Cyclostomes

Giacomini (1902a, b) was the first to describe chromaffin tissue in the veins, arteries and heart of cyclostomes. In the large veins, the chromaffin tissue is separated from the blood by a single layer of endothelial cells. Gaskell (1912) demonstrated the adrenaline-like effects on the blood pressure of cats by extracts of the chromaffin tissue from *Petromyzon*. Later both adrenaline and noradrenaline was detected in heart, kidney and blood vessels of cyclostomes (Table 6.1).

The catecholamine-storing cells in the cyclostome hearts give a weak chromaffin reaction (Augustinsson et al. 1956) and the Falck-Hillarp technique has revealed numerous small intensely fluorescent cells in the hearts of both lampetroids

Table 6.2. Adrenaline and noradrenaline concentration in blood plasma from a few non-mammalian vertebrates before and after disturbance (asphyxia or physical disturbance). Concentrations are expressed in µg/100 ml blood plasma

	"Resting" Adr	Noradr	"Disturbed" Adr	Noradr	Reference
Cyclostomes					
Petromyzon marinus (physical disturbance)	0.19	0.29	1.08	1.36	Mazeaud (1971)
Elasmobranchs					
Scyliorhinus canicula (asphyxia 30 min)	0.81	1.81	8.50	13.84	Mazeaud (1971)
Squalus acanthias (asphyxia 10 min)	0.9	0.5	3.0	5.7	Abrahamsson (1979b)
Teleosts					
Cyprinus carpio (asphyxia 30 min)	0.15	1.6	0.67	3.3	Mazeaud (1971)
Cyprinus carpio (physical disturbance)	0.15	1.6	0.2	7.6	Mazeaud (1971)
Gadus morhua (asphyxia 10 min)	0.55	0.22	5.35	0.54	Wahlqvist and Nilsson (1980)
Dipnoans					
Protopterus aethiopicus (physical disturbance)	2.29	2.36	6.41	29.10	Abrahamsson et al. (1979a)
Amphibians					
Rana esculenta/lessonae (blood sampling in permanently cannulated animal compared to blood sampling by heart puncture after sacrifice)	0.190	0.035	0.370	0.895	Bourgeois et al. (1978)

and myxinoids (Bloom et al. 1961, Dahl et al. 1971, Shibata and Yamamoto 1976, Otsuka et al. 1977). The cells have a diameter of 6–18 µm and processes of up to 70 µm have been described (Bloom et al. 1961). Electron microscope studies revealed granules with diameters of 100–300 nm within the cells (Bloom et al. 1961). In the sea lamprey, *Petromyzon marinus*, the release of catecholamines from the chromaffin tissues during haemhorrage or disturbance is large enough to produce a significant increase in the plasma concentrations of adrenaline and noradrenaline (Table 6.2, Mazeaud 1971).

6.4 Chromaffin Tissue in Elasmobranchs

The chromaffin tissue in elasmobranchs forms segmental bodies associated with the paravertebral autonomic ganglia (Chap. 2.3.1). These "suprarenal bodies" were first described by Leydig (1853) whereafter the chromaffin reaction was demonstrated by Balfour (1877). The "suprarenal bodies" of elasmobranchs are separate from the corticosteroid-secreting "interrenal tissue", which is the equivalent of the mammalian adrenal cortex (Coupland 1965).

The two largest masses of chromaffin tissue are associated with the gastric ganglia within the posterior cardinal sinus to form the "axillary bodies" (Fig. 2.28) (Lutz and Wyman 1927, Gannon et al. 1972). In chimaeroids these chromaffin masses completely surround the subclavian arteries (Pettersson and Nilsson 1979 b).

Noradrenaline predominates over adrenaline in the axillary bodies of all elasmobranchs studied (Table 6.1, Shepherd et al. 1953, von Euler and Fänge 1961). During stimulation of the anterior spinal cord a release of catecholamines into the perfused posterior cardinal vein of *Squalus acanthias* could be demonstrated, and after keeping the animal in air for 10 min a marked increase in the plasma concentrations of adrenaline and noradrenaline occurs (Table 6.2, Abrahamsson 1979 b).

6.5 Chromaffin Tissue in Teleosts

The chromaffin tissue of teleosts is located within the anterior part of the kidney ("head kidney"), often lining the walls of the posterior cardinal veins. In some species the chromaffin tissue is separated from the "interrenal" tissue, while in others there is variable degree of association between the two types of tissue (Giacomini 1902c, Oguri 1960, Nandi 1961, 1964, Mazeaud 1971, Abrahamsson and Nilsson 1976, Nilsson 1976, Nilsson et al. 1976). In most, but not all, teleosts adrenaline is the dominating catecholamine in the chromaffin tissue of the cardinal vein (Table 6.1).

In the cod, *Gadus morhua*, a separate bundle of myelinated nerve fibres extends to the left posterior cardinal vein (Nilsson 1976). The arrangement of the nerve supply allows experimental stimulation of these fibres and Nilsson et al. (1976) were able to demonstrate a release of catecholamines (especially adrenaline) during stimulation of the fibres (Fig. 6.2). The nervous release of catecholamines could be inhibited by hexamethonium, suggesting that the chromaffin tissue of the head kidney of this species is controlled by preganglionic cholinergic fibres similar to the arrangement in mammals.

The catecholamine release during stimulation of the nerve supply to the left cardinal vein is large enough to affect the isolated perfused heart (Holmgren 1977) and

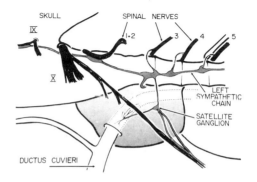

Fig. 6.2. Arrangement of the left sympathetic chain and the nerve to the left cardinal vein in the cod, *Gadus morhua*. Myelinated, presumably preganglionic fibres pass through the sympathetic chain ganglion corresponding to the 3rd spinal nerve and the "satellite ganglion" to enter the posterior cardinal vein within the head kidney. (Reproduced with permission from Nilsson et al. 1976)

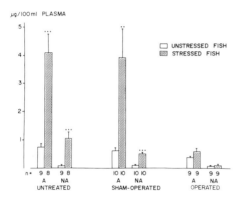

Fig. 6.3. Plasma concentrations of adrenaline (*A*) and noradrenaline (*NA*) from "resting" (*plain bars*) and "stressed" (*hatched bars*) cod. The "stress" was induced by keeping the fish in air for 15 min prior to blood sampling. The three groups represent (*left*) untreated fish, (*middle*) sham-operated fish were incisions in the skin and muscle only were performed and (*right*) fish where the first four spinal nerves had been sectioned 4 days before the blood sampling. Number of fish in each experimental group = n. Levels of significance ("stressed" compared to "resting" fish): ***$p<0.001$; **$p<0.01$. (Reproduced with permission from Nilsson et al. 1976)

the vasculature of the isolated perfused gill (Wahlqvist 1981). Because the release of catecholamines during experimental nerve stimulation increases rapidly, and in a linear manner from 1 to ca. 4–6 Hz, it is not unreasonable to assume a physiological effect of the strategically released catecholamines on the heart and branchial vasculature of the cod (cf. Chap. 7.3).

The catecholamine concentration in cod plasma increases substantially during "stress" induced by keeping the animal in air for ca. 10 min (Nilsson et al. 1976, Wahlqvist and Nilsson 1980). Sectioning of the first four spinal nerves close to their exit from the cranium/spinal column inhibits the release of catecholamines into the plasma during "stress", thereby supporting the idea of a functional autonomic nervous control of the chromaffin tissue in this species (Fig. 6.3).

Small intensely fluorescent cells have been demonstrated in several autonomic ganglia of the cod (Fig. 2.13, Nilsson 1973, 1976, Watson 1980), but the function of these cells in the teleost ganglia is so far unknown.

6.6 Chromaffin Tissue in Ganoids

The arrangement of the chromaffin tissue in chondrosteans and holosteans is similar to that in teleosts. Chromaffin cells have been described in the posterior cardinal vein of the sturgeon (*Acipenser sturio*) (Giacomini 1904) and brightly fluorescent cells were demonstrated by glyoxylic-acid fluorescence histochemistry in the posterior cardinal veins and coeliaco-mesenteric artery of the beluga (*Huso huso*) (Balashov et al. 1981).

In the holostean ganoid, *Lepisosteus platyrhincus*, Nilsson (1981) demonstrated chromaffin cells in the posterior cardinal veins. In both *Huso* and *Lepisosteus* adrenaline is the more abundant catecholamine in the chromaffin tissue of the posterior cardinal vein (Table 6.1), and also, at least in *Lepisosteus*, in blood plasma. It was concluded that the concentration of adrenaline in the plasma is sufficient to affect the heart in both species (Balashov et al. 1981, Nilsson 1981). Granulated cells, resembling the adrenomedullary cells of the higher vertebrates, have also been described in the posterior cardinal and renal veins of the bowfin, *Amia calva* (Youson 1976).

6.7 Chromaffin Tissue in Dipnoans

Giacomini (1906) described chromaffin tissue in the anterior part of the left cardinal vein ("azygos vein") and in the intercostal arteries of *Protopterus*. Holmes (1950) failed to find chromaffin cells in the cardinal veins, but confirmed Giacomini's finding of chromaffin tissue associated with the intercostal arteries and also described an innervation of these cells.

A study of *Protopterus aethiopicus* using Falck-Hillarp fluorescence histochemistry revealed chromaffin cells in the intercostal arteries and also in the most anterior part of the left cardinal vein, thus confirming Giacomini's early descriptions (Abrahamsson et al. 1979a, Fig. 2.27). In addition, large masses of chromaffin tissue were found in the atrium of the heart, where the catecholamine content exceeds that found in cyclostome hearts (Table 6.1). The intracardiac chromaffin cells are thought to be innervated (Scheuermann 1979) and Scheuermann et al. (1981) also concluded that a substantial storage of dopamine is held in these cells.

During disturbance, induced by chasing the animal around its tank for 10–15 min, there is a massive release of catecholamines, particularly noradrenaline, into the blood plasma of *Protopterus* (Table 6.2, Abrahamsson et al. 1979a). It would be of great interest to study the release of catecholamines into the blood during arousal of aestivating of lungfish.

Small, elongated, intensely fluorescent cells have been described associated with the non-fluorescent ganglion cells in sympathetic chain ganglia of *Protopterus*, but the role of these cells in these ganglia is unknown (Abrahamsson et al. 1979b).

The arrangement of the chromaffin tissue in *Protopterus* thus shows striking similarities with the arrangement in the cyclostomes, with venous, arterial and cardiac chromaffin tissue. The physiological significance of the intra-cardiac chromaffin tissue, particularly the role in cardiac control, remains to be studied in both cyclostomes and dipnoans.

6.8 Cromaffin Tissue in Amphibians

In urodele and apodan amphibians, the chromaffin tissue is arranged in segmental bodies associated with the sympathetic chain ganglia along the ventral surface of the kidneys, while in most anurans the chromaffin tissue forms a distinct yellow stand ("adrenal") on the ventral surface of the kidney. In all groups, the chromaffin and interrenal tissue is intermingled, but there appears to be no activation of the PNMT by corticosteroids in amphibians (Leydig 1853, Grynfeltt 1904, Dittus 1936, Nicol 1952, Coupland 1965, Wurtman et al. 1968). In both urodeles and anurans two distinct cell types are found in the adrenal chromaffin tissue, one storing adrenaline and the other storing noradrenaline (Coupland 1971, Kojima et al. 1976, Mastrolia et al. 1976). An innervation of the chromaffin cells by "c-type" nerve endings has also been described (Piezzi 1966, Coupland 1972, Mastrolia et al. 1976).

Extra-adrenal chromaffin cells have been demonstrated in autonomic ganglia of anurans (e.g., McLean and Burnstock 1966, Weight and Weitsen 1977) and Banister et al. (1967) have described catecholamine-storing cells in the carotid labyrinths of *Rana temporaria*.

6.9 Chromaffin Tissue in Reptiles

In reptiles the chromaffin tissue is intermingled with interrenal tissue in a discrete adrenal gland; some chromaffin tissue is also associated with the autonomic ganglia (Coupland 1965, 1972, Unsicker 1976a, b). Extra-adrenal chromaffin tissue is also found in the carotid bodies and in paraganglia near the heart, including the truncus arteriosus (Coupland 1965, 1972, Gabe 1970).

The differential storage of adrenaline and noradrenaline in separate cells has been demonstrated in reptiles and from work on lizards and snakes it has been concluded that the synthesis of adrenaline, and thus the formation of adrenaline storing cells, is dependent on the proximity of the interrenal tissue (Wassermann and Tramezzani 1963, Wurtman et al. 1967, Coupland 1971, Unsicker 1976a, b, Del Conte 1977) (Chap. 3.2.6.1).

6.10 Chromaffin Tissue in Birds

The adrenal glands are discrete organs that are innervated by fibres from the lesser splanchnic nerves passing via the adrenal plexus. Adrenaline- and noradrenaline-storing cells are separate and the chromaffin tissue is closely intermingled with interrenal tissue (Eränkö 1957, Coupland 1965, 1971, 1972, Wells and Weight 1971). The catecholamine content in the adrenal glands of *Columba* and *Gallus* are given in Table 6.1.

Extra-adrenal chromaffin tissue is widespread in birds, including cells in the carotid body and in the autonomic ganglia (Kose 1907, Coupland 1965, 1972, Bennett and Malmfors 1970, Fig. 2.12).

6.11 Conclusions

The chromaffin cells show many features in common with the adrenergic neurons: embryological origin, innervation by preganglionic fibres and synthesis "storage" release of catecholamines. There is a gradual transition from the adreno-medullary chromaffin cells via the paracrine chromaffin cells and interneurons of the autonomic ganglia to the short and long adrenergic neurons. The chromaffin cells are thus part of the autonomic control system.

Since chromaffin tissue is closely accociated with all parts of the cardiovascular function in most vertebrates, it can be argued (Chap. 7) that the adrenergic control of the circulatory system is a phylogenetically old part of the autonomic nervous system. A transition from the intra-cardiac chromaffin cells, via strategically located chromaffin tissue in the large veins to a direct adrenergic innervation of the heart and vasculature can be discerned during vertebrate evolution (cf. Table 7.1). It can be argued that the circulating catecholamines released from chromaffin cell stores mediate a general "humoral adrenergic tonus", which, in fish, may be of major importance for the adrenergic control of different organs, and which reinforces the more direct and restricted control of individual organs by adrenergic nerves in the higher vertebrates.

Chapter 7

The Circulatory System

Profound changes in the demands on the circulatory system have occurred during the evolution of the vertebrates from the aquatic forms to the more advanced terrestrial forms. These changes reflect a variety of anatomical and functional alterations, and also the adjustment from aquatic life at zero gravity to the demands of terrestrial life. The best documented driving force in the evolution of the vertebrate cardiovascular functions is the need for an efficient transport of respiratory gases between the gas exchanger (skin, gill, lung) and the tissues (Johansen and Burggren 1980, Johansen 1982).

Cardiac output is closely related to the breathing pattern of the organism, and undergoes marked changes in conjunction with the respiratory events particularly in the air-breathing vertebrates with periodic ventilation, i.e., periods of ventilation alternating with periods of apnoea (breath holding). Such periodic ventilation occurs in practically all air-breathing vertebrates except birds and mammals (Johansen and Burggren 1980, Johansen 1982). The cardiac output is basically related to two aspects of cardiac performance: the heart rate (chronotropic effect) and stroke volume (inotropic effect). Both effects are controlled by aneural factors, and by the autonomic innervation of the heart.

The aneural control of cardiac performance includes "hormonal" influences, particularly from the catecholamines secreted from chromaffin tissue. The heart rate can also be shown to increase as a result of an elevated venous return, which induces stretching of the pacemaker and therefore an increased excitability. The importance of this effect at the central venous pressures which occur in vivo is, however, doubtful (Jensen 1961, 1965, 1969, 1970, Johansen et al. 1964, Harris and Morton 1968). It should also be noticed that a dramatic increase in heart rate may actually impair cardiac output due to incomplete filling of the heart during the shortened diastole (Johansen et al. 1966, Johansen 1971).

An increased force of beat (positive inotropic effect) is the result of an increased stretching of the ventricular wall induced by an increased filling of the ventricle(s) (Starling's law of the heart). This effect is essential in the control of stroke volume in fish (and also in amphibians and reptiles) in which filling of the ventricle is directly dependent on the contraction of the atrium. The filling of the piscine atrium is, in turn, dependent on the suction of blood from the central venous system due to a negative pericardial pressure during the diastole (*vis a fronte*). This phenomenon is created by the rigidity of the pericardium and the surrounding tissues, especially evident in lampetroids and elasmobranchs. In the avian and mammalian hearts, atrial filling is instead dependent on positive central venous pressure (*vis a tergo*), and the autonomic control of the large veins is therefore imperative in the regulation of cardiac filling (Johansen 1971, Johansen and Burggren 1980). "Hormonal" control, particularly by circulating catecholamines, and an autonomic innervation of the myocardium are also of paramount importance in regulation of stroke volume as will be outlined in the sections below.

The vasculature is controlled by the autonomic nervous system, with the regulation being most developed in the higher vertebrates. As with the heart, "hormonal" control by circulating catecholamines is also essential, particularly in fish in which a vasomotor innervation of some of the vascular beds may be poorly developed. The oxygen tension and locally produced metabolites, e.g., lactic acid, can also affect resistance to blood flow in individual vascular beds (Folkow and Neil 1971).

The vascular resistance, controlled chiefly by the smooth muscles of the arterioles and precapillary sphincters and cardiac output (which are also controlled by the autonomic nervous system) generate the blood pressure necessary for the (1) maintenance of an adequate blood perfusion of the various organs, (2) distribution of blood among the vascular beds, and (3) the control of transcapillary fluid transfer (pre-/postcapillary pressure ratio) (Folkow and Neil 1971).

The density of the vascular innervation varies among the different organs and also varies markedly between the different vertebrate groups. Circulation in the cerebral vasculature, crucial to the survival of the animal is poorly controlled by vasomotor fibres (in mammals) and may depend mainly on local metabolite concentration. On the other hand, the skin vasculature, which is intimately involved in thermoregulation, receives a dense innervation by vasomotor fibres (Folkow and Neil 1971). The vascular beds in mammals are in some cases innervated by both excitatory (vasoconstrictor) and inhibitory (vasodilator) fibres, but the presence and function of vasodilatory fibres in the lower vertebrates is less well documented.

The circulatory system is under surveillance by a variety of receptors that gauge the pressure (baroreceptors) and levels of chemical constituents (chemoreceptors) of the blood. Sensory pathways from these receptors signal any deviations from "homeostasis" to the vasomotor and cardioregulatory centres of the brain (medullary cardiovascular centre). Autonomic nerve fibres from the same centres are then adjusting rate and stroke volume of the heart and the resistance in the vascular beds (Fig. 7.1). There is also a close relationship with the respiratory control centres of the brain, which adjust the respiratory pattern of the animal.

Although cardiovascular reflexes have not been studied extensively in non-mammalian vertebrates, there is good evidence for the existence of cardioinhibitory reflexes triggered by elevated blood pressure in several of the vertebrate groups and also for reflexes affecting the distribution of the cardiac output between the systemic and pulmonary circuits of the vertebrates with periodic breathing (Johansen and Burggren 1980, Johansen 1982).

The autonomic fibres controlling the heart and vasculature are often continuously active, with a low frequency of impulses (Folkow 1952). This continuous influence creates a nervous *tonus* (or *tone*) in the effector organ, e.g., adrenergic (or sometimes "sympathetic") tonus, vagal tonus etc. Inhibition of the heart, for example, can thus be due to either an increased cholinergic (vagal) tonus or a decreased adrenergic tonus or both.

The aim of the present chapter is not to give a comprehensive account of cardiovascular control in the different vertebrate groups, since a detailed treatment of this subject can be obtained by the reader from such reviews and textbooks as those

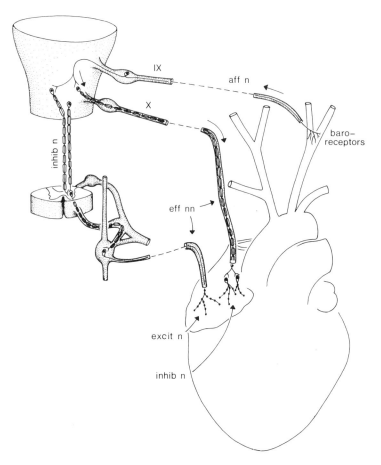

Fig. 7.1. Diagrammatic representation of the pathways involved in the cardio-inhibitory baroreceptor reflex in mammals. Baroreceptive nerve endings, which are sensitive to the stretching of the arterial wall exerted by the blood pressure, are shown in the carotid sinus. Similar fibres (not shown) are also present in the aortic arch. The afferent (sensory) fibres from the carotid sinus run in a branch of the glossopharyngeal nerve to the "medullary cardio-vascular centre", from which cardio-inhibitory vagal fibres run to the heart. In addition to the vagal fibres, inhibitory fibres within the central nervous system impair the transmission in spinal autonomic cardioaccelerator pathways. Abbreviations: *aff n*, afferent nerve; *eff nn*, efferent nerves; *excit n*, excitatory nerve; *inhibit n*, inhibitory nerve; *IX*, glossopharyngeal nerve; *X*, vagus nerve

of Johansen et al. (1970), Folkow and Neil (1971), Johansen (1971, 1982), Eckert and Randall (1978), Johansen and Burggren (1980). Instead the chapter will deal with the role of the autonomic nervous system in the control of cardiac performance and vascular resistance, including a few examples of cardiovascular reflexes in some of the vertebrate groups. A summary of the patterns of cardiac innervation in the different vertebrates is offered in Figs. 7.7 and 7.8.

The typical piscine circulatory system consists of a branchial (gill) and a systemic vasculature coupled in series with the heart. The blood ejected from the heart enters the branchial vasculature directly via the ventral aorta and oxygenation of the blood and ion transfer takes place in the gill lamellae. The branchial vascular bed of fish is thus a high-pressure vasculature compared with the systemic vasculature of the same animal. This is in contrast to the respiratory vasculature of those air-breathing vertebrates with a functionally divided pulmonary and systemic circuit. In these vertebrates the pulmonary vasculature is characterized by comparatively low resistance leading to a lower pulmonary blood pressure (Johansen and Burggren 1980, Johansen 1982).

7.1 Cyclostomes

In cyclostomes, the blood flow through the gills is facilitated by the pumping activity of "branchial hearts" derived from non-vascular tissue. Blood flow in the venous system is enhanced by a "caudal heart" located in the caudal venous sinus, and in the myxinoids there is also an auxiliary portal vein heart (Fänge et al. 1963, Johansen 1960).

7.1.1 The Cyclostome Heart

The primitive vertebrate heart of the type found in cyclostomes, elasmobranchs and actinopterygian fish is formed from a single tube consisting of four sequentially arranged chambers: sinus venosus, atrium, ventricle and bulbus cordis (conus arteriosus). In teleosts the bulbus cordis is absent and there is instead a bulbus arteriosus, which is derived from the proximal portion of the ventral aorta (Johansen and Burggren 1980, Johansen 1982).

One of the most peculiar features of the cyclostome heart, including the myxinoid portal vein heart, is the storage of large quantities of catecholamines (adrenaline and noradrenaline) in specialized myocardial granule-containing cells (see also Chap. 6). These cells are in many ways similar to the adrenomedullary cells of the higher vertebrates. The myocardia of *Lampetra* and *Myxine* show a weak chromaffin reaction (Augustinsson et al. 1956). Falck-Hillarp fluorescence histochemistry has revealed the presence of intensely fluorescent cells embedded in the myocardium (Dahl et al. 1971, Shibata and Yamamoto 1976).

The physiological significance of the catecholamine-storing cells in the cyclostome hearts is poorly understood and there appears to be no extrinsic innervation of these cells in either group of cyclostomes (Caravita and Coscia 1966, Beringer and Hadek 1973). In *Myxine*, a depletion of the catecholamines with reserpine causes bradycardia or cardiac arrest (Bloom et al. 1961), which strongly suggests an involvement of the endogenously stored catecholamines in the normal performance of the cyclostome heart. It is doubtful whether circulating catecholamines play any role in cardiac control in the cyclostomes.

In myxinoids, there is no functional extrinsic innervation of the systemic or portal-vein heart although ganglionic cells and nerve fibres have been described in the heart of *Eptatretus stouti* (Greene 1902, Carlson 1904, Augustinsson et al. 1956, Fänge et al. 1963, Johansen 1971, Hirsch et al. 1964). The heart of *Myxine* is remarkably insensitive to drugs that exert marked cardiac effects in other vertebrates. Thus acetylcholine, catecholamines and tyramine in concentrations known to have strong effects on the heart rate and force in other vertebrates have very weak or no effects on the *Myxine* heart. Negative chronotropic and inotropic effects can be obtained with reserpine or dihydroergotamine, after which adrenaline and noradrenaline will produce distinct positive chronotropic and inotropic effects on the heart of *Myxine* (Fänge and Östlund 1954, Östlund 1954, Augustinsson et al. 1956, Bloom et al. 1961). Positive chronotropic and inotropic effects of 5-hydroxytryptamine have also been described (Augustinsson et al. 1956). An increased venous return causes an acceleration of the portal vein heart in *Myxine*, but there are no accounts of drug effects on this organ (Johansen 1960, Fänge et al. 1963).

In lampetroids, ganglion cells have been described in the heart, and there is a well-developed functional vagal innervation of the heart by branches following the jugular vein (Tretjakoff 1927, Augustinsson et al. 1956, Johnels 1956). The main effect of vagal stimulation in lampetroids is an *acceleration* of the heart, although sometimes a slight bradycardia during or after stimulation is observed (Carlson 1906, Zwaardemaker 1924, Augustinsson et al. 1956). Acetylcholine induces an acceleration of the heart, a unique response among the vertebrates. Other nicotinic cholinoceptor agonists, such as nicotine (even when applied to the isolated preparation in the form of tobacco smoke!) have the same effect (Otorii 1953, Augustinsson et al. 1956, Falck et al. 1966, Lukomskaya and Michelson 1972). Otorii (1953) reports a stimulation of the heart of *Endosphenus japonicus* by muscarinic cholinoceptor agonists (pilocarpine and choline muscarine), but no effects of pilocarpine or muscarine were detected by Falck et al. (1966) for the heart of *Lampetra*. The excitatory effect of vagal stimulation or nicotinic agonists can be blocked by nicotinic cholinoceptor antagonists such as tubocurarine and hexamethonium (Augustinsson et al. 1956, Falck et al. 1966, Lukomskaya and Michelson 1972).

Adrenaline, noradrenaline, isoprenaline and tyramine also stimulate the lampetroid heart, although the effects are less pronounced than that of acetylcholine. The effect of the adrenergic agonists is blocked by pronethalol, suggesting an effect via β-adrenoceptors in lampetroids as in the higher vertebrates (Otorii 1953, Augustinsson et al. 1956, Nayler and Howells 1965, Falck et al. 1966).

7.1.2 The Cyclostome Vasculature

Very little is known about the autonomic innervation of the cyclostome vasculature or the general control of vascular resistance in these animals. In lampetroids, spinal autonomic fibres appear to innervate the blood vessels (Tretjakoff 1927, Johnels 1956); Falck-Hillarp fluorescence histochemistry has shown that at least some of these fibres are adrenergic (Leont'eva 1966, Govyrin 1977).

Catecholamines produce a biphasic response in the perfused branchial vasculature of *Myxine*. Early in an experiment a dilation followed by a constriction of the branchial vascular bed was recorded after adrenaline or noradrenaline, but later during the same experiment the dilatory phase disappeared (Reite 1969). The constrictor response can be antagonized by α-adrenoceptor antagonists, while the dilator response can be abolished by the β-adrenoceptor antagonist propranolol (Reite 1969). A constriction of the branchial vasculature was also obtained by acetylcholine and both the catecholamines and acetylcholine produced an increased systemic vascular resistance in *Myxine* (Reite 1969).

7.2 Elasmobranchs

7.2.1 The Elasmobranch Heart

The heart of the elasmobranchs is supplied by branches from the vagi but contributions to the vagi or direct cardiac nerves from paravertebral ganglia are, with rare exceptions (e.g., *Mustelus*, Pick 1970), absent (Stannius 1849, Izquierdo 1930, Lutz 1930a–c, Young 1933c, Pick 1970, Short et al. 1977). An adrenergic influence on the heart of both selachians and chimaeroids may be exerted by specialized catecholamine-storing endothelial cells in the sinus venosus and atrium. These cells are innervated by cholinergic vagal fibres (Saetersdal et al. 1975, Pettersson and Nilsson 1979b). In addition to the adrenergic stores in the sinus venosus, the anterior chromaffin bodies (axillary bodies) of all elasmobranchs are situated strategically in the venous blood entering the heart, so that an efficient adrenergic control of the heart via catecholamines released from this tissue by the action of preganglionic cholinergic fibres is likely (Satchell 1970, 1971, Johansen 1971, Gannon et al. 1972, Abrahamsson 1979b).

Acetylcholine as well as stimulation of the vagus produce negative chronotropic effects on the elasmobranch heart. The effects are antagonized by atropine, indicating an effect mediated by muscarinic cholinoceptors as in the higher vertebrates (Lutz 1930a–c, Lutz and Wyman 1932, Babkin et al. 1933, Hiatt 1943, Burger and Bradley 1951, Fänge and Östlund 1954, Östlund 1954, Jullien and Ripplinger 1957, Johansen et al. 1966, Butler and Taylor 1971, Capra and Satchell 1977a, Taylor et al. 1977). A variation in the degree of cholinergic vagal tonus on the heart could, in the absence of an adrenergic innervation, serve as an important mode of *nervous* cardioregulation in elasmobranchs (Johansen et al. 1966, Butler and Taylor 1971, Taylor et al. 1977).

The effects of adrenaline and noradrenaline on the elasmobranch heart are somewhat variable. The main response to the catecholamines is a positive chronotropic and, especially, inotropic effect mediated by a β-adrenoceptor mechanism but in many cases an initial bradycardia has been reported after the administration of catecholamines (Lutz 1930a–c, Hiatt 1943, Fänge and Östlund 1954, Capra and Satchell 1977a) The inhibitory effects of the catecholamines could, in some of the experiments, be blocked by atropine, indicating a cholinergic element in the mechanism (Lutz 1930a, b, Fänge and Östlund 1954, Östlund 1954). Capra and Satchell (1977a) after demonstration of a consistent inhibitory effect of noradrenaline, but not by adrenaline or isoprenaline, suggested an α-adrenoceptor mediated effect of noradrenaline since the response could be blocked by phentolamine but not by atropine. Since the antagonists were added in single doses with the concentration in the case of phentolamine reaching 1 mM, the selectivity of this blockade is very doubtful. Regardless of the exact mechanism of action, the possibility of a selective cardiac control via the two naturally occuring catecholamines adrenaline and noradrenaline exists (Capra and Satchell 1977b).

7.2.2 The Elasmobranch Branchial Vasculature

Compared with the number of studies of the teleost branchial vasculature, there are few on the control of branchial vascular resistance in elasmobranchs. Östlund and Fänge (1962) failed to demonstrate vasomotor effects of adrenaline in the isolated perfused gill apparatus of *Squalus acanthias*. In a later study of the same species, however, Capra and Satchell (1977a) demonstrated both an α-adrenoceptor-mediated constriction and a β-adrenoceptor-mediated dilation of the branchial vasculature after administration of catecholamines. Marked dilatory effects of catecholamines have also been demonstrated in perfused gills of *Scyliorhinus canicula* (Davies and Rankin 1973) and a selective vasodilation of the posterior gill arches associated with exercise has been demonstrated in the skate (*Raja rhina*) by Satchell et al. (1970).

Virtually nothing is known about the direct autonomic nervous control of the elasmobranch branchial vasculature. Stimulation of the vagal branches in *Scyliorhinus* produced a small increase in the vascular resistance of the gill (Rankin, personal communication), but it seems that this effect is mainly due to contraction of skeletal muscle associated with the gill arch (Metcalfe, personal communication). It is very likely that the strategically located chromaffin tissue releases catecholamines which control the branchial vascular resistance (Satchell 1971, Gannon et al. 1972). Furthermore it has been demonstrated that the catecholamine levels in blood plasma from "stressed" dogfish are sufficiently high to dilate the branchial vasculature (Davies and Rankin 1973).

7.2.3 The Elasmobranch Systemic Vasculature

Injection of catecholamines (adrenaline and noradrenaline) in vivo produces an elevated arterial blood pressure, which is associated with a slight, possibly reflexive, bradycardia (Wyman and Lutz 1932, Burger and Bradley 1951, Johansen et al. 1966, Capra and Satchell 1977b). It appears that the two catecholamines produce different effects on the central venous pressure and cardiac performance: adrenaline increases and noradrenaline decreases venous pressure and stroke volume (Capra and Satchell 1977b).

The major visceral arteries are contracted by adrenaline via an α-adrenoceptor-mediated effect and slightly relaxed by isoprenaline in low concentrations due to a β-adrenoceptor effect (Nilsson et al. 1975). There is histochemical evidence off an adrenergic innervation of the major systemic arteries in the dogfish, *Squalus acanthias* (Nilsson et al. 1975). The bulk of the literature on the adrenergic control of the circulation in elasmobranchs seems, however, to favour the circulating catecholamines as the most important factor in the control of blood presure (Opdyke et al. 1972, Short et al. 1977, Capra and Satchell 1977b, Butler et al. 1977).

There are few examples of cardiovascular reflexes in elasmobranchs but one well-documented phenomenon is the bradycardia that follows rapidly induced hypoxia in the dogfish (Satchell 1961, Butler and Taylor 1971). The afferent fibres in the reflex arc originate in oxygen receptors which are widely distributed in the gills (Butler et al. 1978). The vagal cholinergic innervation of the heart comprises the efferent cardioinhibitory pathway (cf. Fig. 7.1). During the bradycardia induced by hypoxia, stroke volume of the heart increases. This effect is due to the increased filling of the heart between systoles (Starling's law of the heart) and there is no evidence for an involvement of adrenergic mechanisms in the positive inotropic effect. However, an adrenergic control of the heart during normoxia is present (Short et al. 1977).

7.3 Teleosts

7.3.1 The Teleost Heart

In teleosts, the cardiac branches of the vagi follow the ducts of Cuvier to the sinus venosus and atrium, but vagal fibres do not reach the ventricle. A ganglion in the vagal pathway lies close to the sino-atrial border and appears to consist solely of non-adrenergic cell bodies (Laurent 1962, Yamauchi and Burnstock 1968, Gannon and Burnstock 1969, Santer and Cobb 1972, Santer 1972, Holmgren 1977, 1981).

The vagus is cardioinhibitory as in all vertebrates with the exception of cyclostomes. This inhibitory effect is due to the release of acetylcholine acting via muscarinic cholinoceptors associated with the pacemaker and atrial musculature (Young 1936, Couteux and Lauret 1957, Jullien and Ripplinger 1957, Laurent 1962, Laffont and Labat 1966, Randall 1966, Randall and Stevens 1967, Stevens and Randall 1967, Gannon and Burnstock 1969, Gannon 1971, Johansen 1971, Cobb and Santer 1973, Helgason and Nilsson 1973, Priede 1974, Saito and Tenma 1976, Holmgren 1977, 1981, Jones and Randall 1978, Cameron 1979, Wood et al. 1979). Variations in the vagal tonus will affect heart rate and a vagal, inhibitory, resting tonus has been demonstrated in some species (*Catostomus*, Stevens and Randall 1967; *Prionotus*, Roberts 1968; *Ophiodon*, Stevens et al. 1972; *Salmo*, Priede 1974; *Carassius*, Cameron 1979). There are some indications that the importance of a vagal tonus in cardiac control is temperature-dependent, with adrenergic mechanisms taking over the cardioacceleratory function at higher temperatures (Laffont and Labat 1966, Priede 1974, Wood et al. 1979). Although the negative inotropic influence of the vagi does not reach the ventricle, the cardiac output is greatly affected by the inotropic control of the atrium, which directly regulates the filling of the ventricle (Jones and Randall 1978, Johansen and Burggren 1980).

A great variety of external stimuli produce transient bradycardia or cardiac arrest in teleosts. This "startle response" appears to be vagally mediated since it is readily abolished by atropine (Leivestad et al. 1957, Randall 1968, Stevens et al. 1972, Priede 1974, Wahlqvist and Nilsson 1980).

An adrenergic innervation of the teleost heart was first demonstrated by histochemical and ultrastructural studies (Govyrin and Leont'eva 1965, Yamauchi and Burnstock 1968) and has since been confirmed by a number of physiological and further anatomical studies. Adrenergic fibres have now been demonstrated in all parts of the teleost heart. These fibres reach the heart (1) together with the vagi ("vago-sympathetic trunks"), (2) along the anterior spinal nerves or (3) along the coronary arteries to the ventricle (Gannon and Burnstock 1969, Holmgren 1977). In the cod, *Gadus morhua*, the adrenergic innervation of the heart also includes pericellular networks around the intracardiac ganglion cells (Holmgren 1977).

The density and importance of the adrenergic cardiac innervation seems to vary with species. For example, there is no adrenergic innervation at all of the plaice (*Pleuronectes platessa*) heart, although the heart is sensitive to catecholamines (Falck et al. 1966, Santer 1972, Cobb and Santer 1973). A rebound excitation of the plaice heart following the inhibition caused by vagal stimulation has been demonstrated, but the physiological role of this mechanism awaits further explanation (Cobb and Santer 1973).

An adrenergic tonus on the hearts of cod (*Gadus*) and goldfish (*Carassius*) has been demonstrated, but the relative importance of the neuronal and humoral adrenergic control of the teleost heart remains uncertain (Helgason and Nilsson 1973, Wahlqvist and Nilsson 1977, Cameron 1979, Cameron and Brown 1981). Catecholamines released from the head kidney have marked effects on the perfused cod heart (Holmgren 1977) and the levels of circulating catecholamines in the plasma of cod and several other species are sufficient to produce an adrenergic tonus on the heart. In some species of teleosts, specialized catecholamine-storing endothelial cells have been demonstrated in the atrium, but the physiological significance of these cells is unknown (Saetersdal et al. 1974, Leknes 1980).

The positive chronotropic and inotropic effects on the teleost heart produced by adrenergic agonists and adrenergic nerves are mediated via β-adrenoceptor mechanisms associated with the pacemaker and the myocardial cells (Fänge and Östlund 1954, Östlund 1954, Randall and Stevens 1967, Gannon and Burnstock 1969, Forster 1976a, Holmgren 1977, Wahlqvist and Nilsson 1977, Cameron and Brown 1981). The β-adrenoceptors of the trout (*Salmo gairdneri*) heart resemble the β_2-adrenoceptors of mammals (Ask et al. 1980).

7.3.2 The Teleost Branchial Vasculature

The blood entering the gills via the four pairs of afferent branchial arteries may leave either via the efferent arterial system (the arterio-arterial pathway), or via the venous drainage of the gills (the arterio-venous pathway) (Fig. 7.2). The arterio-arterial pathway may, in a somewhat simplified manner, be looked upon as a respiratory pathway in which deoxygenated blood entering the gills is oxygenated in the gill lamellae and continues into the systemic arterial system. The arterio-venous pathway may, with a similar simplification, be regarded as non-respiratory and particularly important in ionic regulation and nutrition of the gill tissue (Laurent and Dunel 1980). For detailed descriptions of the vascular microanatomy in the

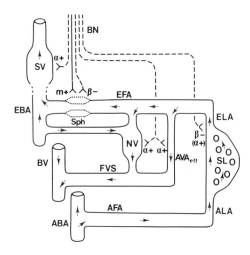

Fig. 7.2. Diagrammatic representation of the arrangement of the branchial vasculature and possible sites of action of the autonomic fibres in the cod, *Gadus morhua*. The blood enters the afferent filamental artery (*AFA*) from the afferent branchial artery (*ABA*) and leaves the filamental circulation either via the efferent branchial artery (*EBA*) (arterio-arterial pathway) or the branchial vein (*BV*) (arterio-venous pathway). The cranial autonomic fibres in the branchial nerve (*BN*) constrict the arterio-arterial pathway possibly by contracting the sphincter (*Sph*) at the base of the efferent filamental artery (*EFA*). The adrenergic fibres originate from the cephalic sympathetic chain ganglia and act chiefly by constricting the efferent arterio-venous anastomoses (AVA_{eff}) and/or the filamental nutritional vasculature (*NV*) which drain into the filamental venous system (*FVS*). A β-adrenoceptor-mediated vasodilatior control by adrenergic fibres innervating the vasculature of the secondary lamellae (*SL*), probably the afferent (*ALA*) or efferent (*ELA*) lamellar arterioles, is also present. A general adrenergic control of the systemic blood pressure by innervation of the systemic vasculature (*SV*) is probably of importance in the control of arterio-venous shunting in the gill filaments. $\alpha+$, $\beta-$ and $m+$ refer to α-adrenoceptors mediating vasoconstriction, β-adrenoceptors mediating vasodilation and muscarinic cholinoceptors mediating vasoconstriction, respectively. [Based mainly on Petterson and Nilsson (1979a), Nilsson and Petterson (1981)]

teleost gill, see Laurent and Dunel (1976, 1980), Vogel et al. (1976), Dunel and Laurent (1977) and Cooke and Campbell (1980).

The general reduction in the branchial vascular resistance of the pike (*Esox lucius*) in response to adrenaline demonstrated by Krawkow (1913), has been confirmed by a large number of studies on different species of teleosts (Keys and Bateman 1932, Kirschner 1969, Rankin and Maetz 1971, Bolis and Rankin 1975, Forster 1976a, Dunel and Laurent 1977: *Anguilla anguilla;* Richards and Fromm 1969, Wood 1974, 1975, Payan and Girard 1977: *Salmo gairdneri;* Pettersson and Nilsson 1979a, Wahlqvist 1980: *Gadus morhua;* Belaud et at. 1971: *Cyprinus carpio, Conger conger;* Reite 1969: *Anguilla anguilla, Gadus morhua;* Östlund and Fänge 1962: *Anguilla anguilla, Zoarces viviparus, Gadus morhua, Labrus berggylta*). In some of these studies, an α-adrenoceptor-mediated vasoconstriction in the arterio-arterial pathway has been demonstrated together with the dilatory effect of adrenaline, which is mediated via β-adrenoceptors (Reite 1969, Beleaud et al. 1971, Wood 1975, Wood and Shelton 1975, 1980a, Dunel and Laurent 1977, Payan and Girard 1977, Wahlqvist 1980). Evidence for a constrictory effect of acetylcholine on the teleost branchial vasculature has also been presented (Östlund and Fänge 1962, Reite 1969, Wood 1975, Dunel and Laurent 1977, Smith 1977).

Since there have been very few studies on the direct autonomic innervation of the vascular effectors in teleost gills, the available knowledge rests mainly on experiments with *Gadus morhua*. The gills receive an autonomic innervation from the branchial nerves, which are branches of the glossopharyngeal (1st pair of gill arches) and vagus (1st to 4th pair of gill arches). Cranial autonomic fibres appear to be absent in the glossopharyngeal (Pettersson and Nilsson, unpublished), but the sympathetic chains carry ganglia in connection with the cranial nerves IX and X and the branchial nerves are therefore mixed "vago/glossopharyngeo-sympathetic trunks" of the type seen elsewhere.

Stimulation of the branchial nerve to the 3rd gill arch produces a constriction of the arterio-arterial pathway (Nilsson 1973, Pettersson and Nilsson 1979a) and this effect is at least partially blocked by atropine, showing a cholinergic component in the constrictory innervation. Fibres joining the branchial nerves from the sympathetic chain ganglia also affect the branchial blood flow, an action that favours the arterio-arterial in comparison with the arterio-venous blood flow (Nilsson and Pettersson 1981). The effects of fibres from the sympathetic chains, which are solely adrenergic, are mediated via both α- and β-adrenoceptors. A summary of the possible sites of action of the adrenergic nerves in the various parts of the complex branchial vasculature is presented in Fig. 7.2. Nerve fibres have been demonstrated by electron microscopy in the efferent lamellar arterioles and in the spincter of the efferent filamental artery (Laurent and Dunel 1980 and unpublished). Whether the sphincter is important in the control of vascular resistance is not clear (see Nilsson and Pettersson 1981). A possible function of the nerve endings as baro- or chemoreceptors needs investigation in view of the difficulties involved in the ultrastructural identification of nerve profiles outlined by Gibbins (1982).

In spite of the clearly demonstrated autonomic innervation of the branchial vasculature, the most important control of at least the arterio-arterial pathway is probably by circulating catecholamines released from the chromaffin tissue in the head kidney (Wahlqvist 1980, 1981, Wahlqvist and Nilsson 1980). The direct adrenergic nervous control of the branchial vasculature may instead be of particular importance in the regulation of shunting between the arterio-arterial and the arterio-venous pathway (Pettersson and Nilsson 1979a, Nilsson and Pettersson 1981, Pettersson and Johansen 1982).

7.3.3 The Teleost Systemic Vasculature

Numerous studies in a variety of teleost species show that injected adrenaline and noradrenaline induce an increased arterial blood pressure due to a constriction of the systemic vasculature by an α-adrenoceptor-mediated mechanism (e.g., Mott 1951, Stevens and Randall 1967, Reite 1969, Helgason and Nilsson 1973, Forster 1976a, b, Wood 1976, Wood and Shelton 1980b, Davie 1981). In some cases injection of isoprenaline has been shown to cause a decrease in blood pressure and a decreased systemic vascular resistance, which has been taken as evidence for the presence of β-adrenoceptors in the systemic vasculature (Helgason and Nilsson

1973, Chan and Chow 1976, Wood 1976, Pettersson and Nilsson 1980, Davie 1981). Also to be borne in mind is the possibility of an α-adrenoceptor *antagonistic* effect of isoprenaline which can be expected to reduce the adrenergic constrictory tonus on the vasculature (Holmgren and Nilsson 1974, Wood 1976).

Fluorescence histochemistry of the catecholamines (Fig. 2.23), together with studies of isolated blood vessels or perfused vascular beds, provide good evidence for a substantial adrenergic innervation of major arteries and resistance vessels (arterioles) in several species of teleosts (Kirby and Burnstock 1969a, Nilsson 1972, 1973, Holmgren and Nilsson 1974, Klaverkamp and Dyer 1974, Nilsson and Grove 1974, Holmgren 1978, Wahlqvist and Nilsson 1981). An interesting question is the extent to which the adrenergic control over visceral and somatic vascular beds (and thus the systemic blood pressure) in teleosts is exerted via adrenergic neurons, compared to the adrenergic tonus created by the circulating adrenaline (and noradrenaline) released from the chromaffin tissue. For the rainbow trout, Wood and Shelton (1975) and Smith (1978a) conclude that the major adrenergic control of the cardio-vascular system is neuronal, while experiments with the cod, *Gadus morhua*, have been interpreted in favour of a mainly humoral adrenergic tonus (Wahlquist and Nilsson 1977). While it seems reasonable to believe that the adrenergic vasodilation of the arterio-arterial pathway in the gills during "stress" is primarily due to circulating amines (Wahlqvist 1980, Wahlqvist and Nilsson 1980), the relative importance of neuronal und humoral catecholamines in the control of the systemic vasculature and the heart is still unclear.

A cholinergic innervation of the major arteries in the eel, trout and other lower vertebrates was postulated by Kirby and Burnstock (1969a) as a complement to the adrenergic vasconstrictory innervation. It has also been suggested that a cholinergic "sympathetic" innervation of the vasculature is a phylogenetically early type of vasoconstrictory innervation, which is taken over in the higher vertebrates by a solely adrenergic vasoconstrictory innervation (Kirby and Burnstock 1969a, Burnstock 1969).

However, more recent experiments on the spinal autonomic innervation of the coeliac artery and the responses to adrenergic and cholinergic agonists of the same vessel of the cod and other species of teleosts have yielded no unequivocal evidence for a cholinergic innervation. In fact, the coeliac artery of the cod and several other gadid species reveal a very poorly developed sensitivity to acetylcholine and carbachol, which is also true for perfused visceral and somatic vascular beds (Reite 1969, Nilsson 1972, Holmgren and Nilsson 1974, Wahlqvist and Nilsson 1981, Holmgren, unpublished). Thus, although it is clear that cholinergic fibres are abundant in the spinal autonomic control of various organs in teleosts, the idea of a primarily cholinergic vascular control is not universally valid among the teleosts.

Information on cardiovascular reflex mechanisms in teleosts, which is somewhat better than for elasmobranchs, includes evidence for both baro- and chemoreceptors in the gill region. A connection between the two types of reflexes at a central level has also been demonstrated (Wood and Shelton 1980b). The available evidence rests primarily on experiments with salmonids, notably *Salmo gairdneri*, in which "oxygen receptors" involved in the hypoxic bradycardia reflex (see elasmobranchs above) appear to be located primarily in the anterior gill arches (Daxboeck and Holeton 1978, Jones and Randall 1978, Smith and Jones 1978).

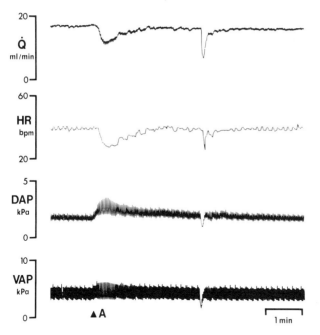

Fig. 7.3. Effects of intravenously injected adrenaline (*A*) on ventral aortic blood flow ($\dot{Q}$), heart rate (*HR*), dorsal (*DAP*) and ventral (*VAP*) aortic blood pressure in the cod. Note the decrease in heart rate and blood flow due to a cardioinhibitory reflex triggered by the elevated systemic blood pressure. The drop in heart rate, flow and blood pressure occurring some time after the adrenaline response is due to a "startle response" by the fish. (Reproduced with permission from Pettersson and Nilsson 1980)

As in the higher vertebrates, injection of adrenaline produces an elevated blood pressure which is accompanied by a bradycardia due to a baroceptor reflex (Fig. 7.3, Mott 1951, Randall and Stevens 1967, Ristori 1970, Helgason and Nilsson 1973, Wood and Shelton 1980a). The baroreceptors that initiate this reflex are, at least in the carp, located in various parts of the gills (Ristori 1970, Wood and Shelton 1980b).

Laurent (1967) demonstrated changes in the impulse frequency in the afferent fibres from the pseudobranch of the tench (*Tinca tinca*) in response to changes in perfusion pressure (baroreceptor function) and chemical composition of the perfusion medium (chemoreceptor function). Although denervation of the pseudobranch in the rainbow trout proved ineffective in changing the ventilatory and cardiac responses to hypoxia (Randall and Jones 1973, Bamford 1974), further studies of the physiological role of the pseudobranch would be of great interest.

7.4 Ganoids

The heart of both chondrostean and holostean ganoids is innervated by branches of the vagi, which have connections with the ganglia of the sympathetic chain (Stannius 1849, Ganfini 1911, cited in Nicol 1952, Allis 1920). In the sturgeon (*Acipenser*: Chondrostei), a vagal cholinergic inhibitory pathway to the heart has been postulated (Kisch 1950); acetylcholine produces negative inotropic effects on both ventricular and, especially, atrial strip preparations from the beluga (*Huso huso*) (Balashov et al. 1981). There appears to be no direct innervation of the beluga heart by adrenergic fibres, but adrenaline produces positive inotropic effects on the atrium and ventricle, and the levels of circulating catecholamines are sufficient to affect the cardiac force of contraction (Balashov et al. 1981).

Acetylcholine also produces negative inotropic effects on the atrium of the gar (*Lepisosteus:* Holostei). The heart of this species is richly innervated by adrenergic fibres in all parts except the ventricle, and a functional adrenergic control of the gar heart, both by adrenergic neurons and circulating catecholamines, has been postulated (Nilsson 1981).

In the beluga (*Huso huso*), chromaffin cells have been demonstrated in the coeliaco-mesenteric artery, but the extent of an adrenergic innervation of the major arteries is unknown (Balashov et al. 1981). Both acetylcholine and catecholamines constrict the arteries of *Huso*.

There is a dense adrenergic innervation of blood vessels in the swimbladder (lung) of *Lepisosteus* but nothing is known about the involvement of the autonomic nervous system in the vascular control in this species (Nilsson 1981).

7.5 Dipnoans

The heart of the lungfish is innervated by the cardiac branches of the vagi, and there is a possibility of a contribution of the vagi from the spinal autonomic nervous system. The sympathetic chains are, however, poorly developed in lungfish and it is doubtful if there are any adrenergic neurons in this system (Jenkin 1928, Abrahamsson et al. 1979 b, Chap. 2.5).

High doses of acetylcholine produce bradycardia, but there is no evidence for a cholinergic tonus on the heart in vivo, since atropine is without effect on the heart rate in the intact animal (Johansen and Reite 1968). Strong negative inotropic effects of carbachol have been demonstrated in the atrium, but not the ventricle, of *Protopterus* (Abrahamsson et al. 1979 b).

As in cyclostomes, there is a large store of catecholamines in the heart of *Protopterus*, where adrenaline and noradrenaline have been demonstrated in chromaffin cells lining the atrial lumen (Abrahamsson et al. 1979 a, Scheuermann 1979). A positive chronotropic effect of adrenaline has been recorded; this effect appears to be most pronounced in aestivating animals (Mohsen et al. 1974). Abrahamsson et al. (1979 b) failed to demonstrate inotropic effects by adrenaline on the *Protopterus* atrium, which is interesting in view of the storage of catecholamines in this part of the heart (cf. cyclostomes, Sect. 7.1.1).

Acetylcholine constricts the branchial, pulmonary and systemic vascular beds of *Protopterus*, whereas adrenaline constricts the pulmonary and systemic vasculature but dilates the branchial vasculature (Johansen and Reite 1968, Reite 1969). In view of the apparently poorly developed adrenergic nervous system in the same species, an adrenergic control mainly via circulating catecholamines seems likely (Abrahamsson et al. 1979 a, b). The complex reflexogenic control of blood flow in the branchial and pulmonary circuits during the change from aquatic to air breathing is of great interest in view of the peculiar arrangement of the cardiovascular system in the lungfish. With the development of the lung as a second respiratory organ, the heart receives both hypoxic blood from the venous system and oxygenated blood from the lung directly into the atrium. A high degree of separation of the two types of blood has been demonstrated, showing the first signs of a functional separation of venous and arterial blood in the heart (Johansen et al. 1968, Johansen 1971, 1982, Johansen and Burggren 1980), but the role of the autonomic nervous system in the shunting of blood in gills and lung is poorly understood.

7.6 Amphibians

7.6.1 The Amphibian Heart

The amphibian heart is divided into two separate atria and a single ventricle. Despite the lack of anatomical division of the ventricle there is, however, a surprisingly high degree of separation of oxygenated and deoxygenated blood leaving the ventricle both in urodele and anuran amphibians. The separation is due chiefly to the trabeculation of the ventricle, which creates to laminar outflow pattern, and the presence of a spiral valve in the bulbus cordis, which separates and directs the blood from the left and right half of the ventricle to the appropriate arteries (Johansen 1963, 1982, Johansen and Hanson 1968, Johansen et al. 1970, Johansen and Burggren 1980).

The oxygenated blood from the lungs passes via the left atrium into the left half of the ventricle and is ejected mainly, but to a variable degree depending on the mode of breathing, into the systemic circuit during systole. Conversely, the deoxygenated venous blood enters the right half of the ventricle via the right atrium and is ejected chiefly into the pulmo-cutaneous artery (Tazawa et al. 1979, Johansen and Burggren 1980). Shunting of the blood into the pulmonary circuit (during lung breathing) and mainly into the systemic circuit (during periods of apnoea) is controlled by autonomic fibres (see below).

The amphibian heart is innervated by vagal fibres and by fibres from the sympathetic chain ganglia that join the vagi near the cranium to form "vagosympathetic trunks." There are no direct spinal autonomic cardiac nerves in amphibians. The cardiac branches of the "vago-sympathetic trunks" enter the heart along the vena cava, and form three pairs of ganglia within the heart: Remak's ganglion in the sinus venosus at the base of the pulmonary veins, and Ludwig's and Bidder's ganglia in the atrial septum (Nicol 1952, Taxi 1976).

The vagal fibres, which are cholinergic, produce negative chronotropic and inotropic effects both in urodeles and anurans (Bidder 1868, Gaskell 1884, McWilliams 1885, Elliott 1905, Loewi 1921, Kirby and Burnstock 1969b, Woods 1970a, b, Campbell et al. 1982). Stimulation of the "vagosympathetic trunk" in amphibians is claimed to produce different responses depending on the season: the vagal (inhibitory) response is most marked during the summer (Schitt cited in Mills 1885, Burnstock 1969). In the toad, *Bufo marinus*, an electron microscope study of the non-adrenergic autonomic fibres in the heart shows nerve profiles with many large granular vesicles typical of "p-type" nerve terminals (see Chap. 3.2.1). There is now good evidence that these fibres contain and release both acetylcholine (the classical "Vagusstoff" of Loewi) and somatostatin (Campbell et al. 1982). All of the intrinsic neurons of the toad heart (i.e., the postganglionic vagal neurons) show somatostatin-like immunoreactivity, and it has been concluded that the hyoscine-resistant cardioinhibition seen during vagal stimulation at frequencies above 3 Hz is due to somatostatin release from the mixed cholinergic/somatostatin-containing neurons (Campbell et al. 1982). The presence of both acetylcholine and somatostatin within the same neurons innervating the toad heart is indicated by a mixed nerve symbol in Fig. 7.7.

The spinal autonomic fibres that enter the "vagosympathetic trunk" of the amphibians are adrenergic and produce positive chronotropic and inotropic effects on the heart (Gaskell 1884, Elliott 1905, Loewi 1921, Woods 1970a, b). In the frog (*Rana* sp.) the sinus venosus, atria and ventricle are densely innervated by adrenergic fibres (Falck et al. 1963, Stene-Larsen and Helle 1978a). The main transmitter substance in the amphibian heart is, as in other tissues, adrenaline, and it has been concluded from agonist potency studies that the adrenoceptors of the heart of the frog (*Rana* spp.) are of the β_2-variety. This is in accordance with the hypothesis of the β-$_2$-adrenoceptor as the "adrenaline-receptor" (Lands et al. 1969, Stene-Larsen and Helle 1978a, Stene-Larsen 1981, Chap. 4.2.1.1).

Another type of adrenoceptor ("junctional adrenoceptor") mediating positive chronotropic and inotropic effects in the toad heart, and which differs from the α- and β-adrenoceptors, has also been described (Morris et al. 1981).

7.6.2 The Amphibian Pulmonary Vasculature

It has been emphatically pointed out that the heart and circulatory system of the lower vertebrates do *not* represent imperfect evolutionary stages, awaiting the perfection of the separate pulmonary and systemic circuits of birds and mammals, but are in fact highly functional systems adapted to the specific needs of the animal (Johansen and Burggren 1980). This statement rings particularly true in the shunting of blood associated with the multiple modes of respiration in amphibians (lungs, skin and in urodeles sometimes also gills), and the circulatory adaptation to periodic breathing in amphibians and reptiles (Johansen et al. 1970, Johansen and Burggren 1980, Johansen 1982). Thus the *anatomically* incomplete ventricular separation, despite its *functional* separation of oxygenated and deoxygenated blood, allows an intracardiac shunting of blood and therefore an individual control of bloodflow in the pulmonary and systemic circuits respectively. Such an individual flow regulation is not possible in birds and mammals, in which pulmonary flow always equals the systemic flow.

During breathing periods, the vascular resistance in the pulmonary circuit of *Xenopus* is about half that of the systemic circuit, but during periods of apnoea a constriction of the pulmonary vasculature increases the pulmonary vascular resistance fivefold (Shelton 1970). The blood is thus directed to a chiefly systemic flow (right-to-left shunt in the heart) and the cardiac work is limited to the perfusion

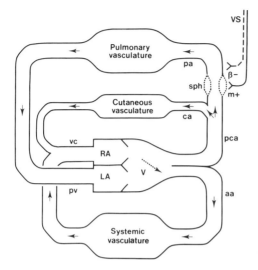

Fig. 7.4. Diagrammatic representation of the cardiovascular system of an amphibian, showing the site of vasoconstrictory innervation of the pulmonary artery. During constriction of the pulmonary artery sphincter (*sph*), the blood is directed into the cutaneous vasculature and a right-to-left shunting of the blood takes place in the ventricle (*broken arrow*). Abbreviations: *aa*, aortic arches; *ca*, cutaneous artery, *LA*, left atrium; *m+*, muscarinic cholinoceptors mediating constriction of the sphincter; *pa*, pulmonary artery; *pca*, pulmo-cutaneous artery; *pv*, pulmonary vein; *RA*, right atrium; *sph*, pulmonary artery sphincter; *V*, ventricle; *vc*, vena cava; *VS*, "vago-sympathetic trunk"; *β-*, β-adrenoceptors mediating dilation of the sphincter and/or pulmonary vasculature generally

of the systemic circuit (Fig. 7.4). The shunting of blood is controlled by an autonomic reflex, triggered by the deflation of the lung.

It has long been known that pulmonary blood flow in anurans can be stopped by stimulation of the "vagosympathetic trunk" (Couvreur 1889), and that this vasoconstrictory nervous influence is tonic (Luckhardt and Carlson 1921c). Campbell (1971b) showed that the vasoconstrictor fibres to the lung of the toad (*Bufo marinus*) are truly vagal in origin and cholinergic. Cholinesterase-positive nerve fibres have been demonstrated in the pulmonary artery of *Rana temporaria* (Leont' eva 1978).

In *Rana temporaria* an anatomically distinct sphincter has been demonstrated in the pulmonary artery close to the point of branching from the pulmocutaneous artery (Fig. 7.4) (de Saint-Aubain and Wingstrand 1979). This sphincter is innervated by cholinergic constrictory vagal fibres and may, at least in some species of amphibians, represent the most important effector unit in the cholinergic constrictory control of the pulmonary blood flow (de Saint-Aubain and Wingstrand 1979).

During constriction of the pulmonary vasculature, the blood in the pulmocutaneous artery is directed primarily towards the skin, which may serve as an important gas exchanger during diving in anurans. The cutaneous circulation is of paramount importance in the excretion of CO_2 during prolonged submersion, and it has been demonstrated that hypercapnia in fact triggers the pulmonary vasoconstrictory reflex in the toad (Smith 1978b, Johansen 1982).

In addition to the important cholinergic constrictory innervation of the pulmonary vasculature, there is an adrenergic innervation that reaches the lung via the "vago-sympathetic trunk". This innervation has been demonstrated both by fluorescent histochemistry (McLean and Burnstock 1967a, b) and physiological experiments (Campbell 1971b, Holmgren and Campbell 1978). As in reptiles, but quite contrary to the effect in mammals, adrenaline and the adrenergic innervation dilate the pulmonary vascular bed, possibly acting via a β-adrenoceptor mechanism (Campbell 1971 b, Holmgren and Campbell 1978). Whether an adrenergic innervation of the pulmonary artery sphincter or a general adrenergic innervation of the pulmonary vasculature plays the major role in the dilatory effect in the pulmonary circuit is not known (Fig. 7.4).

7.6.3 The Amphibian Systemic Vasculature

The amphibian systemic vasculature is supplied by fibres from the sympathetic chain ganglia. It was demonstrated long ago that vasomotor fibres from the 8th and 9th sympathetic chain ganglia to the hind leg of the frog are included in the sciatic nerves (Langley 1911). Adrenaline injected into intact amphibians produces an elevated blood pressure and a constriction of perfused vascular beds is also seen after addition of adrenaline (Erlij et al. 1965, Kirby and Burnstock 1969 b). The adrenergic innervation of the major arteries, as revealed by fluorescence histochemistry, seems to be less dense than in reptiles and mammals but adrenergic fibres are present both in arteries and veins (Banister and Mann 1966, McLean and Burnstock 1966, Kirby and Burnstock 1969 a, Leont'eva 1978, Nilsson 1978). In addition to the adrenergic vasoconstrictor innervation, a cholinergic excitatory innervation of the major arteries in the toad (*Bufo marinus*) has been postulated (Kirby and Burnstock 1969 a). Burnstock and Kirby (1968) were unable to demonstrate an inhibitory effect of isoprenaline on the large arteries of the toad but other investigations show a β-adrenoceptor mediated dilatory effect of isoprenaline in the perfused hind limb of the frog (Erlij et al. 1965) and in the perfused renal vasculature of the toad (Morris 1982).

The reflexogenic control of the amphibian cardiovascular system is obviously closely related to the respiratory events, with the control of the pulmonary vascular resistance being of great functional significance. Baro- and chemoreceptor functions have been ascribed to the amphibian carotid labyrinth, a structure homologous with the carotid body and sinus of mammals (Meyer 1927, Ishii et al. 1966). A baroreceptor function associated with the carotid labyrinth may, however, be unimportant at normal physiological blood pressures (Burnstock 1969, Ishii and Ishii 1978). Moreover the main site of baroreceptors in the toad (*Bufo vulgaris formosa* and *Bufo marinus*) appears to be the pulmocutaneous artery (Ishii and Ishii 1978, Smith et al. 1981).

7.7 Reptiles

7.7.1 The Reptilian Heart

There are two separate atria in the reptilian heart, but the ventricle of the non-crocodilian reptiles is composed of three compartments which are anatomically continuous. In spite of the superficially single ventricle, a very high degree of separation of the oxygenated blood from the lung and the deoxygenated blood from the systemic circulation occurs, and the presence of an anatomical continuity between the ventricular compartments opens the possibility of intracardiac shunting and control of the pulmonary and systemic circuit, respectively. In the non-crocodilian reptiles, blood circulation through the heart takes place according to the following pattern: The oxygenated blood from the pulmonary veins enters the *cavum arteriosum* of the ventricle via the left atrium, and then continues during the diastole and early systole into the larger *c. venosum* which ejects the oxygenated "left" blood preferentially into the left and right systemic aortae. The deoxygenated systemic venous blood enters the right atrium via the sinus venosus and from here also passes into the *c. venosum*. From this compartment the deoxygenated "right" blood passes preferentially into the third of the ventricular compartments, the *c. pulmonale*, and is finally ejected into the common pulmonary artery. During periods of breathing, a left-to-right shunt occurs within the heart, due to the lower pulmonary vascular resistance, and blood passes from the *c. arteriosum* via the *c. venosum* to the *c. pulmonale* (White and Ross 1966, Burggren 1977, Johansen and Burggren 1980, Johansen 1982).

In varanid lizards, the *c. venosum* is reduced and the separation of the left and right part of the ventricle by the muscular ridge (Webb et al. 1971) during systole is sufficiently complete enough to allow a difference between pulmonary and systemic blood pressures (Johansen and Burggren 1980).

In crocodiles, finally, there is a complete anatomical separation of the left and right ventricle, but the possibility of a right-to-left shunting persists due to a connection between the left and right aortic arches (foramen Panizzae) (White 1970, Johansen and Burggren 1980).

The reptilian heart is capable of an enormous variation of the cardiac output between periods of breathing and periods of apnoea. In the turtle, *Pseudemys scripta*, a tachycardia resulting in a 20–30-fold increase in cardiac output has been recorded at the onset of breathing (White and Ross 1966, Shelton and Burggren 1976). The tachycardia appears to be due to the withdrawal of a vagal tonus on the heart (Burggren 1975). As in amphibians, the control of the pulmonary and systemic blood flow, and thus the degree of intracardiac shunting of blood, is determined by the relative vascular resistance of the pulmonary and systemic circuits (Shelton and Burggren 1976, Burggren 1977, Sect. 7.7.2).

The vagal innervation of the reptilian heart is cholinergic and inhibitory. Thus negative chronotropic and inotropic effects occur during electrical stimulation of the vagi in crocodiles, turtles, lizards and snakes, and these responses are mimicked by acetylcholine and blocked by muscarinic cholinoceptor antagonists such as atropine or hyoscine (Gaskell 1883, 1884, Mills 1884, Khalil and Malek 1952, de la Lande et al. 1962, Berger 1971, Hedberg and Nilsson 1976, Berger and Burnstock 1979).

The negative chronotropic effect is due to impaired impulse generation in the pacemaker, which is situated in the sinus venosus, and an innervation of this region by "c-type" nerve terminals was demonstrated by Yamauchi (1969). In addition to the block of impulse generation, a cholinergically mediated blockade of impulse propagation from the pacemaker to the atria can occur (Berger and Burnstock 1979). A "vagal escape" of the type seen in mammals is not apparent in reptiles; continuous vagal stimulation can arrest the turtle heart for very long periods (Mills 1885).

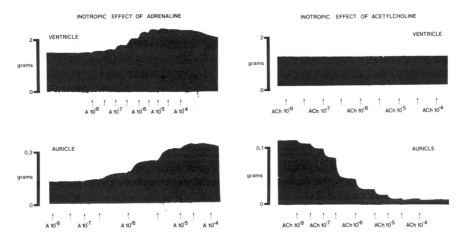

Fig. 7.5. Inotropic effects of adrenaline (*A*, left tracings) and acetylcholine (*ACh*, right tracings) on paced ventricular and atrial (auricular) strip preparations from the heart of the puff adder (*Bitis arietans*). Note the lack of effect of acetylcholine on the ventricular preparation. Concentration of agonists 10^{-8}, 3×10^{-8}, 10^{-7} etc are expressed in mol/l (M). (Reproduced with permission from Hedberg and Nilsson 1976)

A negative inotropic effect of vagal stimulation in the atria is well documented but, although the turtle ventricle is innervated by "c-type" fibres (Yamauchi and Chiba 1973) which may be cholinergic, there is no clear evidence for a negative inotropic effect of these fibres or added cholinergic agonists (Fig. 7.5, Gaskell 1883, Knowlton 1942, Hedberg and Nilsson 1976). The possibility that the "c-type" fibres demonstrated by Yamauchi and Chiba (1973) are non-cholinergic must also be kept in mind in view of the difficulties involved in the interpretation of nerve profiles demonstrated by electron microscopy (Gibbins 1982). In a study of the lizard, *Trachydosaurus rugosus*, Berger and Burnstock (1979) demonstrated a very small negative inotropic effect on the heart by vagal stimulation or carbachol; a possibly more important role of the cholinergic vagal fibres to the ventricle was suggested by Burggren (1978).

In these experiments Burggren (1978) demonstrated two distinct patterns of depolarization of the turtle heart. The first pattern ("apnoea pattern") was recorded during periods of apnoea, and shows a relatively rapid (0.15 m/s) spread of the ventricular depolarization from the left to the right half of the ventricle. During periods of breathing a "ventilatory pattern" evolved, in which the direction of the depolarization wave was reversed, now starting in the right half of the ventricle and moving at a much slower rate (0.09 m/s). It could be shown that this change in depolarization pattern was produced by cholinergic vagal fibres and that the pattern recorded during ventilation helped to improve the separation of the oxygenated and deoxygenated blood in the ventricle (Burggren 1978). The vagal fibres probably innervate the left branch of the strands of specialized conducting cells that run from the atrio-ventricular junction into the two halves of the ventricle (Robb 1953, Burggren 1978). Functionally it appears to be an advantage that the improved cardiac function during ventilation achieved by the vagal control of conduction velocity (dromotropic effects) is not impaired by a negative inotropic effect on the ventricular myocardium.

The reptilian heart is also innervated by adrenergic fibres of spinal autonomic origin, which mediate positive chronotropic and inotropic effects. The fibres either run in a separate nerve to the heart from the stellate complex (Fig. 2.9) or may join the vagi near the cranium (snake: Hedberg and Nilsson 1975, turtles: Gaskell and Gadow 1884, Mills 1885, lizard: Khalil and Malek 1952) or near the heart (crocodiles: Gaskell 1884, Gaskell and Gadow 1884, lizard: Berger 1971).

"a-type" nerve profiles have been demonstrated in the turtle heart (Yamauchi 1969, Yamauchi and Chiba 1973) and there are several fluorescence histochemical demonstrations of adrenergic fibres in all parts of the reptilian heart (Fig. 2.14; turtle, lizards: Govyrin and Leont'eva 1965, lizard: Furness and Moore 1970, snake: Hedberg and Nilsson 1975, crocodile: Berger and Burnstock 1979). In the lizard *Agama causospinosa*, fluorescent nerve terminals have also been observed surrounding ganglion cells in the cardiac ganglion (Nilsson, unpublished).

The effect of the adrenergic nerves is mediated by β-adrenoceptors (Fig. 7.5, Han et al. 1973, Hedberg and Nilsson 1975, 1976). It appears that the heart of some reptiles is under tonic influence by both cholinergic and adrenergic fibres (Hedberg and Nilsson 1975, Lillywhite and Seymour 1978). Thus the heart rate in the anaesthetized puff adder (*Bitis arietans*) increases from 44 to 56 beats per min after injection of atropine, and is reduced from 44 to 21 beats per min after injection of the β-adrenoceptor antagonist propranolol (Hedberg and Nilsson 1975).

7.7.2 The Reptilian Pulmonary Vasculature

Since, as in amphibians, periods of apnoea are accompanied by a marked vasoconstriction of the pulmonary circuit, it is not surprising that the arrangement of the innervation of the pulmonary vasculature in reptiles is very similar to that of am-

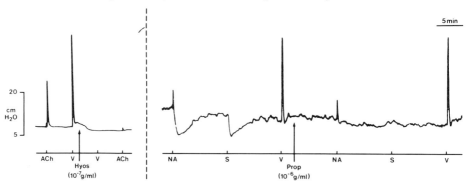

Fig. 7.6. Responses to drugs and to nerve stimulation by the isolated perfused pulmonary artery of the lizard, *Trachydosaurus rugosus*, recorded as pressure changes during constant flow perfusion. The *left part* of the figure shows the effect of acetylcholine (*ACh*) and vagal stimulation (*V*) before and after hyoscine (*Hyos*). The *right part* of the figure shows the effects of noradrenaline (*NA*) and stimulation of spinal autonomic ("sympathetic") (*S*) and vagal (*V*) fibres before and after the addition of the β-adrenoceptor antagonist propranolol (*Prop*). Note that the dilatory effect of noradrenaline, but not the constrictory effect, is abolished by propranolol (*Prop*). (Reproduced with permission from Berger 1972)

phibians. Thus vagal cholinergic constrictory control of the pulmonary vasculature has been demonstrated in a lizard (*Trachydosaurus rugosus:* Berger 1972, 1973; Fig. 7.6), turtles (*Chrysemys scripta, Pseudemys scripta, Testudo graeca, Chelodina longicollis:* Berger 1972, Burggren 1977, Milsom et al. 1977, Berger and Burnstock 1979) and snakes (*Thamnopis spp.:* Smith and Macintyre 1979). With the possible exception of the snakes, the extrinsic pulmonary artery is the major site of vasoconstriction in the pulmonary circuit.

In the lizard *Trachydosaurus rugosus* an additional adrenergic vasodilatory innervation of the extrinsic pulmonary artery has been demonstrated (Berger 1972, Fig. 7.6). A similar arrangement seems to be present also in snakes, in which the adrenergic fibres and exogenous catecholamines dilate the vasculature by a β-adrenoceptor mediated mechanism (Smith and Macintyre 1979). Fluorescence histochemistry reveals an adrenergic innervation of the pulmonary blood vessels (Leont'eva 1966, McLean and Burnstock 1967c, Furness and Moore 1970, Figs. 2.20, 2.21) and the extrinsic pulmonary artery of *Trachydosaurus rugosus* is "by far the most heavily innervated vessel in the lizard" (Furness and Moore 1970).

7.7.3 The Reptilian Systemic Vasculature

A dense innervation by adrenergic fibres of all parts of the systemic vasculature has been demonstrated in the lizard *Trachydosaurus rugosus* (McLean and Burnstock 1967a, Furness and Moore 1970), and both adrenergic and cholinergic vasomotor fibres to the major arteries have been concluded from experiments with isolated strip preparations (Kirby and Burnstock 1969a). Injection of adrenaline produces an elevated blood pressure, as in other vertebrates, and catecholamine-induced constriction of perfused vascular beds has also been recorded (Akers and Peiss 1963, Kirby and Burnstock 1969b, Reite 1970, Berger and Burnstock 1979).

A baroreceptor reflex has been demonstrated in lizards (*Iguana iguana:* Hohnke 1975) and snakes (*Notechis scutatus:* Lillywhite and Seymour 1978); and a careful study of the lizard *Trachydosaurus rugosus* by Berger et al. (1980) has revealed the presence of baroreceptors in the truncus arteriosus.

Although there is some information about the reactions and control of the pulmonary vasculature of reptiles during periods of apnoea (e.g., diving), complementary information about the control of the systemic vasculature is still wanting.

7.8 Birds

7.8.1 The Avian Heart

The avian heart is built much to the same plan as the mammalian heart, with two atria and two completely separated ventricles. The possibility of a right-to-left intracardiac shunt during periods of apnoea is thus absent. Contrary to the lower air-breathing vertebrates, the breathing of birds and mammals is continuous, although prolonged periods of apnoea do occur normally in diving birds and mammals. During diving there are marked cardiovascular adjustments taking place.

The avian heart is innervated by the vagi and by direct nerves from the sympathetic chains (Bolton 1971 a, Bennett 1974). The vagi form dense plexuses in the region of the SA- and AV-node, and the cholinergic vagal fibres mediate negative chronotropic and inotropic effects on the heart (Johansen and Reite 1964, Bolton and Raper 1966, Tummons and Sturkie 1968, 1969, 1970, Yamauchi 1969, Cohen et al. 1970, Bolton 1971 b, Bennett 1974). In contrast to the situation in reptiles and mammals, there is a substantial cholinergic innervation of the ventricles that mediates a negative inotropic effect (Paton 1912, Bolton and Raper 1966, Bolton 1971 b, Kissling et al. 1972). The effect is strong enough to reduce the cardiac output during forced diving in the mallard (*Anas boscas = platyrhynchos*), despite an increased cardiac filling due to an elevated central venous pressure (Johansen and Aakhus 1963, Folkow and Yonce 1967, Folkow et al. 1967).

The cardiac nerves in the domestic fowl (*Gallus domesticus*) leave the sympathetic chains from the first thoracic ganglia. Together with the vagal branches, five cardiac nerve plexuses are formed: anterior, posterior, superior, left and right coronary plexuses (Bolton 1971 a). All parts of the heart, but particularly the SA- and AV-node, are innervated by adrenergic fibres, which mediate positive chronotropic and inotropic effects via a β-adrenoceptor mechanism (Johansen and Reite 1964, Enemar et al. 1965, Govyrin and Leont'eva 1965, Bolton and Raper 1966, Tummons and Sturkie 1968, 1969, 1970, Akester et al. 1969, Yamauchi 1969, Bennett and Malmfors 1970, Bolton 1971 b, Bennett 1974).

The heart rate in birds is high compared to other vertebrates of comparable size. The high demands on the cardiovascular system during flight places special emphasis on the autonomic nervous control of cardiovascular function. In the mallard and domestic fowl the heart is under substantial cholinergic (vagal) and adrenergic tonus (Johansen and Reite 1964, Tummons and Sturkie 1969). For example, bilateral vagotomy in the mallard produces an increase in heart rate from 290 to 480 beats per minute, while subsequent injection of the β-adrenoceptor antagonist pronethalol (Alderlin) reduces the heart rate to only 150 beats per minute (Johansen and Reite 1964).

7.8.2 The Avian Vasculature

An adrenergic innervation of the vasculature in birds is well documented (Enemar et al. 1965, Folkow et al. 1966, Bennett 1969 a, 1971, 1974, Bennett and Malmfors 1970) and there are also indications of a cholinergic innervation (Bell 1969, Bennett 1969 a, 1974). An interesting example of a cholinergic vascular innervation has been demonstrated in the longitudinal muscle of the adventitia in the anterior mesenteric artery of the domestic fowl; this muscle is contracted by acetylcholine or stimulation of the cholinergic nerve supply. Catecholamines produce a β-adrenoceptor-mediated relaxation of the muscle, as does stimulation of the adrenergic nerve supply (Bolton 1968 a, b, 1969, 1971 b, Bell 1969). The circular muscle of the same vessel, however, contracts in response to catecholamines or adrenergic nerve stimulation (Bell 1969). The physiological significance of these findings is not clear, but an involvement of the longitudinal muscle in control of the blood flow through the vessel during emergency responses has been suggested (Bell 1969).

Profound cardiovascular adjustments occurs in birds during *forced* diving, but the role of the autonomic nervous system in the observed haemodynamic responses in only partially understood. The best documented of the cardiovascular adjustments to submersion is a marked bradycardia but the underlying reflexes are complex and so far not clearly understood. The response resembles a "primary chemoreceptor reflex" (Folkow and Neil 1971) in which receptors in the nostrils, in the laryngeal mucosa as well as chemo- and baroreceptors seem to be involved (Jones 1973, Millard et al. 1973, Purves 1975). The baroreceptors of the mallard are located in the ascending aorta and play an important role in the maintenance of a constant systemic blood pressure during a dive (Johansen and Aakhus 1963, Jones 1973, Purves 1975).

A second notable adjustment of the cardiovascular system during diving in birds is the redistribution of blood from the vascular beds of the skeletal muscle to the cerebral and coronary vasculature (Johansen 1964, Purves 1975). The vasoconstriction in skeletal muscle is due to an α-adrenoceptor mediated mechanism, and the possibility of a neurogenic vasodilatory mechanism in the cerebral and coronary vasculature has been suggested (Butler and Jones 1971, Purves 1975).

Although a lot has been learned about the diving reflexes in birds during *forced* submersion of the head of the animal, it has now become clear that the cardiovascular adjustments to *voluntary* diving differ dramatically from those during the forced diving at least in the mallard. Heart rate monitored by radiotelemetry during voluntary dives in the penguin (*Pygoscelis* sp.) is reduced to about 1/3 of the pre-dive level (Millard et al. 1973), while in the mallard a transient *increase* in the heart rate immediately *prior* to the voluntary dive is followed during the dive by a heart rate that is the same as the pre-dive heart rate. The duration of the voluntary dives is relatively short (around 1/2 min), and only by preventing the bird from surfacing (which induces hypoxia) can the bradycardia of forced diving be elicited (Butler 1982, Butler and Jones 1982).

Interestingly, there seems to be a dissimilarity in the distribution of baro- and chemoreceptor afferents between the right and left vagus of birds. During stimulation of the central end of one cut vagus nerve, with the contralateral nerve intact, Johansen and Reite (1964) demonstrated bradycardia and fall in blood pressure during stimulation of the left central vagus stump, while stimulation of the right central vagus stump instead induced tachycardia and a rise in blood pressure.

The autonomic nervous control of the blood flow pattern in the foot, particularly in aquatic birds with webbed feet, plays an important role in thermoregulation. The blood flow in the foot can be directed either through a counter-current heat exchange system, the *rete tibiotarsale*, where heat loss in the foot is reduced by heat transfer from the descending arterial blood to the ascending venous blood, or through an arterial by-pass shunt segment. In the latter case the blood reaches the foot without the pre-cooling in the *rete tibiotarsale*, and heat can be given off to the environment. The bypass shunt artery is controlled by adrenergic vasoconstrictory fibres acting via α-adrenoceptors. Both cholinergic and NANC vasodilator fibres have been implicated in the reflexogenic control of the blood flow in the avian foot (Johansen and Millard 1974, Reite et al. 1977, Midtgård 1980, 1981, Midtgård and Bech 1981).

7.9 Mammals

The autonomic nervous control of the mammalian cardiovascular system has been very well studied and detailed information of the involvement of the autonomic nerves in the general circulatory control in mammals is available in many text-

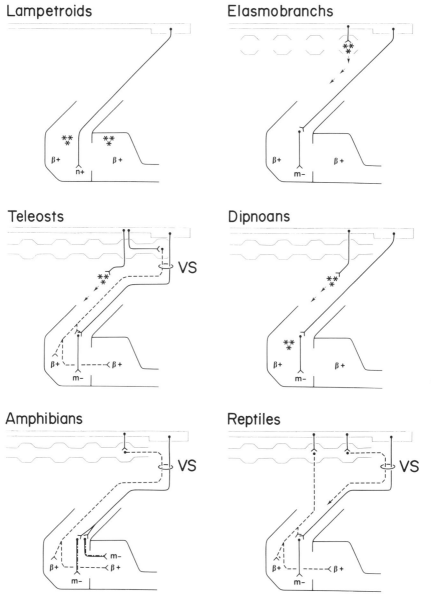

Fig. 7.7. A summary of the pattern of autonomic innervation and control of the heart by circulating catecholamines in some vertebrate groups. For explanation of symbols see Fig. 7.8

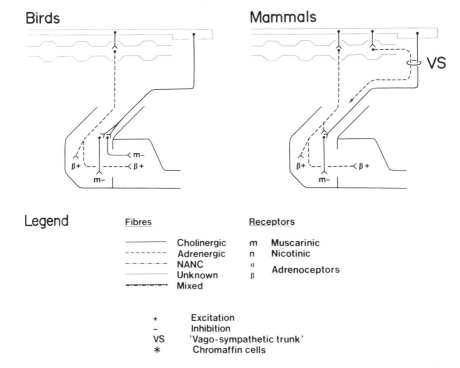

Fig. 7.8. A summary of the pattern of autonomic innervation of the heart in birds and mammals. The legend shows abbreviations and symbols used in the summary of autonomic innervation patterns of several organs. In these summaries, no difference is made between sympathetic chain ganglia and prevertebral ganglia. To avoid unnecessary confusion, the contribution of adrenergic fibres running in the "vagosympathetic trunk" is sometimes shown by an arrow only (cf. mammals above)

books (e.g., Folkow and Neil 1971). The following account is an extensively simplified and reduced description of some of the features of the mammalian cardiovascular control.

7.9.1 The Mammalian Heart

The vagal innervation of the mammalian heart is cholinergic and inhibitory, and cholinergic fibres occur in the pacemaker region (SA-node) and in the atria. An innervation of the ventricle by cholinergic fibres is, however, scarce or absent in most mammals (Bolton and Raper 1966).

There are numerous small ganglia close to, or within the wall of, the heart that are connected to form an extensive cardiac plexus (Jacobowitz et al. 1967, Gabella 1976). The neurons of these ganglia are exclusively non-adrenergic, but extra-adrenal chromaffin cells (SIF-cells) frequently occur within the cardiac and other autonomic ganglia of mammals (Nielsen and Owman 1968).

Adrenergic fibres reach the heart from the stellate ganglia, or in some species the inferior cervical ganglia, and innervate the SA- and AV-nodes, atria and ventricles. The innervation of the ventricle is very rich in some species (e.g., cat), while in other species, particularly the hibernating mammals, it is sparse (Dahlström et al. 1965, Nielsen and Owman 1968, Gabella 1976) It has been proposed that the lack of an adrenergic innervation of the ventricular myocardium explains the resistance of the hearts of these species to ventricular fibrillation induced by low temperature (Nielsen and Owman 1968).

The adrenergic fibres to the mammalian heart mediate positive chronotropic and inotropic responses via both β_1- and β_2-adrenoceptors, which differ in their relative distribution between the different parts of the heart (Hedberg 1980, Hedberg et al. 1980).

Adrenergic fibres enter the heart mainly along the cardiac nerves (*nervi accelerantes*) from the sympathetic chains, but there is also a substantial contribution by fibres entering the vagi from the superior cervical ganglia (Nielsen et al. 1969, Campbell 1970a, Priola et al. 1981). Adrenergic nerve terminals are found surrounding the ganglion cells in the cardiac plexus, at least in some species, and it is possible that an adrenergic control of the cholinergic vagal transmission takes place (Dahlström et al. 1965).

Nerve fibres showing substance P-like immunoreactivity have been demonstrated in the guinea-pig heart, but the function of these fibres is as yet unknown (Wharton et al. 1981).

7.9.2 The Mammalian Vasculature

The integrated cardiovascular control mechanisms in mammals are comparatively well understood. Various aspects of the autonomic nervous control of the vasculature, including the cardio-vascular reflex phenomena and central nervous connections, are discussed by Folkow and Neil (1971).

One particular aspect of the autonomic nervous control of the blood vessels, the systemic vasodilator fibers, has been extensively studied in mammals, but remains largely untouched in the lower vertebrates.

In mammals there are basically two groups of vasodilator fibres that innervate the systemic blood vessels of specific vascular beds. The first group of fibres are the classically "sympathetic" fibres that innervate the resistance vessels of the skeletal muscle. These neurons are active during the "defence reaction" and mediate marked dilation of the skeletal muscle vasculature, while adrenergic discharge to other vascular beds (except the vasculature of the brain and heart) produces vasoconstriction, thereby securing the blood flow in the most crucial circuits (Folkow and Neil 1971). The vasodilatory fibres to the skeletal muscle vasculature are cholinergic, and mediate the effects via muscarinic cholinoceptors (Bülbring and Burn 1935, Folkow and Uvnäs 1948, Folkow et al. 1948, Burnstock 1980b).

The second group of vasodilatory fibres comprises the classically "parasympathetic" fibres to some of the exocrine glands (via cranial autonomic fibres) and erectile tissue of the genitalia (via sacral autonomic fibres). The transmitter mechanisms involved in the control of the vasculature of the genitalia are not fully understood but, as pointed out by Folkow and Neil (1971), it is clear that they are "of paramount importance in the preservation of the species". Available results indicate that NANC transmission plays a dominant role in the vasodilatory responses of the penile vasculature (Dorr and Brody 1967, Klinge and Sjöstrand 1974, Burnstock 1980 b).

The vasodilatory effect of nerve stimulation in the salivary glands has long been known to include a non-cholinergic component (Heidenhain 1872). A detailed study by Lundberg (1981) has provided a very elegant model for the mechanisms involved in the nervous control of the gland cells and vasculature of exocrine glands, including the salivary glands, of the cat. According to this model, VIP stored in the cholinergic nerve terminals is released as a co-transmitter and is responsible for the non-cholinergic part of the vasodilatory response (Fig. 3.11). Acetylcholine, which stimulates the gland cells via a muscarinic cholinoceptor mechanism, is also partially responsible for the vasodilation, and a functional interaction of acetylcholine and VIP at different levels has been concluded (Lundberg 1981). In view of the accumulating evidence for an innervation of gastrointestinal (Chap. 9) and other blood vessels by peptidergic nerve fibers, there is every possibility of local vascular control by a variety of NANC neurons.

7.10 Conclusions

An inhibitory control of the heart by cholinergic vagal fibres is present already in the elasmobranchs. It appears that in many vertebrate groups a variation in the vagal tonus on the heart is an effective means of cardiac control (e.g., in fish, reptiles, diving birds). An anatomical arrangement which allows an adrenergic influence on the transmission in intracardiac vagal ganglia is present in several vertebrate groups including teleosts. Such an arrangement can be expected to amplify the adrenergic stimulation of the heart, e.g., during the "defence reaction", by impairing the inhibitory vagal influence.

The adrenergic control of the vertebrate heart is a phylogenetically old feature. In cyclostomes endogenous adrenergic cells (chromaffin cells) store catecholamines which may be of vital importance for the normal function of the heart. There appears to be no extrinsic nervous control of the catecholamine release from these cells in cyclostomes. In dipnoans, in which a similar intracardiac store of catecholamines exists, an extrinsic innervation of the cells by vagal fibres may be present. Endothelial catecholamine-storing cells have also been demonstrated in chimaeroids and some teleosts, but the role of these cells is unclear.

In addition to the intracardiac chromaffin cells of dipnoans, chromaffin tissue is situated in the posterior cardinal vein outside the heart. A similar arrangement is also found in elasmobranch, ganoid and teleost fish. A cardiac control by circulating catecholamines released from this strategically located chromaffin tissue is probably present in these fish. A direct adrenergic innervation of the heart occurs in holosteans and teleosts and this more advanced, direct, rapid and restricted mode of cardiac control is retained in the tetrapods (Table 7.1).

An adrenergic innervation of the vasculature is evident in some of the cyclostomes and elasmobranchs and well established in the holosteans and teleosts, but

Table 7.1. A speculative summary of the modes of adrenergic cardiac control in the different vertebrate groups. The table summarizes the known or probable *major* adrenergic control mechanism(s) in each group, although species differences may well occur within each group. Circulating plasma catecholamines from chromaffin sources are present in all vertebrates, but their relative importance in cardiac control varies

	Endogenous catecholamine storing cells	Exogenous catecholamine storing cells (circulating amines from chromaffin sources)	Adrenergic nerves
Cyclostomes	×		
Chimaeroids	×	×	
Selachians		×	
Dipnoans	×	×	
Chondrosteans		×	
Holosteans		×	×
Teleosts		×	×
Amphibians			×
Reptiles			×
Birds			×
Mammals			×

not, for unknown reasons, in dipnoans. The conclusion reached by Burnstock (1969) that the vascular innervation in the lower vertebrates is primarily cholinergic, and that the vasoconstrictory control is taken over by adrenergic fibres late in the phylogenetic series, is not supported by later studies. A cholinergic vasoconstrictor innervation is doubtful in vertebrate vasculature other than in the branchial vasculature of elasmobranchs and teleosts and the pulmonary vasculature of amphibians and reptiles. The systemic vascular innervation of the lower vertebrates obviously needs further attention using quantitative pharmacological methods.

The central venous pressure is of little consequence for the atrial filling in the lower vertebrates, where a negative pressure in the atrium prevails during diastole. There appears to be little or no venous innervation in these animals but with an increasingly important central venous pressure control in reptiles, and especially in birds and mammals, there is also an increase in the vasoconstrictory adrenergic innervation of the large veins.

Although a β-adrenoceptor-mediated vasodilation may occur under experimental conditions in both fish and tetrapods, it is difficult to see the functional significance of this mechanism in view of the very dominant α-adrenoceptor-mediated vasoconstriction by the adrenergic fibres or circulating catecholamines. Only in mammals is there clear evidence for a non-adrenergic vasodilatory autonomic innervation of the blood vessels in some organs. The demonstration of peptidergic innervation of blood vessels opens interesting possibilities for NANC control mechanisms. The demonstration of a mixed cholinergic/somatostatinergic innervation of the toad heart likewise points towards an entirely new level of complexity in the autonomic innervation of the vertebrate heart.

Chapter 8

Spleen

The vertebrate spleen is typically a well-defined organ or reddish/brown colour situated in the dorsal mesentery and having a discrete vascular and nervous supply (Romer 1962). In cyclostomes and dipnoans, however, the splenic tissue lies embedded in the gut wall (Parker 1892, Romer 1962). The spleen has several known functions, which may vary in importance among the groups and species of the vertebrates. The functions include formation of lymphocytes in the "white pulp" (Malphigian bodies), erythropoiesis (notably in elasmobranchs; Fänge and Sundell 1969), erythrocyte degradation, phagocytosis and storage and release of erythrocytes. The most important role of the autonomic innervation of the spleen involves the control of erythrocyte release.

Our view on the nervous control of the spleen is based primarily on the studies of mammalian spleens. The effects of drugs on mammalian spleens have been extensively reviewed by Davies and Withrington (1973). The discrete vascular and nervous supply and the large and easily recordable changes in blood flow and volume of especially the spleen of dogs and cats have made it a preferred preparation for the study of adrenergic nerve functions.

In mammals, the smooth muscle of the spleen forms two effector units. One is the capsular and trabecular smooth muscle, forming the splenic capsule and the three-dimensional network of trabeculae (formed from the inner layer of the capsule), which divides the organ into lobules. The development of the capsular smooth muscle, which varies among species, is particularly prominent in cats and dogs (Davies and Withrington 1973). The second effector unit is the vascular smooth muscle. In the cat and dog, maximal contraction of the capsular and trabecular smooth muscle occurs at a stimulation frequency of about 4 Hz, while the maximal vascular contraction is not obtained until the stimulation frequency is raised to about 10 Hz (Celander 1954, Blakely 1968, Davies et al. 1973).

The arrangement of the effector units in non-mammalian spleens is similar, although early histological studies denied the existence of smooth muscle in the spleens of some amphibians and fish (see Zwillenberg 1964, Tischendorf 1969).

The knowledge of the autonomic innervation of spleens from non-mammalian vertebrates is sparse, but there appears to be general pattern of postganglionic adrenergic innervation in all species studied. Thus fibres from the coeliac ganglion reach the spleen in the splanchnic (elasmobranchs and teleosts) or splenic (tetrapods) nerves. Dipnoans are exceptional in that a separate nerve to the spleen has not been described. Instead a nervous supply to the dipnoan spleen could run with the vagus along the gut (Jenkin 1928, Abrahamsson et al. 1979b).

There are no conclusive reports of a vagal innervation of the vertebrate spleens (Utterback 1944). In mammals, peripheral stimulation of the vagus causes a passive reduction in the spleen volume due to a decrease in blood pressure, while stim-

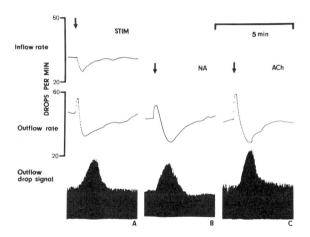

Fig. 8.1. Polygraph recordings of changes in perfusion flow through the spleen of the cod (*Gadus morhua*) caused by electrical stimulation of the splanchnic nerve (*A*), addition of noradrenaline (*NA*) 1 nmol (*B*) or addition of acetylcholine (*ACh*) 1 nmol (*C*). Note the close similarities in the responses to the different stimuli. The increased amplitude in the "outflow drop signal" reflects an increased optical density of the effluent due to expulsion of red blood cells. (Reproduced with permission from Nilsson and Grove 1974)

ulation of afferent fibres in the cranial vagus will induce a reflex contraction of the spleen (Schäfer and Moore 1896, Strasser and Wolf 1905, Davies and Withrington 1973).

The magnitude of the change in spleen volume varies considerably between species, and the release of sequestered erythrocytes during splenic contraction is very marked in some mammals (e.g., cat and dog) but may be insignificant in others (e.g., man). Among the non-mammalian vertebrates studied, nerve stimulation produces a pronounced volume change only in the small spleen of the toad (*Bufo marinus*) (Nilsson 1978). Release of sequestered blood cells from the spleen during nerve stimulation is also evident in elasmobranchs and teleosts, although the quantitiative importance judged from hematocrit measurements in the dogfish (*Squalus*) may be small (Bonnet 1929, Opdyke and Opdyke 1971, Nilsson and Grove 1974, Nilsson et al. 1975, Yamamoto et al. 1980) (Fig. 8.1).

8.1 Adrenergic Control

Falck-Hillarp flurorescence histochemistry has revealed adrenergic fibres in the spleens from a number of vertebrates. In the spiny dogfish (*Squalus acanthias*) catecholamine-containing fibres were observed mainly associated with blood vessels and sparsely in the trabeculae (Nilsson et al. 1975). A similar pattern was observed in the cod (*Gadus morhua*; Nilsson and Grove 1974). In the toad (*Bufo marinus*)

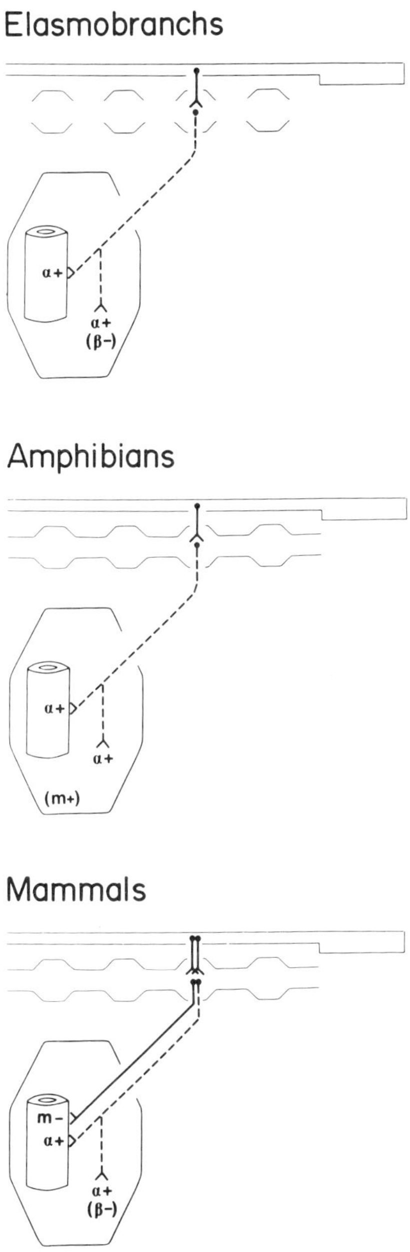
Elasmobranchs
α+
α+
(β-)
Teleosts
α+
α+m+
Amphibians
α+
α+
(m+)
Reptiles
?
Mammals
m-
α+
α+
(β-)

Fig. 8.2. A summary of the pattern of autonomic innervation of the spleen in some vertebrate groups. For explanation of symbols see Fig. 7.8

and a lizard (*Trachydosaurus rugosus*), a dense adrenergic innervation of the capsule and the blood vessels has been demonstrated (Furness and Moore 1970, Nilsson 1978). Mammalian studies indicate a variation in the density of the adrenergic innervation of the splenic smooth muscles, with a sparse innervation in the cat and a more dense innervation in the dog (Dahlström and Zetterström 1965, Gillespie and Kirpekar 1966, Fillenz 1970). In many of the vertebrates species studied, the intense autofluorescence emanating from porphyrine derivatives in the spleen could have obscured the weaker catecholamine fluorescence of fine nerve terminals (Abrahamsson et al. 1979b, Balashov et al. 1981, Nilsson 1981).

The order of potency for the adrenergic agonists studied and the effects of adrenoceptor antagonists indicate a uniform adrenergic control via α-adrenoceptors. In the majority of studies on non-mammalian species, the relative involvement of the capsular and trabecular and the vascular smooth muscle in the response has not been resolved. α-Adrenoceptor agonists constrict the vasculature and contract the capsular and trabecular smooth muscle to a variable degree, mimicking the effect of electrical stimulation of the nervous supply to the spleen (*Squalus acanthias, Scyliorhinus canicula*: Opdyke and Opdyke 1971, Nilsson et al. 1975; *Huso huso*, Balashov et al. 1981; *Lepisosteus platyrhincus*, Nilsson 1981; *Tinca tinca*, Vairel 1933; *Gadus morhua*, Nilsson and Grove 1974, Fig. 8.1, Holmgren and Nilsson 1975; *Bufo marinus*, Nilsson 1978).

A splenodilatory effect involving β-adrenoceptors has been shown in the mouse (Ignarro and Titus 1968) but the physiological significance of this effect and whether the capsular and trabecular or the vascular smooth muscle (or both) are responsible for the effect is obscure. The dilatory effect of isoprenaline in elasmobranchs, teleosts and amphibans suggests a similar mechanism in these species (Opdyke and Opdyke 1971, Nilsson and Grove 1974, Nilsson et al. 1975, Nilsson 1978). However, when the concentration of isoprenaline is increased, the response of the toad spleen is reversed to an α-adrenoceptor mediated constriction, with a remarkably high intrinsic activity ($\alpha = 0.88$) for isoprenaline (Nilsson 1978).

Since circulating catecholamines affect the splenic smooth muscles in a number of species, it is possible that in the dogfish (*Scyliorhinus canicula*) and perhaps also in the lungfish (*Protopterus aethiopicus*) the only adrenergic control of the spleen is via the circulating amines (Nilsson et al. 1975, Abrahamsson et al. 1979b).

8.2 Cholinergic Control

In their review of drug effects on the mammalian spleen, Davies and Withrington (1973) conclude that "the action of acetylcholine on the spleen is extremely complex." This is certainly also true for a number of non-mammalian vertebrate species.

Cholinoceptor agonists affect the splenic smooth muscle in elasmobranchs, dipnoans, teleosts, amphibians and mammals but not in ganoids. With the exception of teleosts and dipnoans, the responses are often weak and inconsistent; there is also a marked desensitization to acetylcholine (Vairel 1933, Davies and Withrington 1973, Nilsson and Grove 1974, Holmgren and Nilsson 1975, Nilsson et al. 1975, Nilsson 1978, 1981, Abrahamsson et al. 1979b, Balashov et al. 1981).

The contraction of the spleen produced by acetylcholine observed in teleosts, dipnoans and amphibians is mediated by muscarinic cholinoceptors (Nilsson and Grove 1974, Holmgren and Nilsson 1975, Nilsson 1978, Abrahamsson et al. 1979

b) but only in the cod are there strong indications of a cholinergic component in the splanchnic nervous control of the spleen (Nilsson and Grove 1974, Winberg et al. 1981).

In the cod the effects of catecholamines and acetylcholine are identical and mimic closely the response to splanchnic nerve stimulation (Fig. 8.1). A functional cholinergic innervation of the splenic vasculature seems unlikely since the blood vessels and vascular beds are very insensitive to acetylcholine (Reite 1969, Holmgren and Nilsson 1974, Wahlqvist and Nilsson 1981). The main cholinergic control of the spleen is therefore probably of the capsular and trabecular smooth muscle.

Both catecholamines and the "cholinergic neuron marker," choline acetyltransferase (ChAT) are associated with neurons of the cod spleen and there is a substantial reduction in the catecholamine levels and ChAT activity after surgical denervation or treatment with 6-hydroxydopamine treatment (Chap. 5.2.3, Abrahamsson and Nilsson 1975, Winberg et al. 1981). The possibility that both types of transmitter are stored in and released from the same neuron has been discussed previously (Chap. 3.2.5, Holmgren and Nilsson 1976).

The spleen of *Protopterus aethiopicus* is atypical in contracting more strongly in response to cholinoceptor agonists than to the catecholamines. The position of the spleen, which is embedded in the gut wall, could mean that the splenic smooth muscle is like the gastrointestinal muscle rather than the specific splenic musculature of the other vertebrates (Abrahamsson et al. 1979b).

The mammalian spleen contains acetylcholine. For the spleens of cats and dogs there is also some evidence that nervous control involves a cholinergic component which mediates vasodilation via muscarinic cholinoceptors (Burn and Rand 1960, Daly and Scott 1961).

The splenic innervation in some of the vertebrate groups is summarized in Fig. 8.2.

Chapter 9

The Alimentary Canal

The functions of the alimentary canal of the vertebrates are basically to (1) receive and act as a reservoir for ingested food and fluid; (2) process the food chemically and physically; (3) absorb water and nutrients and (4) dispose of the wastes. These functions of the gut are controlled by a multitude of systems which regulate the muscular activity of the gut wall, the secretion of digestive juices into the lumen of the gut, and the blood flow in the mucosa. The control systems include the coordinated reflexes of the enteric nervous system and some reflexes that also include extrinsic autonomic pathways (vagal, splanchnic and pelvic). In addition paracrine cells release substances that affect the adjacent cells, and endogenous endocrine systems exert control over the gut motility and, especially, the secretory activity of the gastric and intestinal glands (Larsson 1980).

The present comparative account is focused on the nervous control of the gastro-intestinal smooth muscle, but it must be strongly emphasized that the enteric nervous system is also involved in the control of many other functions, some of which are poorly understood. The muscle of the gut is capable of several types of functional activities, e.g., "receptive relaxation", "pendular mixing movements" and peristalsis, all apparently controlled by enteric neurons. Three types of gastro-intestinal reflexes are considered as examples of the numerous types of reflexes that involve the autonomic nervous system: (1) gastric receptive relaxation (Cannon and Lieb 1911), (2) the peristaltic reflex (Bayliss and Starling 1899, 1900), and (3) the intestinal vasodilatory reflex (Biber 1974).

Since information about the gastro-intestinal reflexes is almost exclusively based on investigations with mammals, it is practical to first consider this group before going on to discuss the nervous control of the gut in the other vertebrate groups. Attempts to summarize the patterns of the *extrinsic* autonomic innervation of the vertebrate stomach and intestine are made in Figs. 9.7 and 9.8 respectively.

The degree of differentiation among the different segments of the gut varies considerably with species among the vertebrates, often reflecting the different food habits of the animal. A distinct stomach is absent in many fish, e.g., cyclostomes, chimaeroids, dipnoans and some teleosts (Romer 1962). The arrangement of the intestine is also variable, with the greatest specialization of the different segments found in birds and mammals.

9.1 Mammals

9.1.1 Structure of the Mammalian Alimentary Canal

In most mammals the histological appearance of the gut wall makes it possible to subdivide the various segments of the gut further. In the stomach, there are generally three parts – cardia, fundus and antrum (pylorus). The small intestine can also be subdivided into three parts – duodenum, jejunum and ileum. At the junction between the ileum and the main part of the large intestine, the colon, there is a blind sac, the caecum, which, in herbivorous mammals, can be large. The most posterior part of the large intestine, the rectum, is closed off to the outside by the anal sphincters, an internal smooth muscle sphincter and an external striated muscle sphincter (Adams and Eddy 1951, Romer 1962).

The general structure of the gut wall is very uniform throughout the entire tube, and also among the different groups of vertebrates. The lumen of the gut is lined by the mucosa whose many folds and villi greatly increase the surface area, thereby enhancing the processes of digestion and absorption. The mucosa contains numerous endocrine and paracrine cells, which produce and release substances involved in the control of various gut functions, especially the secretion of digestive juices (Larsson 1980). The mucosa, including its smooth muscle layer, the muscularis mucosa, rests on a layer of connective tissue, the submucosa. The major trunk blood vessels supplying the mucosa, and other parts of the gut wall, occur in the submucosa (see also Fig. 2.5). Outside the submucosa is the muscularis externa which consists of an inner, relatively thicker circular layer of smooth muscle, and an outer, thinner layer of longitudinally arranged smooth muscle. The longitudinal smooth muscle layer is covered by an outer connective tissue coat, the serous membrane or tunica externa.

In the guinea-pig the longitudinal muscle of the caecum is arranged in three distinct bands, superficially resembling the taenia of the human colon. The guinea-pig "taenia coli", as these caecal smooth muscle bands are often incorrectly called (see Burnstock et al. 1966), are (together with the isolated ileum from the same animal) very widely used preparations for the study of the physiology and pharmacology of gastrointestinal smooth muscle.

A phenomenon peculiar to the smooth muscle of the gut and some other organs (ureters, uterus), is the electrotonic spread of a stimulus between smooth muscle cells and cell bundles, which makes the smooth muscle act as a "single unit" often showing myogenic, rhythmic contractions. In this type of system not all the smooth muscle cells are innervated, but the stimulus travels between the cells and cell bundles in specific low resistance pathways (Bennett and Burnstock 1968, Burnstock 1970, 1977, Bennett 1972).

The smooth muscle of the gut is innervated by intrinsic neurons of the enteric plexuses, and by extrinsic neurons of both cranial and spinal origin. There are two major nerve plexuses in the gut (Chap. 2.3.1, Fig. 2.5). One of these is situated in the submucosa, bordering the circular smooth muscle layer (the submucous plexus or Meissner's plexus) and the other between the circular and longitudinal smooth muscle layers (the myenteric plexus or Auerbach's plexus).

The extent of the extrinsic control of the gut varies among the different levels. In general it appears that the most anterior and posterior parts of the alimentary canal receive a comparatively greater control from extrinsic, traditionally "parasympathetic" nerves (vagus and pelvic nerves, respectively), while the activities in the middle part of the gut (small intestine and anterior part of large intestine) are relatively more dependent on the integrated intrinsic reflex activities of the enteric nervous system. Innervation by extrinsic adrenergic fibres is received by all parts of the gut; the fibres emanate from postganglionic nerve cell bodies of exclusively spinal autonomic pathways. The cell bodies are located in sympathetic chain ganglia (e.g., superior cervical ganglia which send fibres into the vagi) or prevertebral ganglia (coeliac and mesenteric ganglia) (Gabella 1979).

9.1.2 Gastro-Intestinal Reflexes in Mammals

9.1.2.1 *Gastric Receptive Relaxation*

Although there is a substantial secretion of digestive juices from the mucosa of the stomach, it is probably true to say that a major function of this organ is to act as a reservoir for the food ingested during a meal. It has long been known that an increase in stomach volume during filling is not accompanied by any major increase in the intragastric pressure. It was also observed that swallowing or mechanical stimulation of the pharynx and oesophagus, as well as experimental filling of the stomach, elicits a reflex relaxation of the muscularis of the stomach (Cannon and Lieb 1911, Jansson 1969, Abrahamsson 1973). This reflex has been called "gastric receptive relaxation" and obviously serves to increase the receptive capacity of the stomach during a meal. In mammals, the reflex involves two separate, but synergistic, reflex arcs which both involve extrinsic nerves. The two arcs have afferent and efferent fibres in the vagus (vago-vagal reflex) and splanchnic nerve respectively (Abrahamsson 1973, Fig. 9.1).

The vago-vagal inhibitory reflex is activated by distension of the oesophagus and stomach, which means that the afferent (sensory) fibres of the reflex arc are tension receptors (Fig. 9.1). The efferent vagal postganglionic fibres are of NANC type, with the postganglionic neurons within the stomach wall (Martinson and Muren 1963, Martinsson 1964, 1965, Campbell 1966a, Campbell and Burnstock 1968, Jansson 1969, Abrahamsson 1973). VIP has been suggested as the inhibitory NANC transmitter (Fahrenkrug 1979).

A second reflex arc of the gastric receptive relaxation consists of afferent and efferent fibres in the splanchnic nerve. The afferent fibres are again tension receptors, since the reflex is triggered by distension of the stomach. The efferent pathways, which are of spinal autonomic origin, have adrenergic postganglionic fibres. As in the other parts of the mammalian gut, it appears that the main function of the adrenergic neurons is inhibition of the vagal cholinergic excitatory pathways, rather than direct innervation of the smooth muscle (Jansson 1969, Abrahamsson 1973, Fig. 9.1).

In addition to the inhibitory reflexes, there is an excitatory reflex controlling stomach peristalsis. Vagal neurons are normally involved also in this case, but after

Fig. 9.1. Arrangement of the extrinsic neurons in the vagal and splanchnic pathways involved in the gastric receptive relaxation reflex of mammals. Distension of the stomach activates afferent (sensory) fibres, and the efferent parts of the reflex arcs consist of splanchnic adrenergic fibres which inhibit the transmission in vagal cholinergic excitatory pathways, and vagal NANC inhibitory fibres to the smooth muscle. Abbreviation: *m*+, muscarinic cholinoceptors mediating excitation. (Based primarily on Abrahamsson 1973)

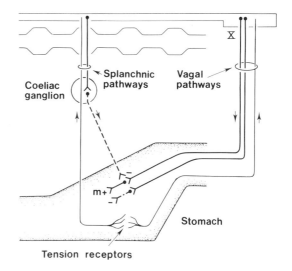

sectioning of the vagi a certain amount of intrinsic reflex activity develops with time (Thomas and Baldwin 1968). Part of the contractile response of the cat stomach to vagal stimulation is insensitive to hexamethonium, which would be expected to block the ganglionic transmission in efferent vagal autonomic pathways. The response is, however, sensitive to atropine, and Delbro (1981) proposed a model in which afferent (sensory) vagal and splanchnic fibres are responsible for the hexamethonium-resistant excitatory effects. In this model, collaterals of thin afferent axons are thought to release substance P which stimulates intrinsic cholinergic neurons innervating the smooth muscle (Delbro 1981).

9.1.2.2 The Peristaltic Reflex

The function of peristalsis is to move the contents of the gut in a "conveyor-belt" fashion from the anterior to the posterior end of the alimentary canal. In the intestine, the peristaltic propulsion of the gut content is coordinated entirely by the intrinsic enteric neurons, but the extrinsic innervation can sometimes modulate the activity in the enteric nervous system and thereby accelerate or decelerate the velocity of the peristaltic wave.

Bayliss and Starling (1899, 1900) in describing the effects of stimulating the intestinal wall, noted a contraction of the intestine on the oral side of the stimulus and a relaxation anally to the stimulus. The enteric reflexes responsible for the two types of reaction have since been studied in experimental preparations in which the peristaltic movements induced by distension or other stimuli have been recorded (Trendelenburg 1917). The terminology used to describe these phenomena has become very complex (Kottegoda 1970), but it is clear that the excitation of the circular muscle oral to the point of distension of the intestine and the inhibition on the anal side are due to two different components of the reflex, an *ascending excitatory reflex* and a *descending inhibitory reflex* (Costa and Furness 1976).

Distension of the intestine first produces a contraction of the longitudinal muscle anally to the stimulus and a subsequent sustained relaxation of the circular muscle of the same segment, "preparatory phase" of Trendelenburg (Bayliss and Starling 1899, 1900, Kosterlitz and Lees 1964, Costa and Furness 1976). The final link in the relaxation of the circular smooth muscle (descending inhibitory reflex) is mediated by NANC inhibitory neurons, which innervate the circular muscle. It has been suggested that somatostatin-releasing and possibly also cholinergic interneurons are involved in the reflex (Hirst et al. 1975, Furness et al. 1980a, Furness and Costa 1980). The transmitter in the final NANC inhibitory neurons is unknown although both ATP and VIP have been suggested (Fig. 9.2; Burnstock 1972, Furness and Costa 1980, Furness et al. 1980a).

Together with the relaxation of the circular muscle anally to the distension, there is a marked transient contraction of the circular muscle orally to the point of stimulation. This "ring" of contracted circular muscle travels anally, thus moving along the contents of the intestine, "emptying phase" of Trendelenburg. The final link in the part of the peristaltic reflex that produces the contraction (the ascending excitatory reflex) involves cholinergic excitatory neurons that innervate the smooth muscle. These neurons are part of the enteric nervous system, with their cell bodies mainly located in the myenteric plexus (Cannon 1912, Costa and Furness 1976). There is also some evidence for the presence of serotonergic interneurons in the ascending reflex, but such neurons probably do not innervate the muscle directly in mammals. The ascending excitation is blocked by nicotinic

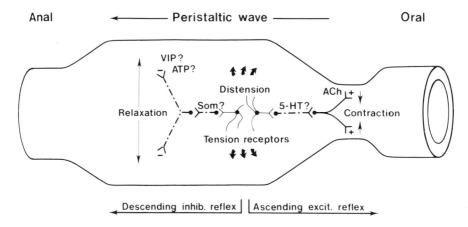

Fig. 9.2. A simplified representation of the possible arrangement of the ascending and descending pathways involved in the peristaltic reflex of mammals. Distension of the gut wall activates intrinsic sensory neurons. The ascending pathway possibly involves serotonergic neurons (*5-HT*)(?), and the final link in this pathway consists of cholinergic excitatory neurons which innervate the smooth muscle. The descending pathway may involve a chain of somatostatin-releasing neurons (*Som*), which stimulate the final NANC inhibitory neurons. The transmitter of these neurons could be *VIP* or *ATP* (?). Both the ascending and the descending pathway may also involve cholinergic (nicotinic) synapses (not shown). The number of neuronal links in each pathway may be greater than shown in the figure. (Based primarily on Furness and Costa 1980)

cholinoceptor antagonists, which suggests the presence of cholinergic synapses in the pathways (Bülbring and Crema 1958, Bülbring and Lin 1958, Kosterlitz and Lees 1964, Costa and Furness 1976, Fig. 9.2).

9.1.2.3 The Vasodilatory Reflex

Gentle stimulation of the intestinal mucosa, e.g., by the presence of food in the gut, elicits a local vasodilation ("functional hyperemia") in the intestine. The reflex underlying this response has been studied by Biber and co-workers in a series of experiments on isolated segments from cat intestine (Biber et al. 1971, 1973 a b, Biber 1974). The reflex vasodilation obtained by mechanical stimulation of the intestinal mucosa in these experiments could not be blocked by extrinsic denervation of the intestine, which shows that the reflex is intrinsic. The vasodilatory reflex in the most posterior part of the large intestine (distal colon), however, involves extrinsic pathways in the pelvic nerves (Fasth et al. 1977).

The intestinal vasodilatory reflex could not be blocked by adrenergic or cholinergic antagonists, but it was impaired by the 5-HT antagonist bromo-lysergic acid diethylamide (Biber et al. 1971, Biber 1974). Biber (1974) proposed a model in which 5-HT, released from enterochromaffin cells or serotonergic nerve fibres, serves as an intermediary step in the vasodilatory reflex. Later studies suggest that VIP may be the final mediator in the reflex arc (Fahrenkrug 1979, Furness and Costa 1980, Fig. 9.3). Local action of plasma kinins, e.g., bradykinin, has also been proposed (Fasth and Hultén 1973).

A functional advantage of a local vasodilation of the type described, compared to the more generalized vasodilatory effect of hormones such as secretin and gastrin/cholecystokinin, could be that the hyperemia is triggered only in the segments of the intestine where food is present rather than generally in the entire gut (Fara et al. 1972, Biber 1974).

9.1.3 "Rebound Excitation"

A phenomenon associated with the smooth muscle of the gut is the contraction that follows *at the end* of a period of inhibition produced, for example, by stimulation

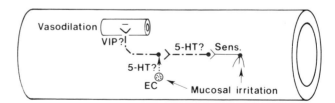

Fig. 9.3. Possible arrangement of the intrinsic neurons and enterochromaffin cells involved in the local vasodilator reflex in the mammalian gut. Irritation of the mucosa, e.g., by the presence of food, causes release of *5-HT* either from serotonergic neurons or from enterochromaffin cells. The *5-HT* then stimulates vasodilator fibres to the blood vessels. It is possible that the transmitter in these NANC vasodilator neurons is *VIP*. (Based primarily on Biber 1974)

of inhibitory nerves. This phenomenon, first described in the guinea-pig taenia coli, and called "rebound excitation", occurs in a number of preparations from vertebrate gut (Burnstock et al. 1963a, Campbell and Burnstock 1968, Fig. 9.5). Little can be surmised about the physiological significance of the "rebound excitation" phenomenon, but a "rebound contraction" of circular muscle in the gut following a distension/relaxation could facilitate the ascending excitatory part of the peristaltic reflex.

The presence of a rebound contraction has caused some difficulties in the interpretation of the responses of the gut to nerve stimulation since it is often difficult to judge, particularly from older records, whether a contraction recorded is primary (occurring during stimulation) or rebound (following cessation of the stimulus). There is, therefore, every reason to be careful in the evaluation of results that indicate excitatory effects caused by stimulation of autonomic nerves (Campbell and Burnstock 1968, Burnstock 1969).

Initially, the rebound excitation was thought to be a non-specific over-shoot of the membrane potential of the smooth muscle cells due to adaptation during the inhibitory phase (Bennett 1966, Campbell and Burnstock 1968). Later, other explanations for the phenomenon have also been suggested. Thus Burnstock et al. (1975) showed that the rebound contraction of the guinea-pig taenia coli, which follows electrical stimulation of the NANC inhibitory fibres, was blocked by indomethacin. Since indomethacin is known to block prostaglandin synthesis, it was suggested that prostaglandins are responsible for the excitation following the cessation of the inhibitory stimulus. An alternative explanation is that the rebound excitation is neurogenic. Thus it has been suggested that substance P, released together with the inhibitory transmitter and with more long-lasting (excitatory) effects, produces the rebound excitation (Nakazato et al. 1970, Brodin et al. 1981). The rebound excitation of the guinea-pig taenia coli is in fact strongly reduced by the competitive substance P antagonist (D-Pro^2, D-Trp^{7,9})-substance P (Leander et al. 1981 b).

9.1.4 Neurotransmitters in the Mammalian Enteric Nervous System

The different types of nerves in the enteric nervous system of mammals may be classified according to several groups of criteria. Thus, (1) the ultrastructural appearance of the nerve terminals, (2) the electrophysiological behaviour of the neurons, (3) their biochemical and histochemical properties, and (4) their pharmacological and functional differences, have been used to characterize more than ten

Fig. 9.4. An attempt to summarize a few of the neuron types of the enteric nervous system in mammals. *Enteric neurons to smooth muscle:* The type of neuron (*Chol*, cholinergic; *Adr*, adrenergic; *NANC*, non-adrenergic, non-cholinergic) and its effect on the smooth muscle (+, contraction; −, relaxation) are shown together with the possible transmitter substance of the neuron (*ACh*, acetylcholine; *ATP*, adenosine triphosphate; *VIP*, vasoactive intestinal polypeptide; *SP*, substance P; 5-HT, 5-hydroxytryptamine; *NA*, noradrenaline). Adrenergic nerve cell bodies are only exceptionally present in the en-

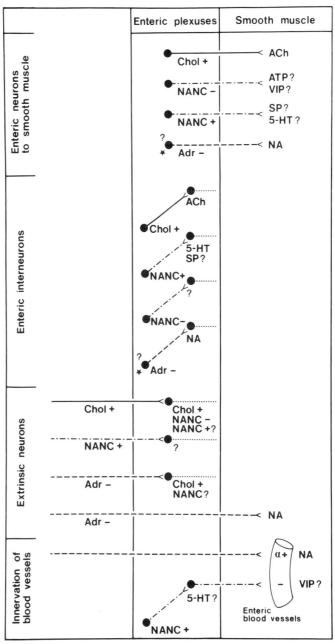

teric plexuses (*). *Enteric interneurons:* The type of neuron, its effect on other enteric ganglion cells and the possible transmitter of the interneuron are shown as above. *Extrinsic neurons:* The type of neuron and its effect on the intrinsic ganglion cells or smooth muscle is shown as above. The possible types of intrinsic neurons innervated by the extrinsic fibres are also shown. *Innervation of blood vessels:* extrinsic adrenergic fibres and instrinsic VIP-ergic fibres are shown. The VIP-releasing neurons are part of the local vasodilatory reflex arc, which probably also involves serotonergic neurons or 5-HT from enterochromaffin cells (see also Fig. 9.3). (Based primarily on Furness and Costa 1980)

different types of neurons in the enteric nervous system (including extrinsic neurons of the gut) (Furness and Costa 1980, Sundler et al. 1980, Gershon 1981, Wood 1981). However, there are still great difficulties when the neurons characterized by one set of criteria are to be compared to those classified by another set of criteria: e.g., the electrophysiological behaviour of a nerve cell is often impossible to relate to the type of transmitter of that particular neuron, or to its ultrastructural appearance.

There appears to be a general agreement that at least acetylcholine and noradrenaline fulfill the five criteria cited above (Chap. 3.1) to establish them as neurotransmitters in the mammalian gut (Furness and Costa 1980, Gershon 1981). There is also good evidence in favour of several of the putative NANC transmitters, although there is as yet no general agreement about their functions as neurotransmitters in the gut. This group includes primarily ATP, 5-HT and several peptides (Burnstock 1972, 1979, 1981, Gershon et al. 1979, Furness and Costa 1980, Sundler et al. 1980, Gershon 1981). The present short account of this very dynamic field is focused upon the presence and possible function of some of the putative transmitters in the enteric nervous system (Fig. 9.4). Several excellent reviews on the subject can be consulted for details (Furness and Costa 1980, Sundler et al. 1980, Gershon 1981).

9.1.4.1 Noradrenaline

Adrenergic neurons reach the gut mainly along blood vessels in the mesenteries but also in the vagus and pelvic nerves. As far as is known, all the pathways containing postganglionic adrenergic neurons are of spinal autonomic ("sympathetic") origin. The presence of adrenergic neurons in the vagus and pelvic nerves is due to contributions from the sympathetic chain ganglia or prevertebral ganglia. With few known exceptions in mammals (e.g., the anterior part of the guinea-pig colon), all the adrenergic neurons of the gut are extrinsic, i.e., with their cell bodies outside the gut wall and thus by definition are not part of the enteric nervous system (Costa et al. 1971, Furness and Costa 1971, 1974, Gabella 1976, 1979, Furness et al. 1979).

Ultrastructurally the noradrenaline-releasing neurons are identified by their small granular vesicles (Chap. 3.2.1) and also by their reactions to 5- and 6-hydroxydopamine (Chap. 5.2.3) (Furness and Costa 1980). Furthermore, the Falck-Hillarp fluorescence histochemical technique and immunohistochemical techniques (identification of dopamine-β-hydroxylase and tyrosine hydroxylase) can be used to localize these neurons (see Chap. 3.2.6.4).

The adrenergic fibres in the mammalian gut occur chiefly in the enteric plexuses, often forming pericellular networks about the enteric ganglion cells. It is known that the adrenergic fibres can decrease the outflow of acetylcholine from the enteric system and that the main role of the adrenergic neurons may be regulation of enteric ganglia, rather than a direct effect on the smooth muscle. Noradrenaline probably acts mainly presynaptically by reducing the release of acetylcholine in the synapse by presynaptic heteromodulation of the cholinergic preganglionic nerve terminals (Chap. 3.2.4, Norberg and Sjöqvist 1966, Paton and Vizi 1969, Burnstock and Costa 1973, Gershon et al. 1979). Numerous adrenergic terminals innervate the smooth muscle of the sphincters, and to some extent also the circular muscle, and the longitudinal muscle of the guinea-pig taenia coli (Åberg and Eränkö 1967, Jenkinson and Morton 1967, Read and Burnstock 1969a, Furness and Costa 1974, 1980).

In addition to the adrenergic innervation of the enteric neurons and the smooth muscle, adrenergic neurons in the vagus and splanchnic nerve have been shown to induce release of 5-HT from the enterochromaffin cells of the gut mucosa. The adrenergic control of these cells is mediated chiefly by a β-adrenoceptor mechanism (Ahlman et al. 1976, Pettersson 1979, Larsson 1981).

9.1.4.2 Acetylcholine

The role of acetylcholine as a transmitter in the gut was first proposed by Dale (1937a, b). Later it was shown that acetylcholine is in fact synthesized in and released from the myenteric nerve plexus of the mammalian gut (Paton and Zar 1968, Wikberg 1977).

The cholinergic fibres in the mammalian gut mediate a final step in the excitatory control of both longitudinal and circular smooth muscle (Kosterlitz and Lees 1964, Bennett and Burnstock 1968, Costa and Furness 1976, Furness and Costa 1980). In addition, both intrinsic and extrinsic cholinergic preganglionic fibres innervate neurons of the enteric plexuses. Intrinsic cholinergic preganglionic neurons are involved in the peristaltic reflex, and extrinsic cholinergic neurons are present in the vagus and in the pelvic nerves (Kosterlitz and Lees 1964, Abrahamsson 1973, Gonella 1978, Furness and Costa 1980, Gershon 1981).

9.1.4.3 Adenosine Triphosphate (ATP)

Since the initial discovery of NANC effects following stimulation of enteric neurons (Chap. 3.2.8), evidence has been obtained in support of the hypothesis that ATP is the inhibitory NANC transmitter in the autonomic neurons of the gut (Burnstock 1972, 1981). Although there appears to be little doubt about systems of synthesis, storage, release and degradation of ATP, or the similarities in the responses to ATP compared to NANC nerve stimulation in many cases (Fig. 3.8), there is a lack of truly selective pharmacological tools to manipulate this type of transmission (Chap. 5.4). Nor is there a specific histochemical method that can be used to visualize the purinergic neurons. The result of these shortcomings is that ATP has not been generally accepted as a transmitter in the autonomic nervous system.

Inhibitory neurons, which could be puringergic, include the postganglionic vagal neurons in the oesophagus and stomach, which are responsible for the efferent part of the gastric receptive relaxation reflex, and also inhibitory neurons in the posterior part of the large intestine and internal anal sphincter. Inhibitory NANC fibres are also involved in the descending inhibitory phase in the peristaltic reflex (Burnstock 1972, 1981, Abrahamsson 1973, Campbell and Gibbins 1979, Furness and Costa 1980).

9.1.4.4 5-Hydroxytryptamine (Serotonin)

There is good fluorescence histochemical evidence for the presence of serotonergic neurons in the enteric nervous system of fish (Chap. 9.2 and 9.4) but the demonstration of such neurons by the Falck-Hillarp technique in the mammalian gut is less convincing. A group of intrinsic neurons with a high-affinity uptake of 5-HT and related amines has been called "intrinsic amine-handling neurons" (Furness and Costa 1978) in analogy with the APUD system of Pearse and co-workers (Pearse 1969, 1976). It is possible that the intrinsic amine-handling neurons and the serotonergic neurons are related, although their content of 5-HT is too low to allow detection by the fluorescence histochemical method (Gershon 1981). Immunohistochemical demonstration of 5-HT, using specific antisera against this amine, will be of interest in solving this problem (Consolazione et al. 1981).

In the enteric system, the serotonergic neurons are thought to be interneurons in the ascending phase of the peristaltic reflex (Fig. 9.2) and in the vasodilatory reflex (Fig. 9.3), but the role of the enterochromaffin cells as a source of 5-HT must also be kept in mind. There is also some evidence for a direct innervation of smooth muscle, at least in some parts of the gut (Furness and Costa 1980, Gershon 1981).

Serotonergic neurons may also innervate inhibitory, intrinsic NANC neurons of the gut (Bülbring and Gershon 1967, Furness and Costa 1973, Costa and Furness 1979 a). Slow excitatory postsynaptic potentials (EPSP) in the enteric neurons can be induced by 5-HT (Wood and Mayer 1978, 1979, Wood et al. 1980, Gonella 1981), supporting the idea that this substance acts as an excitatory transmitter in the enteric nervous system.

9.1.4.5 Vasoactive Intestinal Polypeptide (VIP)

The presence of VIP in the mammalian gut was first demonstrated by Said and Mutt (1970), working with pigs. Investigations of the amino acid sequence revealed it to be a polypeptide with 28 amino acids (Said and Mutt 1972, Mutt and Said 1974). VIP-like immunoreactivity has since been demonstrated in the gut of several vertebrates. The amino acid sequence of VIP from the domestic fowl (chicken) has also been determined (Nilsson 1975, Fig. 5.8).

Although VIP was initially thought of as a "classical" gut hormone, immunohistochemical localization has now clearly shown that VIP in mammals is stored exclusively in neurons. The VIP-containing neurons occur in all layers of the various parts of the mammalian gut and in many other organs. In the enteric system, VIP-containing neurons are particularly abundant in the enteric plexuses, especially in the submucous plexus of some regions of the gut, where VIP-like immunoreactivity has been demonstrated in as many as 45% of the cell bodies. VIP fibres are also found surrounding blood vessels of the gut, and in the smooth muscle layers, especially the circular layer (Fahrenkrug 1979, Costa et al. 1980b, Furness and Costa 1980, Furness et al. 1980a, Larsson 1980, Sundler et al. 1980, Schultzberg et al. 1980, Jessen et al. 1980, Leander et al. 1981a).

VIP relaxes the smooth muscle of the gut. Release of VIP following vagal stimulation or distension of the pig, cat and rat gut has also been demonstrated. A possible role of VIP as a neurotransmitter in the vagal inhibitory innervation of the stomach was therefore suggested (Fahrenkrug et al. 1978a, b, Pederson et al. 1981). Since some of the VIP-immunoreactive neurons in the enteric plexuses of the intestine project anally to innervate the circular muscle layer, VIP could play a role as the neurotransmitter in the final step in the descending inhibitory part of the peristaltic reflex (Costa and Furness 1976, Campbell and Gibbins 1979, Fahrenkrug 1979, Furness and Costa 1979a, 1980, Furness et al. 1980a, Gershon 1981). Excitatory effects of VIP on about 45% of the enteric neurons studied have been recorded by Williams and North (1979b).

In addition to the functions of VIP as a neurotransmitter in "purely VIP-ergic" neurons, the possibility of VIP as a co-transmitter and/or neuromodulator in other types of neurons should be kept in mind. Such co-existence has been shown in the "classically" cholinergic neurons of some exocrine glands of mammals (Lundberg 1981).

9.1.4.6 Substance P

Substance P was demonstrated in extracts from the brain and gut by von Euler and Gaddum (1931), and is thus the first described "brain-gut peptide". Analysis of the primary structure shows a peptide with 11 amino acids (Chang et al. 1971) and modern immunohistochemical techniques have confirmed the localization of the peptide to both endocrine cells of the intestinal mucosa and neurons in the brain and gut of mammals and other vertebrates (Sundler et al. 1980).

Substance P-immunoreactive neurons form very dense networks in the submucous and, especially, the myenteric plexuses in all parts of the gut of rat and guinea-pig. They also occur in the mucosa, muscle layers and surrounding blood vessels of the submucosa (Franco et al. 1979b, Schultzberg et al. 1980, Sundler et al. 1980, Costa et al. 1980a, 1981, Leander et al. 1981a). Most of the substance P-containing neurons in the enteric system are intrinsic, but some of the neurons to the submucous ganglia and the blood vessels of the submucosa disappear after extrinsic denervation of the gut, and are thus of extrinsic origin (Franco et al. 1979a, b, Costa et al. 1980a). The intrinsic substance P neurons of the submucous plexus innervate the mucosa whilst the substance P neurons of the myenteric plexus innervate ganglion cells of both plexuses, the mucosa and the muscle layers (Costa et al. 1981). The proportion of substance-P-immunoreactive cell bodies in the two plexuses of the rat and guinea-pig gut varies among the different parts of the gut, being completely absent in the stomach of both species (Schultzberg et al. 1980, Costa et al. 1980a).

Substance P is excitatory both on smooth muscle and on the enteric neurons. The neurons show slow excitatory postsynaptic potentials (Katayama and North 1978, Katayama et al. 1979, North et al. 1980, Furness and Costa 1980, Leander et al. 1981a). It has also been shown that substance P can act via excitation of cholinergic neurons in the cat gut (Edin et al. 1980, Delbro 1981). A preganglionic cholinergic input to substance-P-releasing neurons has also been demonstrated (Furness and Costa 1980).

The most likely role of substance P-releasing neurons of the enteric system is therefore an excitatory control of smooth muscle and, especially, of ganglia in the enteric plexuses. Immunohistochemical evidence also suggests that substance P could contribute to the control of blood flow in the gut (Furness and Costa 1980, Schultzberg et al. 1980, Sundler et al. 1980, Gershon 1981, Holzer et al. 1981).

9.1.4.7 Somatostatin

Somatostatin is a peptide of 14 amino acids, originally demonstrated in the brain by Brazeau et al. (1973). Immunohistochemical localization of somatostatin-like immunoreactivity in the mammalian gut shows nerve cell bodies in both the submucous and myenteric plexuses in the small and large intestine, but not in the stomach (Schultzberg et al. 1980, Furness and Costa 1980). There are no reports of an innervation of blood vessels by somatostatin-immunoreactive nerve fibres; an innervation of the muscle layers is sparse or absent (Schultzberg et al. 1980).

Somatostatin has either a small excitatory (Cocks and Burnstock 1979, Fig. 3.8) or no effect on the gut muscle, but an inhibitory action on both adrenergic and cholinergic nerve function has been described (Guillemin 1976, Williams and North 1978, Cohen et al. 1978, Furness and Costa 1979b). It is possible that somatostatin-releasing neurons inhibit excitatory neurons in the enteric nervous system and excite inhibitory neurons. The distribution and effects of the somatostatin-releasing neurons are compatible with the view that they act as interneurons in the descending inhibitory pathways of the intestine (Fig. 9.2, Furness and Costa 1979b, 1980, Furness et al. 1980a, b).

Somatostatin-like immunoreactivity has been demonstrated in neurons, which in addition show gastrin/cholecystokinin-like immunoreactivity in the guinea-pig colon, and also in noradrenergic neurons that project to the gut (Hökfelt et al. 1977a, Schultzberg et al. 1980).

9.1.4.8 Enkephalin

Met- and leu-enkephalin are a pair of pentapeptides that differ in the COOH-terminal amino acid (methionine or leucine). The enkephalins are the endogenous agonists for the so called opiate receptors and were first isolated from brain extracts (Hughes 1975). Immunohistochemical techniques have demonstrated enkephalins also in preganglionic vagal and pelvic neurons, preganglionic neurons to the sympathetic chain and prevertebral ganglia, and in the enteric nervous system (Alumets et al. 1978, Lundberg et al. 1980 b).

In the enteric nervous system, enkephalin-like immunoreactivity has been demonstrated in nerve cell bodies of the myenteric, but not submucous, plexus and in nerve fibres in the myenteric plexus and muscle layers (Furness and Costa 1980, Schultzberg et al. 1980, Leander et al. 1981a). Synthesis of enkephalins in the enteric nervous system, and also release in response to electrical stimulation of enteric nerves has been demonstrated (Sosa et al. 1977, Schultz et al. 1977). Enkephalin produces a rapid, reversible and dose-dependent inhibition of the electrical activity in enteric neurons (Williams and North 1979 a). Also the release of acetylcholine from enteric neurons is reduced by enkephalin. Enteric enkephalin-releasing neurons may thus be involved in an inhibitory control of enteric ganglia (Furness and Costa 1980, Leander et al. 1981a). No direct effect of enkephalin on the smooth muscle could be demonstrated in the guinea-pig taenia coli (Cocks and Burnstock 1979, Fig. 3.8).

9.1.4.9 Other Peptides

Other peptides have been demonstrated in enteric neurons by immunohistochemical techniques, but their function and possible role as transmitter substances is even less known than for the peptides mentioned above.

Gastrin/cholecystokinin-like immunoreactivity has been demonstrated in a sparse network of fibres in the stomach and intestine, and also in cell bodies in the submucous and myenteric plexuses of the large intestine (Schultzberg et al. 1980).

Neurotensin-like immunoreactivity has been demonstrated in nerve fibres, but not in cell bodies, of the rat gut (Schultzberg et al. 1980). Neurotensin contracts the taenia coli of the guinea-pig (Cocks and Burnstock 1979, Fig. 3.8).

Bombesin-like immunoreactivity has been demonstrated in nerve fibres in the mucosa of the rat gut (Dockray et al. 1979, Sundler et al. 1980).

Pancreatic polypeptide-like immunoreactivity has been demonstrated in neurons of the enteric plexuses, perivascular nerve fibres and in other layers of the gut (Sundler et al. 1980).

9.2 Cyclostomes

The knowledge of the autonomic innervation of the cyclostome gut is fragmentary. The gut of both lampetroids and myxinoids is a straight tube, without obvious regional specializations of the different parts of the gut. The smooth muscle coat of the gut in both groups of cyclostomes is poorly developed, except in the most posterior part ("rectum"), which is also densely innervated in both *Lampetra* and *Myxine* (Brandt 1922, Johnels 1956, Fänge et al. 1963).

The gut is innervated by the intestinal branches of the vagi, which in both myxinoids and lampetroids fuse to form a single strand, the *ramus intestinalis impar*, which extends along the dorsal surface of the intestine. Spinal autonomic fibres from the anterior part of the animal may enter the vagi near the cardiac plexus (Marcus 1910) and it is also possible, at least in lampetroids, that a large part of the innervation of the "rectum" is of spinal autonomic origin (Brandt 1922, Johnels 1956, Fänge et al. 1963).

The enteric plexus of the gut is rich in ganglion cells both in *Myxine* (less so in the posterior part of the gut) and *Lampetra*. Ganglion cells are numerous also in the vagus and in the spinal autonomic nerves to the "rectum" (Brandt 1922, Johnels 1956, Fänge et al. 1963, Baumgarten et al. 1973). There is no information about the role of the enteric nervous system in gastrointestinal reflexes of cyclostomes.

Stimulation of the vagus is either without effect (Fänge 1948, Fänge and Johnels 1958, *Myxine*), or produces a weak relaxation of the gut (Patterson and Fair 1933, *Bdellostoma*). Cholinesterase-positive cells have been demonstrated histochemically in the gut (Govyrin and Popova 1969, Hallbäck 1973), but it is unlikely that acetylcholine is the vagal transmitter since this substance contracts all parts of the gut (Fänge 1948, von Euler and Östlund 1957, Johnels and Östlund 1958, Holmgren and Fänge 1981). It should be remembered that the presence of cholinesterase in a neuron is not a reliable proof that the neuron is cholinergic (Chap. 3.2.7.1). The variable effects of adrenaline, including excitatory effects, on the *Myxine* gut (Holmgren and Fänge 1981) makes it doubtful that catecholamines are responsible for an inhibitory vagal control, if such a control is indeed present at all. It is clear that the vagus of *Myxine* contains small quantities of catecholamines (von Euler and Fänge 1961) and there is also evidence for the existence of neurons (cell bodies and processes) containing primary catecholamines (dopamine and noradrenaline) in the gut of the lamprey (Govyrin and Popova 1969, Baumgarten et al. 1973, Sakharov and Salimova 1980).

In contrast to the situation in mammals, there is very good fluorescence histochemical evidence for characteristically yellow fluorescent serotonergic nerve fibres in the gut of both *Myxine* and *Lampetra*, and 5-HT has also been demonstrated chemically in gut extracts (Baumgarten et al. 1973, Goodrich et al. 1980, Sakharov and Salimova 1980). Catecholamine-storing chromaffin cells are present in the *Lampetra* gut but there are no 5-HT-storing enterochromaffin cells (Baumgarten et al. 1973, Sakharov and Salimova 1980).

Autoradiography at the ultrastructural level shows a specific uptake of ^{3}H-5-HT into neurons of the enteric plexus of *Myxine* (Goodrich et al. 1980). Also in *Lampetra* injection of 5,6- or 5,7-dihydroxytryptamine has been shown to cause cytotoxic effects in the "amine-handling neurons". These results support the idea of serotonergic neurons in the cyclostome gut (Baumgarten et al. 1973). In the electron microscope, these neurons show a few small granular vesicles and a dominance of large granular vesicles with a diameter of 75–160 nm (Baumgarten et al. 1973).

5-HT induces a very weak contraction of the *Myxine* gut and is strongly excitatory on the rectum of *Lampetra* (von Euler and Östlund 1957, Johnels and Östlund 1958).

9.3 Elasmobranchs

The alimentary canal of the elasmobranchs shows some elaboration of the straight tube of the cyclostomes. A wide oesophagus opens into the cardiac portion of the stomach, which makes a U-turn and tapers off into the forward-running pyloric portion of the stomach. The pyloric stomach makes a second U-turn and opens through the pyloric sphincter into the wide intestine with its peculiar spiral valve which greatly increases the internal surface area of the short intestine (Fig. 2.6). The spiral intestine extends posteriorly to the short rectum. In chimaeroids, a stomach is absent and the spiral intestine is directly attached to the oesophagus (Nicol 1952, Romer 1962, Young 1980a).

The understanding of the autonomic nervous control of the alimentary canal of elasmobranchs is very unsatisfactory. The available information is often contradictory regarding the types of effect produced by the extrinsic nerves. The enteric nervous system is represented by both a submucous and a myenteric plexus, with the ganglion cells in the myenteric plexus diffusely arranged (Kirtisinghe 1940, Nicol 1952). Nothing is known about the function of the intrinsic neurons, or their role in enteric reflexes, but an involvement in peristalsis seems likely (Young 1933c).

9.3.1 The Elasmobranch Stomach

Electrical stimulation of the intracranial roots of the vagi of MS 222 anaesthetized *Scyliorhinus canicula* produces a contraction of the striated muscle of the oesophagus, but an inhibition of the spontaneous rhythmic activity of the stomach as recorded by an intragastric balloon (Campbell 1975). Recording the tension in longitudinal muscle of the cardiac and pyloric portions of the stomach from pithed specimens of the same species and of *Raja clavata*, however, failed to provide any conclusive evidence in favour of an inhibitory vagal control of the stomach (Young 1980a). It is possible, as pointed out by Young (1980a), that the vagal effects recorded by Campbell (1975) are due to inhibition of the circular muscle layer, the responses of which were not recorded in Young's preparations. In view of the findings by Campbell and Duxson (1978) in a toad-lung preparation (see Chap. 10.3.1), it is also possible that the pretreatment of the animal (MS 222 anaesthesia and pithing, respectively) could affect the response to vagal stimulation.

The extent to which recorded excitatory effects of vagal stimulation in selachians is due to rebound contractions (Bottazzi 1902, Müller and Liljestrand

1918, Lutz 1931, Babkin et al. 1935, Campbell and Burnstock 1968), is not at all clear. In spite of one description of a primary contraction during vagal stimulation (Nicholls 1934), Campbell and Burnstock (1968) concluded that the evidence favours an inhibitory vagal innervation of the elasmobranch stomach. It is also clear that species differences may exist.

Stimulation of the anterior splanchnic nerve with low frequency (1 Hz) causes an inhibition of the pyloric stomach musculature of *Scyliorhinus*, which is followed by a marked rebound contraction at the end of the stimulation period (Young 1980 a). Stimulation with higher frequencies produces a contraction, a response seen also in other experiments (Müller and Liljestrand 1918, Nicholls 1934, Babkin et al. 1935, Young 1980a). Thus both excitatory and inhibitory fibres appear to be present in the splanchnic nerve supply to the stomach of elasmobranchs, and the lack of effects of nicotine and hexamethonium on the responses of the stomach to splanchnic nerve stimulation suggests that these fibres are postganglionic (Young 1980a).

Acetylcholine, at least in high concentrations, induces contractions of the musculature of the elasmobranch stomach, a response that can be blocked by atropine and hyoscine. The excitatory effect of splanchnic nerve stimulation is, however, insensitive to muscarinic cholinoceptor blockade (Nicholls 1934, Babkin et al. 1935, von Euler and Östlund 1957, Young 1980a) which is incompatible with the view that acetylcholine is the splanchnic excitatory transmitter.

Adrenaline and noradrenaline also induce contraction of the elasmobranch stomach (Young 1933c, 1980a, Nicholls 1934, Babkin et al. 1935, Dreyer 1949, Moore and Hiatt 1967). The excitatory splanchnic transmitter could thus be a catecholamine and Young (1980a) showed that the contraction caused by splanchnic nerve stimulation in the *Scyliorhinus* stomach could be abolished by phentolamine. However, since the concentrations of phentolamine used were high (6×10^{-5}–1.8×10^{-4} M) a non-selective blockade cannot be excluded. A lower concentration of phentolamine (10^{-6} M), which did not inhibit the effects of 5-HT or cholinoceptor agonists, effectively blocked the responses of splanchnic stimulation in the *Squalus acanthias* stomach (Nilsson and Holmgren, to be published). The idea of an adrenergic splanchnic excitatory control of the dogfish stomach is further supported by the demonstration of a dense network of adrenergic nerve fibres in the enteric plexuses of the *Squalus* stomach (Fig. 2.22) (Holmgren and Nilsson, to be published). VIP-, gastrin/CCK-, somatostatin- and bombesin-like immunoreactivity has been demonstrated in neurons of the enteric plexuses of *Squalus*, but so far nothing is known about the effects of these peptides on the gastric function of this species (Fig. 2.24) (Holmgren and Nilsson, to be published).

9.3.2 The Elasmobranch Intestine

Very little is known about the innervation of the elasmobranch intestine. As in the stomach, the middle and posterior splanchnic nerves are excitatory, but nothing is known about the transmitter substance(s) involved (Bottazzi 1902, Lutz 1931, Young 1933c, Nicholls 1934). The intestinal musculature of *Raja* is contracted by acetylcholine, acting via muscarinic cholinoceptors, and also by substance P and

5-HT (von Euler and Östlund 1957, Young 1980a). Both excitatory and inhibitory effects of adrenaline have been demonstrated in the elasmobranch rectum and intestine (Lutz 1931, Young 1933c, von Euler and Östlund 1957).

Of the putative neurotransmitters among the peptides, both substance P and VIP-like peptides have been demonstrated in extracts from the elasmobranch intestine but their cellular localization and possible functions remain unknown (von Euler and Östlund 1956, Fouchereau-Peron et al. 1980).

9.4 Teleosts

The alimentary canal in teleosts shows great anatomical variation, largely reflecting the differences in food habits. In most cases, an oesophagus, stomach and intestine can be discerned, and in some cases there may also be an anatomically defined rectum. In many teleosts, e.g., the cyprinids, a stomach is absent and the intestine is a simple tube, more or less elaborately folded, which runs from the oesophagus to the anus. A useful review of the anatomy and physiology of the alimentary canal in fish is that of Fänge and Grove (1979).

The arrangement of the various layers in the wall of the teleost gut is essentially the same as in the mammals (Sect. 9.1.1). In some species, particularly the stomachless tench (*Tinca tinca*), there is, in addition to the smooth muscle layers of the gut, an outer layer of striated muscle throughout the length of the gut. Vagally innervated striated muscle is present in the oesophagus of many fish and invades the upper part of the stomach (Burnstock 1959, Fänge and Grove 1979, Holmgren et al. 1982).

The enteric plexuses in teleosts are well developed but there are at most a few ganglion cells in the submucous plexus. The ganglion cells of the myenteric plexus lie scattered in the nodes and internodal connectives of the plexus (Kirtisinghe 1940, Burnstock 1959). The cell bodies of the neurons in the myenteric plexus of *Salmo trutta* are uni-, bi- and multipolar with diameters of 10–15 µm, or larger multipolar cells with a diameter of 30–60 µm. In the stomach there are additional uni- or bipolar cells (diameter 30–70 µm), which are thought to represent the postganglionic vagal neurons (Burnstock 1959).

In 1952 Nicol concluded that, apart from the studies of colour change, the physiology of the autonomic nervous system of teleosts has received less attention than that of the elasmobranchs. This is no longer true, and while the understanding of the autonomic control of the gut in elasmobranchs is still very poor, there are now numerous studies that contribute to an understanding of the autonomic nervous control of the teleost gut.

The earlier research on extrinsic nervous control of the teleost gut was hampered by the lack of knowledge of the rebound excitation phenomenon and the large spinal autonomic contribution to the vagus ("vago-sympathetic trunk"), which is present in all teleosts studied. Since the early literature has been well covered by reviews, notably Nicol (1952), Campbell and Burnstock (1968) and Burnstock (1969), the present account will focus mainly on the more recent work.

9.4.1 The Teleost Stomach

The teleost stomach is supplied by cranial autonomic fibres running in the intestinal branches of the vagi (*rami intestinales vagi*). There are also contributions of spinal autonomic fibres both directly via the anterior splanchnic nerve(s), and indirectly in the "vagosympathetic trunk" (Fig. 2.7). The vagal influence does not extend beyond the stomach or upper part of small intestine ("duodenum") in most teleosts, but in the stomach-less cyprinids, such as *Tinca*, the vagus influences the entire gut (Ohnesorge and Rehberg 1963). In other stomachless fish (*Ammotretis* and *Rhombosolea*), however, the vagal influence does not reach further than the striated muscle of the oesophagus (Grove and Campbell 1979 a).

The vagal influence on the teleost stomach may be purely or almost purely inhibitory as in *Salmo* and *Conger*, and in these fish the period of inhibition produced by electrical stimulation of the vagus is followed by a rebound contraction at the end of the stimulation period (Campbell 1975, Campbell and Burnstock 1968, Holmgren and Nilsson 1981).

In another group of teleosts, both inhibitory and excitatory fibres have been demonstrated in the vagus (Edwards 1972a, Stevenson and Grove 1977: *Pleuronectes;* Young 1980 b: *Lophius;* Fig. 9.5).

Finally, in a third group of teleosts one observes only primary contractions of the stomach during electrical stimulation of the vagus, an effect that is due to cholinergic vagal fibres. Although it is possible that cholinergic fibres from the sympathetic chain ganglia enter the vagus extracranially, it is clear at least in cod (*Ga-*

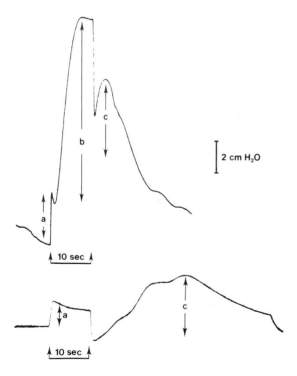

Fig. 9.5. Intragastric pressure responses to vagal stimulation in the plaice (*Pleuronectes platessa*) before (*upper trace*) and after (*lower trace*) addition of atropine (10^{-5} M). Note the rapid contraction of the oesophageal striated muscle (*a*), the primary contraction by atropin sensitive (cholinergic) fibres (*b*) and the "rebound excitation" following cessation of the stimulus (*c*). The vertical calibration bar shows an intragastric pressure increase of 2 cm H_2O (ca. 0.2 kPa). (Reproduced with permission from Stevenson and Grove 1977)

dus morhua) and flathead (*Platycephalus bassensis*) that intracranial stimulation of truly vagal fibres produces primary excitation of the stomach (Nilsson and Fänge 1969, Grove and Campbell 1979b, Nilsson, unpublished).

Acetylcholine and other muscarinic cholinoceptor agonists induce contraction of gastric musculature in all species studied thus far. There is a general agreement that the vagal excitatory fibres are cholinergic (Nilsson and Fänge 1969, Edwards 1972a, b, Stevenson and Grove 1977, Fänge and Grove 1979, Grove and Campbell 1979b, Young 1980b). The nature of the inhibitory transmitter substance is not known and, in view of the lack of effects of the blocking agents for adrenergic neurotransmission used in the studies, it seems likely that the vagal inhibitory fibres of the teleosts are akin to the NANC fibres of the tetrapods. In one study of the rainbow trout stomach, it has been shown that a substance that mimicks the inhibitory response to vagal stimulation is released from the stomach upon vagal stimulation in quantities sufficient to affect a second preparation, i.e. a superfused strip of stomach wall (Holmgren and Nilsson 1981, Fig. 9.6). The cell bodies of the postganglionic neurons of the vagal pathways are enteric (*Salmo:* Holmgren and Nilsson 1981), or may lie in the vagal ganglion (nodose ganglion) complex (*Pleuronectes:* Stevenson and Grove 1977).

The effect of splanchnic nerve stimulation is an excitation of gastric musculature in *Salmo* and *Platycephalus* (Campbell and Gannon 1976, Grove and Campbell 1979b), the response in both cases being due to adrenergic splanchnic fibres acting via α-adrenoceptors. In *Pleuronectes*, on the other hand, the excitatory component in the dual response to splanchnic nerve stimulation is cholinergic. There is also an inhibitory innervation that may be adrenergic since the response is inhibited by the β_2-adrenoceptor antagonist butoxamine and by reserpine (Stevenson and Grove 1978). Adrenaline and noradrenaline induce contraction of the stomach of *Salmo*, *Anguilla* and *Gadus*, whereas responses are variable in *Pleuronectes* and inhibitory in *Lophius* and *Uranoscopus* (Young 1936, 1980b, Burnstock 1958a, b, Nilsson and Fänge 1967, 1969, Edwards 1972b, Campbell and Gannon 1976, Fänge and Grove 1979).

Fluorescence histochemistry has revealed adrenergic fibres in the myenteric plexus, smooth muscle (especially in the circular layer), and surrounding blood vessels in the stomach of *Platicthys*, *Salmo* (Fig. 2.23), *Anguilla*, *Pleuronectes* and *Platycephalus* (Campbell and Gannon 1976, Stevenson and Grove 1978, Fänge and Grove 1979, Grove and Campbell 1979b, Santer and Holmgren unpublished). Fluorescent varicose fibres surrounding the cell bodies of enteric neurons have been described in the stomach of *Salmo* (Santer and Holmgren unpublished). This feature has previously been thought to be unique to the higher tetrapods (Burnstock 1969).

5-HT causes contractions of the stomach of *Pleuronectes* and *Salmo;* at least in *Pleuronectes* this response is thought to involve release of acetylcholine from enteric neurons. A direct effect on the smooth muscle is also present (Grove et al. 1974). ATP induces contractions of the stomach of *Pleuronectes* (Fänge and Grove 1979), but has dual actions in *Salmo* (Holmgren 1982) and therefore remains a putative inhibitory vagal transmitter.

Both VIP- and substance P-immunoreactivity have been demonstrated in neurons of the stomach and intestine of the rainbow trout (*Salmo gairdneri*) and

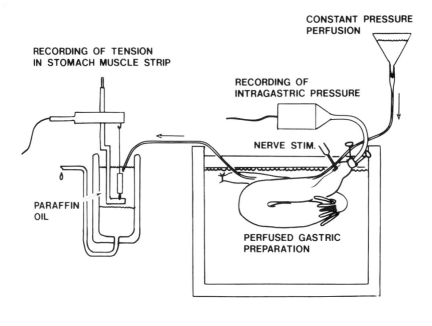

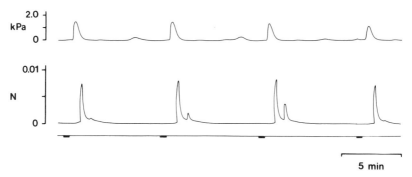

Fig. 9.6. Experimental arrangement of the "Loewi-preparation" used to study the tension responses of a superfused strip of stomach muscle to the venous effluent from the stomach of the rainbow trout (*Salmo gairdneri*). The stomach is perfused with Ringer's solution at a constant pressure, and the venous effluent from the stomach passed over an isolated stomach wall muscle strip suspended in paraffin oil to avoid dilution of the perfusate. The tracings show the "rebound" contraction of the perfused stomach *after* each period of vagus stimulation recorded as an elevation of the intragastric pressure (*upper trace*) and, after a delay depending mainly on the length of the venous catheter, a similar tension response is recorded in the superfused strip (*lower trace*). (Reproduced with permission from Holmgren and Nilsson 1980, 1981)

other teleost species (Langer et al. 1979, van Noorden and Patent 1980, Holmgren et al. 1982). The presence of VIP and substance P has been confirmed by radioimmunoassay of extracts from the gut of *Salmo* (Holmgren et al. 1982) and VIP has also been demonstrated in the gut of *Labrus berggylta* and *Trisopterus luscus* (Fouchereau-Peron et al. 1980).

The VIP-immunoreactive fibres have been demonstrated in all layers of the stomach and intestine of the rainbow trout, particularly in the myenteric plexus (Fig. 2.26) and also in the smooth muscle layers and surrounding blood vessels (Holmgren et al. 1982). A similar distribution of VIP-immunoreactive fibres has also been described in some other species of teleosts, but there appear to be some species differences in the pattern of distribution of these fibres (Langer et al. 1979).

The effect of VIP on the smooth muscle from the rainbow trout is dual, including both excitatory and inhibitory components (Holmgren 1982).

Substance P immunoreactivity has been demonstrated in nerve terminals of the myenteric plexus of *Salmo gairdneri* and *Gillichthys mirabilis* and the substance P-immunoreactive neurons occur also in the muscle layers of the stomach and intestine (van Noorden and Patent 1980, Holmgren et al. 1982). Substance P causes contraction of smooth muscle of the rainbow trout stomach. This effect in not due to excitation of cholinergic neurons in the enteric plexuses, since atropine is without effect on the response to substance P (Holmgren 1982).

In addition to nerve fibres showing to VIP-like and substance P immunoreactivity, Langer et al. (1979) demonstrated nerve fibres with immunoreactivity to enkephalin, neurotensin and bombesin in several species of teleosts. In a study of the rainbow trout gut, bombesin-immunoreactive endocrine cells were seen but neither bombesin nor enkephalin could be demonstrated in neurons (Holmgren et al. 1982). In the stomach and intestine of the ganoid, *Lepisosteus platyrhincus*, VIP-, neurotensin- (Fig. 2.25) and bombesin like immunoreactivity has been demonstrated in nerve fibres within the myenteric plexus (Holmgren and Nilsson, to be published).

9.4.2 The Teleost Intestine

A vagal control of the intestine occurs in the stomachless cyprinids, in which it reaches all parts of the gut. The fibres contract the striated muscle coat of *Tinca* but inhibit the smooth muscle with a rebound contraction following the cessation of the stimulus. Not to be excluded is a possible role of cholinergic fibres in the rebound excitation in the gut of *Carassius* (Ito and Kuriyama 1971, Saito 1973).

The splanchnic innervation of the teleost intestine is mainly excitatory and, since the catecholamines relax this organ, it is unlikely that adrenaline or noradrenaline are the splanchnic transmitters. The best study of the intestinal splanchnic innervation in teleosts has been performed in two stomachless fish, *Ammotretis rostrata* and *Rhombosolea tapirina*, with results that indicate the presence of inhibitory adrenergic, and excitatory NANC and cholinergic fibres in the splanchnic innervation of the intestine (Grove and Campbell 1979a). The innervation of the rectum of *Salmo trutta* by posterior splanchnic fibres is a mixture of excitatory and inhibitory fibres with the prevailing response depending on the stimulus parameters (Burnstock 1958a, b, 1959). Nothing is known about the transmitters involved.

Fluorescence histochemistry by the Falck-Hillarp method reveals adrenergic fibres in the myenteric plexus, smooth muscle and surrounding blood vessels in the teleost intestine (Baumgarten 1967, Read and Burnstock 1968, 1969a, Saito 1973,

Grove and Campbell 1979a, Watson 1979). The results of some studies suggest that the catecholamine present in the neurons is dopamine, rather than adrenaline and noradrenaline. There is also histochemical evidence for yellow-fluorescent serotonergic neurons in the intestine of *Tinca*, *Myoxocephalus* and *Pleuronectes* (Baumgarten 1967, Watson 1979).

ATP induces contractions of the intestine of *Carassius*, *Ammotretis* and *Rhombosolea* (Burnstock et al. 1972, Grove and Campbell 1979a), but relaxation of the intestine of *Pleuronectes* (Fänge and Grove 1979). 5-HT is excitatory on all preparations from teleost intestine studied and appears to exert its effects chiefly directly on the smooth muscle cells (von Euler and Östlund 1957, Burnstock 1958a, b, Baumgarten 1967).

The distribution of peptidergic neurons in the teleost gut is described above (Sect. 9.4.1). It has also been shown that substance P induces contraction of the intestine of *Labrus berggylta* and *Pleuronectes platessa* (Euler and Östlund 1957). A substance P-like peptide has been demonstrated in the gut of *Squalus* and *Gadus* (von Euler and Östlund 1956).

9.5 Amphibians

The alimentary canal of amphibians, at least in the urodeles and anurans, is similar to that of mammals. It is possible to distinguish an oesophagus, a stomach, a long small intestine (duodenum and ileum), a short large intestine (colon) and a rectum. There are well-developed enteric plexuses, but ganglion cells are few or absent in the submucous plexus (Gunn 1951, Campbell and Burnstock 1968).

9.5.1 The Amphibian Stomach

The amphibian stomach is innervated by cranial autonomic fibres in the vagus nerve, and by spinal autonomic fibres both in "vagosympathetic" and splanchnic pathways (Fig. 2.8). The presence of "sympathetic" fibres in the vagus trunk makes it difficult to interpret early investigations of the innervation of the amphibian stomach, since the mixture of spinal and cranial pathways in the vagal trunk has not always been taken into consideration. More recent experiments do, however, give a relatively uniform picture of the mode of innervation of the anuran stomach.

Since vagal fibres to the stomach of urodeles (Patterson 1928) and anurans (Dixon 1902, Campbell 1969) are purely inhibitory, it is possible that the excitatory effects of vagal stimulation observed in many cases are due to the presence in the vagus of fibres from the sympathetic chain ganglia (Contejean 1892, Steinach and Weiner 1895, Hopf 1911, Aikawa 1931, Yüh 1931, Hato 1958b, Campbell and Burnstock 1968, Campbell 1969). The possibility of rebound excitation should also be kept in mind in the interpretation of excitatory effects of nerve stimulation (see Sect. 9.1.2).

In a thorough examination of the innervation of the stomach of *Bufo marinus*, Campbell (1969) showed that cholinergic excitatory fibres from the cervical sympathetic chain enter the vagus and run to the stomach. The effect of the ganglionic blocking agent pentolinium on the excitatory effect of vagosympathetic trunk stimulation suggests that most of the excitatory pathways have ganglionic synapses within the stomach. The cell bodies of the postganglionic neurons are situated either in the myenteric plexus or in the distal part of the vagosympathetic trunk which is rich in ganglion cells (Jullien and Ripplinger 1954, McLean and Burnstock 1967b, Campbell 1969).

In addition to the spinal autonomic fibres in the "vagosympathetic trunk", there is a direct innervation by splanchnic pathways. Stimulation of these fibres either relaxes (Campbell 1969) or has mixed effects (contraction/relaxation) (Yüh 1931, Semba and Hiraoka 1957, Hato 1958a) on the stomach. There are some indications that these differences in effect can be explained in part by seasonal variations. In the summer, the effect is predominantly, if not solely inhibitory and mediated by adrenergic fibres, while the excitatory effects seen in other cases are mediated by cholinergic fibres (Campbell 1969). In urodeles, stimulation of the splanchnic nerve produces a strong contraction of the stomach but the nature of the transmitter involved is obscure (Patterson 1928).

Adrenaline usually relaxes the anuran stomach (Hopf 1911, Carlson and Luckhard 1921, Yüh 1931) but contracts the circular muscle layer in the stomach of the urodele *Necturus* (Friedman 1935). Adrenergic fibres are found in circular muscle and perivascular plexuses of the stomach (Read and Burnstock 1969a).

The nature of the inhibitory vagal NANC transmitter in the anuran stomach is not known, but ATP has been proposed (Burnstock 1975). It is also clear that VIP is present in nerve fibres in the stomach of both urodeles and anurans (Buchan et al. 1980, 1981).

9.5.2 The Amphibian Intestine

Splanchnic nerve fibres innervate the small and large intestine of the amphibians and, in addition, there is a rich innervation of the colon and rectum by pelvic (classically "parasympathetic") fibres from the 9th and 10th spinal nerves (Langley and Orbeli 1911, Boyd et al. 1964). The pelvic innervation of the hindgut is excitatory; the postganglionic cholinergic neurons of these pathways are part of the myenteric plexus (Boyd et al. 1964).

The splanchnic innervation is both excitatory and inhibitory. The excitatory fibres could be cholinergic, since the response to splanchnic nerve stimulation is at least partly blocked by muscarinic cholinoceptor antagonists. The presence of excitatory NANC fibres in the innervation of the intestine of *Bufo marinus* has also been postulated; the effects of these fibres are mimicked by ATP (Boyd et al. 1964, Sneddon et al. 1973).

The inhibitory splanchnic fibres are likely to be adrenergic and at least in the rectum mediated by a β-adrenoceptor mechanism (Boyd et al. 1964, Shiozu 1965). Adrenergic fibres revealed by the Falck-Hillarp technique are more plentiful in the large intestine than in the stomach and small intestine and innervate mainly the circular muscle layer and blood vessels. Adrenergic fibres are also present in the myenteric plexus, but there are no intrinsic fluorescent ganglion cells or pericellular networks of fluorescent fibres (Read and Burnstock 1968, 1969a).

VIP has been demonstrated both in neurons and endocrine cells of the intestine, but as in the higher tetrapods, the presence of VIP-immunoreactivity in endocrine cells may be due to different forms of VIP. The use of alternative antisera in amphibians as in birds would therefore be of interest (Sundler et al. 1979, Sect. 9.7).

9.6 Reptiles

The alimentary canal of reptiles resembles that of mammals and in most cases it is possible to discern an oesophagus, stomach, small and large intestine. Little is known about the physiology of the intrinsic neurons in the reptilian gut, but there is some information of the extrinsic control of the stomach and intestine. This summary of the patterns of innervation of the reptilian gut is based mainly on knowledge from turtles and lizards with the available information condensed to describe a "type reptile".

9.6.1 The Reptilian Stomach

The reptilian stomach is innervated by the vagi and splanchnic nerves. Stimulation of the vagus of turtles and lizards produces both inhibitory and excitatory effects, depending on the frequency of stimulation (Berkovitz and Rogers 1921, Veach 1925, Berger and Burnstock 1979). The excitatory effect can be abolished by muscarinic cholinoceptor antagonists and is therefore cholinergic as in mammals (Sect. 9.1). The postganglionic cell bodies of the excitatory vagal pathways are probably part of the enteric nervous system of the stomach (Noon 1900, Veach 1925), but there is also some evidence from work on crocodiles that suggests that at least some of the postganglionic nerve cell bodies lie outside the gut wall, possibly associated with the vagus trunk (nodose) ganglion (Gaskell 1886).

In addition to the excitatory vagal pathways, there is an inhibitory vagal NANC innervation of the stomach of lizards and turtles and also of the lizard oesophagus (Carlson and Luckhardt 1921). The position of the postganglionic cell bodies in these pathways is not clear, but they may be part of the enteric nervous system as in mammals (Berkovitz and Rogers 1921, Veach 1925).

The splanchnic innervation of the stomach is adrenergic, but there are no specific descriptions of an adrenergic innervation of smooth muscle (except blood vessels) or pericellular networks of adrenergic fibres of the type that occurs in the lizard intestine (Read and Burnstock 1968, 1969a, Berger and Burnstock 1979).

9.6.2 The Reptilian Intestine

The extent to which the vagus affects the gut behind the stomach in reptiles is not known. Physiological information on the splanchnic innervation is also scarce. Stimulation of the splanchnic nerve supply to the intestine of the lizard (*Trachy-*

dosaurus rugosus) produces a contraction, which is insensitive to adrenergic and cholinergic antagonists and therefore probably of the NANC type (Burnstock et al. 1972, Sneddon et al. 1973, Berger and Burnstock 1979). It has been suggested that the excitatory NANC neurons are purinergic on the grounds that quinidine acts as an antagonist to both the excitatory nerve response and to exogenously applied ATP (Sneddon et al. 1973).

There is evidence from fluorescence histochemistry for an adrenergic innervation of both fluorescent (= adrenergic) and non-fluorescent cell bodies in the myenteric plexus of *Trachydosaurus*, and a more sparse supply to the circular muscle layer of the large, but not the small, intestine and to blood vessels throughout the gut (Read and Burnstock 1968, 1969a).

Little is known about the enteric reflexes in reptiles. Peristalsis cannot be triggered by distension alone in the turtle intestine, but addition of 5-HT or acetylcholine initiates peristalsis (Yung et al. 1965). There is no information about the distribution of putative peptide transmitters in the reptilian gut.

9.7 Birds

The oesophagus of some species of birds is peculiar when compared with that of the other vertebrates in having a large swelling, the crop, which serves a food-storing function. Also the stomach proper differs in its construction from other vertebrate stomachs. A swelling of the lower part of the oesophagus, the proventriculus, carries the enzyme-secreting mucosa, while the main stomach or gizzard is usually an extremely muscular organ with a horny lining that is important in the grinding of the ingested food. The small intestine can usually be divided into a duodenum, jejunum and ileum as in mammals. In some species there is a pair of caeca (rectal caeca) at the junction of the ileum and large intestine (Adams and Eddy 1951, Romer 1962).

The layers of the avian gut are the same as in mammals but there is no outer smooth muscle coat in the gizzard (Bennett and Cobb 1969a). Interconnected myenteric and submucous plexuses are present in most parts of the gut, but a submucosal plexus is absent in the gizzard (Bennett and Cobb 1969a, b, Bennett 1974). In addition to the enteric plexuses, the intestine is supplied by nerve fibres along its length by the ganglionated nerve of Remak (Chap. 2.8, Bolton 1971a, 1976, Bennett 1974).

9.7.1 The Avian Oesophagus, Crop and Gizzard

The upper part of the oesophagus is supplied by the glossopharyngeal nerve, which forms a plexus with the vagus that innervates the crop and lower part of the oesophagus. Vagal fibres also supply the proventriculus and gizzard, and run in the nerve of Remak to the rest of the gut.

The vagal fibres provide a cholinergic excitatory innervation of the oesophagus and crop, and in the presence of atropine or hyoscine a NANC inhibitory response to vagal stimulation is unmasked (Everett 1966, Sato 1969, Sato et al. 1970, Bolton 1971 b, Bennett 1974). There appears to be no control of the oesophagus and crop by perivascular ("sympathetic") fibres (Sato et al. 1970, Bennett 1974).

Perivascular fibres from the coeliac and mesenteric plexuses and vagal fibres innervate the gizzard. The vagal fibres are both cholinergic-excitatory and NANC-inhibitory, similar to the situation in other tetrapods. The synapses in the vagal excitatory cholinergic pathways lie within the gizzard or along the vagus. It should be noted that the nodose ganglion lies distally in the vagus trunk in birds as in reptiles, and a situation like that in crocodiles, where postganglionic vagal nerve cell bodies may be present in the nodose ganglion, is possible (Sato et al. 1970, Bennett 1969 b, c, 1974).

The vagal NANC-inhibitory fibres are probably intrinsic, since hexamethonium blocks the NANC effects of vagal stimulation (Sato et al. 1970).

Adrenaline, or stimulation of the perivascular ("sympathetic") fibres to the proventriculus and gizzard is mainly excitatory, the effect being mediated by α-adrenoceptors (Bennett 1969 b, Bolton 1971 b, 1976). Fluorescence histochemistry reveals an extrinsic adrenergic innervation in the myenteric plexus, and perivascular nerves in the gizzard are also adrenergic. Fluorescent, probably adrenergic, nerve cell bodies are also present within the enteric nervous system (Enemar et al. 1965, Bennett and Malmfors 1970, Bennett et al. 1973, Bennett 1974). In addition to the adrenergic part of the response to perivascular nerve stimulation, there is an excitatory cholinergic component (Sato 1969, Sato et al. 1970).

9.7.2 The Avian Intestine

The intestine is supplied by an extrinsic innervation along its length by Remak's nerve. Through these pathways, vagal fibres may be able to reach all parts of the gut; it has been concluded that such fibres are mainly cholinergic and excitatory (Nolf 1934, Everett 1968, Bennett 1974). Stimulation of the perivascular nerves to the avian intestine produces biphasic responses. The excitatory component is cholinergic, while the inhibitory part is adrenergic acting via both α- and β-adrenoceptors. The β-adrenoceptors of at least the rectal caeca appear to be of the β_2-variety (Lands et al. 1969). There is also a NANC inhibitory component which probably involves enteric neurons (Everrett 1968, Bolton 1971 b). The presence of serotonergic neurons in the chicken gut has been postulated (Epstein and Gershon 1978), but the function of such neurons is not clear.

Substance P has been demonstrated in fibres in all layers of the avian gut, particularly in the enteric plexuses where nerve cell bodies are also present. Substance P induces contractions of the rectal caeca, and its involvement in the rebound excitation phenomenon has been proposed (Sect. 9.1.3) (Sundler et al. 1980, Brodin et al. 1981, Fontaine-Perus et al. 1981, Saffrey et al. 1982).

VIP, with an amino acid sequence slightly different from mammalian VIP (Fig. 3.6, Nilsson 1975), has also been demonstrated in the avian gut. The VIP-immunoreactive neurons innervate glands in the proventriculus, gizzard and small intestine, and there is a prominent innervation of the smooth muscle of the large intestine (Sundler et al. 1979, Vaillant et al. 1980, Fontaine-Perus et al. 1981, Saffrey et al. 1982). Some antisera demonstrate VIP in endocrine cells (Sundler et al. 1979) and different immunoreactive forms of VIP are present (Vaillant et al. 1980). VIP-, substance P-, enkephalin- and neurotensin-immunoreactive neurons in the gut of the domestic fowl appear to be more abundant in young chickens and embryos than in adults (Saffrey et al. 1982).

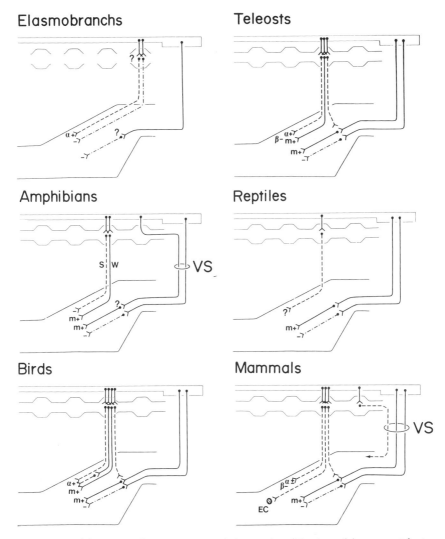

Fig. 9.7. A summary of the pattern of *extrinsic* autonomic innervation of the stomach in some vertebrate groups. The *S* and *W* beside the spinal autonomic fibres in the amphibian refers to possible dominance of inhibitory or excitatory responses depending on the season (Summer and Winter respectively). An adrenergic innervation of the 5-HT storing enterochromaffin cells in the mammalian gut is also indicated. Note that the intrinsic neurons apart from postganglionic neurons of extrinsic pathways are not shown in this figure. (For explanation of symbols see Fig. 7.8)

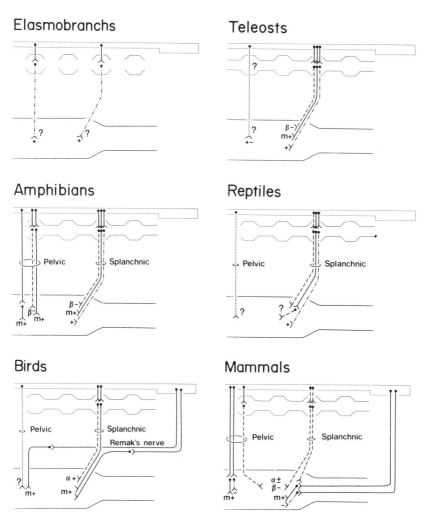

Fig. 9.8. A summary of the pattern of *extrinsic* autonomic innervation of the intestine in some vertebrate groups. No differentiation between small and large intestine has been made. Numerous pathways of pelvic, splanchnic and vagal origin my run in either direction in Remak's nerve of birds (not shown in the figure). Note that the intrinsic neurons apart from the postganglionic neurons of extrinsic pathways are not shown in this figure. For explanation of symbols see Fig. 7.8

9.8 Conclusions

In view of the still very fragmentary knowledge of the complex functions of the enteric nervous system in all vertebrates, and particularly in the non-mammalian groups, it is difficult to draw any conclusions regarding possible phylogenetic relationships. Although it is mainly a matter of semantic interest whether to recognize the enteric nervous system as a separate division of the autonomic nervous system, it is quite clear that the intrinsic neurons of the gut are capable of a high degree of independently coordinated activity, and that these functions are little affected by total extrinsic denervation of the gut. In spite of this, and mainly due to the lack of appropriate techniques to study the enteric nerve function, the knowledge of the autonomic nervous control of the gut is almost entirely dedicated to the extrinsic control. This is particularly true in the non-mammalian vertebrates, for which practically nothing is known about the enteric nerve functions. Attempts to summarize the pattern of extrinsic innervation of the vertebrate stomach and intestine are made in Figs. 9.7 and 9.8 respectively.

Previous reviewers (Campbell and Burnstock 1968, Burnstock 1969) concluded that the vagal control of the vertebrate stomach is primitively inhibitory, involving NANC fibres. Much of the older work is of limited value, since neither the phenomenon of "rebound excitation" nor the presence of spinal autonomic fibres in the "vagosympathetic trunks" were considered. It is possible that an inhibitory vagal innervation of the stomach is a primitive condition, and that such fibres may be part in a gastric receptive relaxation reflex of the type seen in mammals. It must be remembered, however, that excitatory vagal fibres are present in some fish, and that inhibitory vagal fibres may be absent.

The splanchnic fibres to the stomach and intestine may be cholinergic, adrenergic or NANC, and again there are no obvious phylogenetic "trends" in the evolution of this innervation. As with the extrinsic vagal pathways, little is known about the actual functions in vivo, despite a certain knowledge of potential effects (excitation or inhibition of gut motility). Further studies of the functions of the intrinsic neurons, and also of the involvement of intrinsic and extrinsic neurons in gastrointestinal reflexes of the non-mammalian vertebrates, would be most welcome.

There are several fundamental questions regarding enteric nerve functions to which studies of the lower vertebrates could add valuable information. One obvious example is the function of serotonergic neurons, which have been clearly demonstrated in some fish, but which are not universally accepted as parts of the mammalian enteric nervous system. Anatomical and functional aspects of the peptidergic neurons, and also the relationships between enteric neurons, paracrine and endocrine cells, could be elucidated given the advantage of a probably simpler arrangement of the enteric nervous system of the lower vertebrates.

Chapter 10

Swimbladder and Lung

The swimbladder or lung(s) of fish and tetrapods develop from single or paired outgrowths from the anterior part of the gut wall. The available information indicates that the tetrapod lung and the teleost swimbladder represent two separate lines of evolution from a primitive fish lung similar to that found in *Polypterus* (Romer 1962). There are large morphological and functional differences between the respiratory tetrapod lung and the teleost swimbladder which, apart from a respiratory function in some species, may be involved in buoyancy adjustment, pressure and sound perception and sound production (Harden Jones and Marshall 1953, Fänge 1953, 1976, Steen 1970).

10.1 The Teleost Swimbladder

The different functions of the swimbladder are reflected in the large structural variations found. An important function of the swimbladder in many teleost species is that of a buoyancy regulator, and elaborate mechanisms for the transport of gases into and from the swimbladder have evolved. The role of the autonomic nervous system in the control of functions of the swimbladder has undergone a surprising degree of refinement. In order to appreciate fully the beautiful complexity of the autonomic innervation of the various effector tissues in the swimbladder, some general comments on the functions of the swimbladder are in order. Extensive reviews on the subject have been written by Harden Jones and Marshall (1953), Fänge (1953, 1966, 1973, 1976) and Steen (1970).

The most detailed studies of the autonomic innervation of the swimbladder have been made in the eel (*Anguilla anguilla*), cod (*Gadus morhua*) and goldsinny wrasse (*Ctenolabrus rupestris*). The eel belongs to the group of teleosts with the swimbladder open to the anterior part of the digestive tract via a pneumatic duct (physostomes), while the cod and wrasse have completely closed swimbladders (physoclists). The general appearance of the swimbladder of these three species are shown in Fig. 10.1.

If the fish swims downwards, the swimbladder will be compressed so that gas transfer into the swimbladder must be initiated to restore its volume and thus maintain neutral buoyancy. The inflation of the swimbladder is initiated by an "inflatory reflex" (Fänge 1953), which is triggered by afferent fibres from stretch receptors in the swimbladder wall or by the balance organs (Tytler and Blaxter 1973, Fänge 1976). Conversely, if the fish ascends in the water, the swimbladder will expand and gas must be removed to maintain neutral buoyancy. In the physostomes,

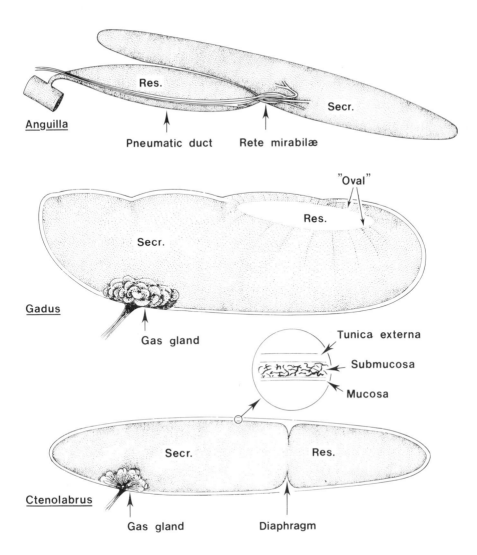

Fig. 10.1. General arrangement of the swimbladder of the eel (*Anguilla anguilla*), cod (*Gadus morhua*) and goldsinny wrasse (*Ctenolabrus rupestris*), showing the division of the swimbladder mucosa into a secretory (*Secr.*) and resorbent (*Res.*) part. In the eel, the gas gland cells are diffusely spread in the secretory part of the swimbladder. The blood to this part passes through a paired system of counter current capillaries, the *rete mirabile*. The resorbent part of the eel swimbladder is equivalent to the pneumatic duct, which opens via a sphincter to the anterior part of the gut. In the cod and wrasse, the gas gland cells from a distinct gas gland, within which there are several *retia mirabilia*. In the cod the secretory and resorbent parts of the mucosa are divided by a dorsally situated opening, the "oval". The size of the "oval" determines the contact surface between the swimbladder gas and the resorbent vasculature in the dorsal part of the swimbladder. Opening of the "oval" takes place by simultaneous contraction of radial smooth muscles and relaxation of circular smooth muscles of the oval edge. In the wrasse, the equivalent of the "oval" is present in the form of a diaphragm with a central hole. This diaphragm marks the division of the anterior secretory mucosa from the posterior resorbent mucosa. During gas resorption, the secretory mucosa contracts, the resorbent mucosa relaxes, and the diaphragm moves forward. The mucosal movements can be very large thanks to the "lubricant" action of the loose connective tissue in the submucosa between the mucosa and the outer tunica externa

this can be achieved simply by releasing gas through the pneumatic duct sphincter in the pharynx (Fig. 10.1), ("Gasspuckreflex"), but in the physoclists other mechanisms are responsible for the removal of gas.

10.1.1 Inflation of the Swimbladder

Transfer of gases from the blood into the swimbladder takes place by a process known as "gas secretion." To achieve a gas transfer from the blood against a partial pressure gradient into the swimbladder, metabolic activity is started in specialized epithelial cells in the swimbladder wall: the gas gland cells. In some species (e.g., cod and wrasse), these cells form a distinct "gas gland" in the ventral part of the swimbladder, while in others (e.g., eel; Fig. 10.1) the cells are diffusely spread in the secretory part of the swimbladder (Fänge 1953, 1976, Fahlén 1971).

The metabolic activity in the gas gland cells produces three effects of interest for the process of gas secretion: Firstly, CO_2 is formed during catabolism of glucose via the so-called "hexose monophosphate shunt." This CO_2, and CO_2 produced from plasma HCO_3^- by the action of the enzyme carbonic anhydrase, can reach a partial pressure high enough to make diffusion into the swimbladder possible.

Secondly, glycolysis in the gas gland cells, which occurs despite the high surrounding partial pressure of oxygen, produces lactic acid. This acidifies the blood

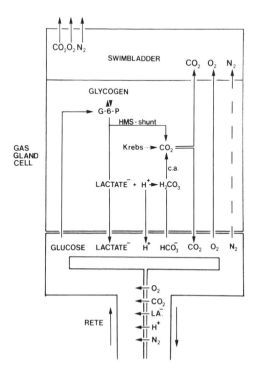

Fig. 10.2. Metabolic events in the gas gland cells, which are of importance for the processes of gas transfer from blood to swimbladder (gas secretion). Glucose from the blood, or stored glycogen, can be catabolized by glycolysis to form lactic acid, or via the "hexose monophosphate shunt" (*HMS-shunt*), to form CO_2. Some metabolic CO_2 is also obtained from the tricarboxylic acid cycle (*Krebs*) and from plasma HCO_3^- by the action of carbonic anhydrase (*c.a.*). Oxygen is released from the haemoglobin by a combination of Bohr and Root effects and diffuses into the swimbladder. Nitrogen is "salted out" of solution due to the increased osmolarity of the plasma that occurs during metabolic activity in the gas gland cells. The high concentrations of O_2, CO_2, N_2, lactic acid (LA^- and H^+) are maintained in the blood in contact with the gas gland cells due to the countercurrent multiplication system of the *rete mirabile*. [Redrawn with permission from Fänge (1973)]

and releases O_2 from the haemoglobin by a combination of Bohr and Root effects. Thirdly, the metabolic activity of the gas gland cells creates an increased osmolarity of the blood. This will decrease the solubility of all gases in the plasma (i.e., also nitrogen), which will be forced out of solution by the "salting out" effect (Fänge 1953, 1973, 1976, Steen 1970) (Fig. 10.2).

The high partial pressures of the gases created by the metabolic activities of the gas gland cells are further enhanced within the gas gland by the counter-current multiplication system of the *rete mirabile*. In the capillaries of the *rete*, O_2, CO_2, H^+ etc. will diffuse from the blood leaving the gas gland, into the arterial blood entering in the afferent capillaries (Fig. 10.2). Detailed accounts on functions of the *rete* can be found in Steen (1963, 1970) and Fänge (1976).

Two effector units involved in the process of gas secretion are controlled by the autonomic nervous system: the gas gland cells themselves, and the vasculature of the gas gland tissue.

Inflation of the swimbladder is prevented by vagotomy. It has been postulated that autonomic fibres in the vagi control directly the metabolic activity in the gas

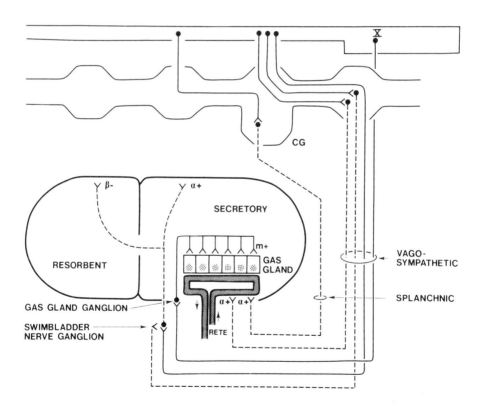

Fig. 10.3. A summary of the autonomic innervation of the teleost swimbladder via "vagosympathetic" and splanchnic pathways. The "wiring diagram" is a combination of the information from the swimbladders of cod (*Gadus*) and wrasse (*Ctenolabrus*). CG, coeliac ganglion. Other symbols as in Fig. 7.8

gland cells (Bohr 1894, Fänge 1953, 1976, Fänge et al. 1976). Because injection of cholinoceptor antagonists (atropine, ganglionic blocking agents) into the cod and wrasse prevents inflation, it has been concluded that the vagal fibres innervating the gas gland cells are cholinergic (Fänge 1953, Fänge et al. 1976). Further indirect evidence for a cholinergic innervation of the gas gland cells comes from the work of Dreser (1892) who described an increased oxygen content in the swimbladder of the pike (*Esox lucius*) following injections of pilocarpine. Nerve fibres that stain for acetylcholinesterase have been demonstrated in the gas gland of the cod, in which the cell bodies of these neurons form a "gas gland ganglion" at the base of the *rete mirabile* (Deineka 1905, Fänge 1953, McLean and Nilsson 1981). Fänge and Holmgren (1982) also demonstrated choline acetyltransferase (ChAT) in the gas gland of the cod. Vagotomy produced a decrease in the activity of this enzyme in the gas gland to about 30% of the control value, which shows that cholinergic neurons to the gas gland are vagal.

The vasculature of the gas gland is controlled by adrenergic fibres that reach the swimbladder in the "vagosympathetic trunks" and in the splanchnic nerve. These fibres have been observed by Falck-Hillarp fluorescence histochemistry to be associated with both pre- and post-*rete* blood vessels in the gas gland (Fahlén et al. 1965, Fänge et al. 1976, McLean and Nilsson 1981). The adrenergic fibres are vasoconstrictor in nature and act via α-adrenoceptors (Fänge 1953, Nilsson 1972). In the cod, most of the splanchnic adrenergic fibres to the gas gland vasculature have their cell bodies in the coeliac ganglion, while the adrenergic fibres of the "vagosympathetic trunk" have their cell bodies mainly in the sympathetic chain ganglion associated with the vagus (Fig. 10.3).

10.1.2 Deflation of the Swimbladder

Gas transfer from the swimbladder takes place by diffusion of the gases into the blood of the "resorbent vasculature." The regulation of gas resorption takes place mainly by mucosal movements that alter the relative surface area of the secretory compared to the resorbent mucosa of the swimbladders. The secretory mucosa is slightly thicker than the resorbent mucosa and is reinforced with relatively gas-impermeable guanine crystals which give the secretory part of the swimbladder a shiny appearance (Denton et al. 1972, Fänge 1976). The mucosal movements are controlled by the autonomic innervation of the muscularis mucosae. In the cod and wrasse extensive movements of the oval edge or diaphragm are facilitated by the "lubricating" effect of the loose connective tissue of the submucosa (Fig. 10.1).

Adrenergic fibres in the mucosa of the eel, cod and wrasse have been demonstrated by Falck-Hillarp fluorescence histochemistry (Fahlén et al. 1965, Fahlén 1967, Nilsson 1971, Fänge et al. 1976, McLean and Nilsson 1981). In the swimbladder of the wrasse, yellow fluorescent cells resembling enterochromaffin(5-HT storing) cells have been described but the function of these cells is not known (Fänge et al. 1976). VIP-immunoreactive nerve fibres have also been demonstrated in the swimbladder mucosa of the cod, *Gadus morhua*, but the function of these fibres is so far unknown (Lundin and Holmgren, unpublished).

In the eel, as in other physostomes, gas can be released from the swimbladder through the pneumatic duct sphincter, but a significant resorption of gas takes place in the pneumatic duct (Steen 1963, 1970, Stray-Pedersen 1970, Fänge 1976). During deflation of the eel swimbladder, the adrenergic innervation produces an α-adrenoceptor mediated contraction of the secretory part, which forces gas into the pneumatic duct. This in turn relaxes, due to a β-adrenoceptor mediated effect, and a direct or indirect dilation of the pneumatic duct vasculature takes place (Fänge 1953, Nilsson and Fänge 1967, Stray-Pedersen 1970).

In the cod, the rate of gas resorption varies with the size of the "oval" opening in the dorsal part of the swimbladder (Woodland 1910, Fänge 1953, Nilsson 1971). Activity in the adrenergic fibres that innervate the muscularis mucosae causes contraction of the radial smooth muscles of the oval opening (an α-adrenoceptor effect) and simultaneously relaxes the circular smooth muscle of the oval edge (a β-adrenoceptor effect) (Fänge 1953, Nilsson 1971, Ross 1978: *Pollachius virens*). Although acetylcholine induces contraction of the circular smooth muscles (Fänge 1953, Nilsson 1971, Ross 1978), there is no evidence for a cholinergic innervation of the oval edge in the cod (Nilsson 1971).

In the wrasse, the rate of gas resorption is controlled by movements of the secretory and resorbent mucosa, which displace forward or backward the diaphragm dividing the two parts of the swimbladder (Fänge 1953). During gas resorption, the secretory part is caused to contract by the adrenergic fibres acting via α-adrenoceptors, and the resorbent part of the mucosa relaxes due to a β-adrenoceptor effect (Fig. 10.3, Fänge 1953, Fänge et al. 1976). In the wrasse acetylcholine induces contraction of the resorbent part of the swimbladder mucosa but there is no evidence for a cholinergic innervation (Fänge et al. 1976).

The available evidence thus favours the view of a solely adrenergic control of mucosal movements in the teleost swimbladder. Generally speaking, the adrenergic nerves exert their action by contracting the secretory mucosa (an α-adrenoceptor effect) and relaxation of the resorbent mucosa (a β-adrenoceptor effect). An attempt to summarize the autonomic influence on the swimbladder is made in Fig. 10.3 and Table 10.1.

Table 10.1. A summary of the autonomic nervous control of the different effector units in the teleost swimbladder during inflation and deflation

	Inflation (= gas secretion)	Deflation (= gas resorption)
Cholinergic fibres:		
Gas gland cells	Stimulation	–
Gas gland vasculature	Dilation (indirect?)	–
Secretory mucosa	–	–
Resorbent mucosa	–	–
Adrenergic fibres:		
Gas gland cells	–	–
Gas gland vasculature	–	Constriction (α-adrenoceptors)
Secretory mucosa	–	Contraction (α-adrenoceptors)
Resorbent mucosa	–	Relaxation (β-adrenoceptors)

10.2 Fish Lungs

There are several examples of actinopterygian and sarcopterygian fish with swimbladders or lungs adapted for air breathing. The most obvious examples are the dipnoans, lungfish, which belong to the sarcopterygians. On the actinopterygian branch, some of the ganoids (*Polypterus, Amia, Lepisosteus*) and several teleost species, notably tropical freshwater forms, have a "lung" that is used for air breathing. However, a true lung with a separate pulmonary vein that empties directly into the atrium of the heart, is present only in the dipnoans (Harden Jones and Marshall 1953, Johansen et al. 1968, Fänge 1976).

In the holostean *Lepisosteus platyrhincus* the lung is innervated by the vagi, which carry excitatory cholinergic fibres to the musculature of the lung wall and trabeculae. Falck-Hillarp fluorescence histochemistry of the lung wall has revealed a dense adrenergic innervation of blood vessels and also a sparse innervation by adrenergic fibres of smooth muscle bundles. The function of the adrenergic innervation of the vasculature and muscles of the lung wall is unknown (Potter 1927, Nilsson 1981).

In the dipnoans, such as *Protopterus*, it seems probable that a similar cholinergic excitatory vagal innervation of the smooth muscle in the lung wall, is involved during expiration. There is no conclusive evidence for an adrenergic innervation of any part of the *Protopterus* lung (Parker 1892, Giacomini 1906, Jenkin 1928, Johansen and Reite 1967, Abrahamsson et al. 1979b).

10.3 Tetrapod Lungs

The amphibian and reptilian lungs are essentially muscular sacs, in which the inner surface is increased by the presence of alveolar septa that protrude into the pulmonary cavity (Smith and Campbell 1976). Autonomic nerve fibres innervate the smooth muscles of the lung wall, the septal edges and the vasculature (see also Chap. 7.6.2 and 7.7.2).

The bird lung is completely different from the lung of the other tetrapods. Air is mainly forced from the posterior air sacs of the bird, through a system of bronchi in the rigid bird lung. As will be seen below, the innervation of the avian lung by autonomic fibres is poorly understood.

In the mammalian lung, there is an elaborate system of branching bronchi and bronchioli ending in the alveoli. There is no distinct muscular lung wall. The lung wall musculature of amphibians and reptiles is homologous with the interstitial muscle of the mammalian lung (Bronkhorst and Dijkstra 1940). Although there is good information about the autonomic innervation of bronchial smooth muscle in mammals, very little is known about the innervation of the interstitial muscle of the lung. As pointed out by Burnstock (1969) and Berger and Burnstock (1979), a certain caution is in order when comparing the modes of innervation of the different tetrapod lungs because of the different embryological origin of the innervated smooth muscle. A summary of the autonomic innervation of the lung in different vertebrates, including the fish lungs, is presented in Fig. 10.4.

10.3.1 The Amphibian Lung

While only scattered information about the lung innervation in urodeles exists (Luckhardt and Carlson 1920, 1921 b), there is a massive amount of information about the autonomic innervation of the anuran lung. Amphibians fill their lungs by a buccal force pump under positive pressure. There are tonically active autonomic nerve fibres that relax the lung wall (Carlson and Luckhardt 1920a, Luckhardt and Carlson 1920, 1921 b). This inhibitory innervation may be important to reduce the force needed to inflate the lungs ("receptive relaxation"). Excitatory fibres to the lung wall musculature are also present, and it is reasonable to believe that an effective contraction of the lung wall during expiration is of advantage in reducing the respiratory dead space. Contraction of the lung wall may also be valuable during sound production ("croaking").

Autonomic fibres reach the lungs in the "vagosympathetic trunks". The inhibitory fibres to the smooth muscle of the lung wall are mainly truly vagal in origin and belong to the NANC category (Campbell 1971a, Campbell et al. 1978, Campbell and Duxson 1978, Holmgren and Campbell 1978).

Cholinergic excitatory fibres have been demonstrated in the "vagosympathetic trunks" of anurans, but there appear to be no excitatory fibres to the urodele lung (Luckhardt and Carlson 1920, 1921 b). The origin of the cholinergic fibres is obscure and an origin from both the vagus (Wood and Burnstock 1967, Campbell and Duxson 1978) and from the sympathetic chains (Shimada and Kobayasi 1966, Wood and Burnstock 1967, Campbell 1971a) has been proposed. In their study of the pulmonary innervation of the toad (*Bufo marinus*), Campbell and Duxson (1978) were unable to demonstrate cholinergic fibres originating in the sympathetic chains, and concluded that the previous reports of such fibres must be due to a spread of current during stimulation. Additional evidence implies that the cholinergic neurons in the "vagosympathetic trunks" lack a central nervous connection, and may thus be part of local reflex arcs (Campbell and Duxson 1978, Holmgren and Campbell unpublished).

In the toad both the smooth muscle of the lung wall and the vasculature of the lung are innervated by adrenergic fibres (Shimada and Kobayasi 1966, McLean and Burnstock 1967b, Wood and Burnstock 1967, Campbell 1971a, b, Campbell et al. 1978, Campbell and Duxson 1978, Holmgren and Campbell 1978). Both a contraction and a relaxation of the lung of the toad (*Bufo marinus*) can be obtained by stimulation of the cervical sympathetic chain. There appears to be a puzzling difference in the nature of the response depending on the pretreatment of the toad: in pithed toads only a contraction is obtained, while in animals which have been anaesthetized in MS 222 (tricaine) a contraction of the lung is followed by a relaxation (Campbell and Duxson 1978).

Holmgren and Campbell (1978) were able to demonstrate a regional difference in the responses of the lung to catecholamines and sympathetic chain stimulation. Thus the smooth muscle of the septal edges contracted due to an α-adrenoceptor effect, while the smooth muscle of the lung wall simultaneously relaxed due to both an α- (sic!) and a β-adrenoceptor effect. The dual responses recorded from the whole lung preparation could in part be explained by these regional differences. The differences in the responses to adrenergic stimulation may be of functional sig-

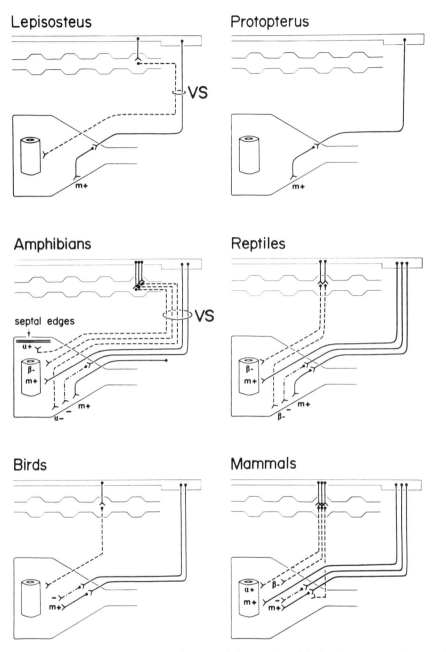

Fig. 10.4. A summary of the pattern of autonomic innervation of the lung in some vertebrate species and groups. For explanation of symbols see Fig. 7.8

nificance, since a simultaneous relaxation of the lung wall and contraction of the septal edges will "lift" the septa into the lung cavity and thus increase the inner surface of the lung.

10.3.2 The Reptilian Lung

In reptiles, the lungs are filled by aspiration (negative pressure). The inhibitory fibres that innervate the smooth muscles of the lung wall are not tonically active (Carlson and Luckhardt 1920b, Luckhardt and Carlson 1921a).

Cholinergic excitatory and NANC inhibitory fibres to the lung wall musculature run in the vagi in all reptiles studied (Burnstock and Wood 1967b, Berger 1973, Berger and Burnstock 1979, Smith and Macintyre 1979). Ganglia in the vagal pathways have been demonstrated at the base of the lungs, but the nature of the neurons in these ganglia is uncertain (Jones 1926, McLean and Burnstock 1967c).

There is no conclusive evidence for an adrenergic control of the reptilian lung by fibres running in the "vagosympathetic trunks". Instead the blood vessels and smooth muscles of the lung wall are directly innervated by spinal autonomic fibres from the stellate complex (Figs. 2.20, 2.21; Jackson and Pelz 1918, McLean and Burnstock 1967c, Berger 1973, Smith and Macintyre 1979). The effect of the adrenergic innervation of the smooth muscle of the lung wall in the lizard *Trachydosaurus rugosus* is a β-adrenoceptor mediated relaxation. A small α-adrenoceptor mediated contraction of the lung wall could also be demonstrated after blockade of the β-adrenoceptors (Berger 1973). It is not known whether regional differences in the adrenergic effects on the smooth muscle of lung wall and septal edges occur in reptiles.

10.3.3 The Avian Lung

The autonomic nervous control of the avian lung is poorly understood. The pulmonary branches of the vagi unite with spinal autonomic nerves from the sympathetic chains to form a pulmonary plexus which extends over the ventral surface of the lung (McLelland 1969, Bennett 1974). Fibres from this plexus innervate the pulmonary vasculature and the smooth muscle of the primary, secondary and tertiary bronchi, with the innervation of the primary bronchi being more dense than of the other two types (Cook and King 1970, Bennett 1971, 1974).

There is evidence for both cholinergic excitatory (King and Cowie 1969, Cowie and King 1969) and NANC inhibitory (Richardson 1979) fibres innervating the bronchial smooth muscle of the avain lung. A sparse innervation of the bronchial muscle by adrenergic fibres has also been demonstrated by fluorescence histochemistry and electron microscopy, but the function of these fibres remains unknown (Akester and Mann 1969, Bennett and Malmfors 1970, Cook and King 1970, Bennett 1971, 1974).

In addition to the autonomic innervation of the bronchial smooth muscle, there is evidence for a cholinergic control of mucus secretion in the airways of the goose (Phipps and Richardson 1976).

Histochemical evidence for both an adrenergic and a non-adrenergic innervation of the avian air sacs have been presented, but no functional studies of this innervation have been made (Groth 1972).

10.3.4 The Mammalian Lung

Vagal fibres and fibres from the stellate ganglion unite to form pulmonary plexuses in the peribronchial area of the mammalian lung. The autonomic fibres innervate bronchial and vascular smooth muscle. The autonomic innervation of the mammalian lung has been reviewed by Richardson (1979).

The diameter of the bronchi is controlled by cholinergic constrictor (Cabezas et al. 1971, Gold et al. 1972, Hahn et al. 1978), adrenergic dilator (Ehinger et al. 1970, Cabezas et al. 1971) and NANC dilator fibres (Middendorf and Russel 1978, Richardson 1979), which innervate the bronchial smooth muscle. The adrenergic inhibitory effect is mediated by β_2-adrenoceptors (Lands et al. 1967a, b).

As in birds, a cholinergic control of mucus secretion in the bronchi has been demonstrated (Brody et al. 1972).

Chapter 11

Urinary Bladder

The urinary bladder is a hollow organ that receives and temporarily stores urine. In amphibians and some reptiles the urinary bladder also plays a major role in the regulation of salt and water balance of the animal.

The control of the urinary bladder is exerted by autonomic nerve fibres innervating the smooth muscle of the bladder wall. The autonomic reflexes involved in bladder control during filling ("receptive relaxation") and emptying have been studied in mammals only. The "receptive relaxation" reflex seen during filling of the bladder is comparable to "gastric receptive relaxation" (Chap. 9.1.2.1) and facilitates a volume increase without a simultaneous pressure increase. The inhibitory fibres responsible for the "receptive relaxation" of the mammalian bladder run in the hypogastric nerve (Downie 1981).

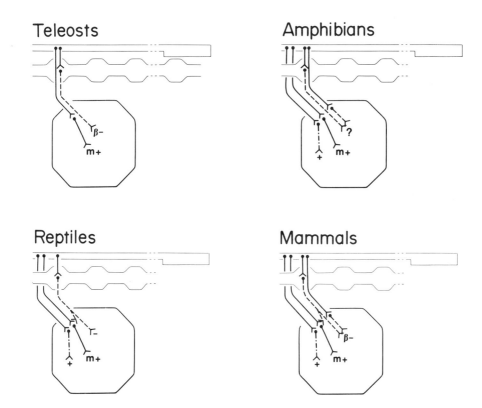

Fig. 11.1. A summary of the pattern of autonomic innervation of the urinary bladder in some vertebrate groups. For explanation of symbols see Fig. 7.8

The emptying reflex (micturition) is induced by activation of stretch receptors in the bladder wall. Pathways in the CNS mediate a sacral-pontine-sacral reflex, where excitatory fibres innervating the bladder run in the pelvic nerves (Downie 1981).

A urinary bladder is absent in cyclostomes, elasmobranchs, some reptiles and all birds (except *Rhea americana*, Skadhauge 1981). In elasmobranchs the urinary ducts may sometimes be enlarged at the distal end (particularly in females). The rich innervation of these urinary sinuses has been described by Young (1933c). Nothing is, however, known about the function or nature of this innervation.

The urinary bladder of most teleosts consists of the fused and enlarged distal parts of the ureters, and is thus of mesodermal origin (Young 1931b, Burnstock 1969). This should be kept in mind when making comparisons with the endodermal urinary bladders of the tetrapods. An attempt to summarize the patterns of autonomic innervation of the urinary bladder of teleosts, amphibians, reptiles and mammals is made in Fig. 11.1.

11.1 The Teleost Urinary Bladder

The mesodermal urinary bladder and gonads of teleosts are innervated by the posterior splanchnic nerve (vesicular nerve; Fig. 2.7), which leaves the posterior part of the sympathetic chains. The nerve runs along the ureters with a more or less distinct "vesicular nerve ganglion" along its course (Young 1931b, Nilsson 1970, 1976). A fluorescence histochemical study of this ganglion in the cod, *Gadus morhua*, revealed mainly non-fluorescent nerve cell bodies within the ganglion, while fluoresent cell bodies are abundant in the ganglia of the sympathetic chains (Nilsson 1976, Holmgren and Nilsson 1982).

Stimulation of the posterior splanchnic nerve to the urinary bladder of *Uranoscopus*, *Lophius* and *Gadus* produces a strong contraction of the bladder, a response that can be mimicked by acetylcholine (Young 1936, Nilsson 1970). Since neither the response to nerve stimulation nor the rhythmical contractions of the bladder induced by acetylcholine in *Lophius* could be blocked completely by atropine, Young (1936) concluded that the effect of acetylcholine is "nicotinic" rather than "muscarinic." In view of the new information about a NANC innervation of the tetrapod urinary bladder, the "atropine resistance" observed in Young's experiments could be interpreted in favour of a NANC excitatory innervation of the bladder of *Lophius* (see Sect. 11.4).

In the cod, *Gadus morhua*, however, the excitatory response was abolished by atropine unmasking a smaller inhibitory response to nerve stimulation (Nilsson 1970). The excitatory, but not the inhibitory, nerve response was abolished by hexamethonium provided that the stimulating electrodes were placed proximal to the vesicular nerve ganglion, suggesting that the non-fluorescent nerve cell bodies in the ganglion represent postganglionic cholinergic neurons of the excitatory pathway (Nilsson 1970, 1976, Holmgren and Nilsson 1982).

The inhibitory response to nerve stimulation was abolished by propranolol, which suggests an adrenergic inhibitory innervation acting via β-adrenoceptors of the bladder smooth muscle (Nilsson 1970). This idea was further supported by a fluorescence histochemical study of the urinary bladder, in which an innervation by adrenergic fibres of smooth muscle bundles in the bladder wall was demonstrated (Nilsson 1973).

11.2 The Amphibian Urinary Bladder

The urinary bladder of frogs and toads is derived from the cloacal wall and is thus of endodermal origin. The autonomic innervation of the urinary bladders of *Rana temporaria* and *Bufo marinus*, which have been studied with both histochemical and physiological methods, runs in the pelvic nerve complex, in *Bufo* the outflow from the spinal cord takes place mainly in the 9th spinal nerve but fibres are also found in the 7th, 8th and 10th spinal nerves (Burnstock et al. 1963 b). Cholinesterase-containing nerve fibres were demonstrated both in the frog (McLean et al. 1967) and toad (Bell 1967) urinary bladder. Fluorescence histochemistry of stretch preparations from the bladder wall revealed a dense (frog) or moderate (toad) innervation of smooth muscle bundles in the bladder wall. Extra-adrenal chromaffin cells (SIF cells) associated with the nerve bundles and ganglia are abundant, particularly in the toad. Non-fluorescent ganglion cells (in the frog also fluorescent) and ganglion cells containing cholinesterase were also described (Falck et al. 1963, McLean and Burnstock 1966, McLean et al. 1967).

In both the frog and the toad stimulation of the pelvic nerve produces a contraction of the bladder, which in the frog could be enhanced by neostigmine. In the toad, the response to nerve stimulation was reduced by atropine, showing that part of the excitatory innervation is cholinergic. Also demonstrated was an "atropine-resistant" excitatory response which is probably due to NANC fibres. Part of the excitatory response of the toad bladder could be blocked by ganglionic blocking agents, showing the presence of intrinsic ganglion cells (Burnstock et al. 1963 b).

Exogenous catecholamines induce contraction of the frog bladder and relaxation of the toad bladder, but there is no conclusive evicence for the function of the extensive adrenergic innervation of the amphibian urinary bladder (Burnstock et al. 1963 b, McLean et al. 1967).

11.3 The Reptilian Urinary Bladder

The knowledge of the autonomic nervous control of the reptilian bladder is based on studies of a single species, *Trachydosaurus rugosus*. In this lizard, the bladder is innervated by adrenergic fibres which form a varicose network in the smooth muscle bundles of the bladder wall. There are no intrinsic fluorescent cell bodies, but pericellular networks of varicose fibres around non-fluorescent ganglion cells in the bladder have been described and the presence of small, intensely fluorescent cells (SIF cells) was also demonstrated (McLean and Burnstock 1967 a).

The bladder of *Trachydosaurus* is innervated by the pelvic nerve, stimulation of which causes contraction of the bladder. This contraction is partially blocked by ganglionic blocking agents, showing that some of the postganglionic neurons in the excitatory pathways are intrinsic. Atropine likewise produces a partial blockade of the excitatory response, revealing the presence of an "atropine-resistant", probably NANC, excitatory innervation. The frequency-response curve for the excitatory response of the lizard bladder has two peaks, one at 20 Hz and one at 35 Hz. The peak at 20 Hz was most sensitive to blockade by atropine, indicating the possibility that this peak represents cholinergic fibres. The peak at higher frequency may represent NANC excitatory fibres (Burnstock and Wood 1967a, Burnstock 1969, Berger and Burnstock 1979).

In about 70% of the lizard bladder preparations studied by Burnstock and Wood (1967a), it was possible to demonstrate an inhibitory component which could be abolished by guanethidine, suggesting the possibility that the inhibitory innervation is adrenergic.

11.4 The Mammalian Urinary Bladder

Contrary to the arrangement in amphibians and reptiles, the urine enters the mammalian bladder directly from the ureters rather than via a cloaca. The region of the mammalian bladder where the ureters enter (the trigone region) has a particularly dense adrenergic innervation and an abundance of intrinsic adrenergic ganglion cells (Norberg and Hamberger 1964). The adrenergic nerve cell bodies represent the short adrenergic neurons of the (classically "sympathetic") hypogastric pathways which enter the pelvic plexus (Fig. 2.3).

The early work of Elliott (1907) described an excitatory pelvic and an inhibitory hypogastric innervation of the mammalian bladder. In the cat, the excitatory effect on the detrusor muscle of the bladder is accompanied by a dilatory effect on the urethral sphincter, and the inhibitory effect of hypogastric nerve stimulation is accompanied by a constrictor effect on the sphincter (Elliot 1907).

The excitatory effect of stimulation of the pelvic nerve is enhanced by cholinesterase inhibitors, such as physostigmine, and partially blocked by atropine, showing that some of the excitatory fibres are cholinergic. The "atropine resistance" of the remaining pelvic nerve response, first described by Langley and Anderson (1895) was initially explained by the presence of a cholinergic junction inaccessible to atropine (Ursillo and Clark 1965, Huković et al. 1965, Vanov 1965). Other studies recognize a NANC-excitatory transmitter (Henderson and Roepke 1934, Chesher and Thorp 1965, Ambache and Zar 1970, Dumsday 1971, Burnstock et al. 1972, 1978a, b, Dean and Downie 1978, Campbell and Gibbins 1979). A similar "atropine-resistant" response has also been described in the bladder of a marsupial mammal, the ringtail possum (*Pseudocheirus peregrinus*) by Burnstock and Campbell (1963). ATP may be involved in the NANC excitatory innervation of the mammalian urinary bladder either as a transmitter (Dumsday 1971, Burnstock et al. 1978b) or as a modulator of the nervous activity (Dahlén and Hedqvist 1980). Substance P causes contraction of the urinary bladder, but the nervous response is insensitive to the substance P antagonist (D-Pro^2, D-Trp^{7,9})-substance P (Leander et al. 1981 b).

The inhibitory pathways in the hypogastric nerve are adrenergic and the inhibitory effect on the detrusor of the bladder is mediated by β-adrenoceptors. An additional inhibitory effect on the transmission in ganglia of the excitatory pathways has also been postulated (Downie 1981).

Chapter 12

Iris

The basic function of the vertebrate iris, which is aptly compared to the aperture diaphragm of a camera, is control of the intensity of light reaching the retina. The pupillary aperture is controlled by two sets of antagonistically acting muscles in the iris: the circular muscle of the *sphincter pupillae* (iris sphincter), which causes narrowing of the pupil (*miosis*), and the radially arranged muscles of the *dilator pupillae* (iris dilator), which causes widening of the pupil (*mydriasis*). In most groups the iris sphincter and dilator are composed of smooth muscle, but in reptiles and birds the iris muscles are, at least mainly, striated. In some groups (fish, amphibians and reptiles) the iris sphincter contains a light-sensitive pigment which causes a direct constriction of the pupil in response to incident light, without involvement of nervous mechanisms (Brown-Séquard 1847, Steinach 1890, 1892, Guth 1901). The pupillary diameter is controlled mainly by the sphincter and, in some species of teleosts, amphibians and reptiles dilator muscles may be absent (see below). It is of interest to note that the muscles of the iris are derived from the embryonic retina (light-sensitive pigments!) and are thus of ectodermal origin. In spite of this unorthodox origin, the muscles of the iris show both structural and functional similarities with other muscle tissue (Romer 1962).

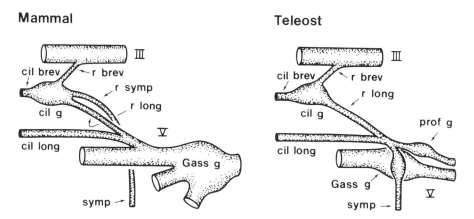

Fig. 12.1. Arrangement of the ciliary ganglion and ciliary nerves in a mammal and a teleost. Note the great similarity in the general pattern. In teleosts the profundus nerve is separate from the trigeminal, and carries its own ganglion (profundus ganglion) which is separate from the Gasserian ganglion of the trigeminal nerve. Abbreviations: *cil brev, ciliaris brevis*, short ciliary nerve; *cil g*, ciliary ganglion; *cil long, ciliaris longa*, long ciliary nerve; *Gass g*, Gasserian ganglion of the trigeminal nerve; *prof g*, profundus ganglion; *r brev, radix brevis*, short ciliary root, *r long, radix longa*, long ciliary root; *r symp, radix sympathica; symp*, nerve fibres of spinal autonomic origin: fibres from superior cervical ganglion (mammal) or fibres in the cephalic sympathetic chain (teleost); *III*, oculomotor nerve; *V*, trigeminal nerve

The muscles of the vertebrate iris are controlled by autonomic nerve fibres. In the birds and mammals a direct effect of light on the iris sphincter is absent. The arrangement of the autonomic innervation of the iris (ciliary roots, nerves and ganglion) is very similar in the different groups of vertebrates (Fig. 12.1). Cranial autonomic fibres to the iris, when present, run in the oculomotor nerve and synapse with postganglionic neurons in the ciliary ganglion. Fibres from the sympathetic chain ganglia enter the ciliary nerves either as a separate root of the ciliary ganglion (*radix sympathica*) or as contributions in the long ciliary nerve or short ciliary root (Fig. 12.1).

12.1 The Elasmobranch Iris

Young (1933a) presented the first thorough description of the autonomic nervous control of the elasmobranch iris. In studies of *Scyllium catulus*, *S. canicula*, *Mustelus laevis* and *Trygon violaceus*, he demonstrated a direct effect of light on the iris sphincter, but no autonomic innervation of the sphincter could be seen. The iris sphincter of these fish is also insensitive to adrenaline and acetylcholine (Young 1933a).

Stimulation of the oculomotor nerve dilates the elasmobranch pupil by contracting the radially arranged dilator muscles. The oculomotor fibres are probably cholinergic, since atropine caused a narrowing of the pupil. There is no evidence for an innervation of the iris by fibres from the paravertebral ganglia (Young 1933a).

The pupillary opening in elasmobranchs is thus controlled by the direct constrictor effect of light on the iris sphincter and the antagonistically acting dilator muscles which are innervated by cholinergic oculomotor fibres (Fig. 12.2).

12.2 The Teleost Iris

A direct effect of light on the iris sphincter of teleost fish and amphibians was early described by several authors (Brown-Séquard 1847, Steinach 1890, 1892, Guth 1901). The first detailed account of the autonomic nervous control of the teleostean iris comes from the work of Young (1931a, 1933b) on *Uranoscopus scaber* and *Lophius piscatorius* (cf. Fig. 12.1). A direct stimulating action by light on the iris sphincter appears to be absent in these species, but constriction of the pupil can be achieved by stimulation of the ipsilateral cephalic sympathetic chain. The fibres to the sphincter muscles are cholinergic, since the effect of sympathetic chain stimulation can be abolished by atropine (Young 1933b).

An innervation of the dilator muscles by oculomotor fibres was demonstrated in *Uranoscopus* but no effect on the iris was seen during stimulation of the oculomotor nerve in *Lophius*. It is likely that the oculomotor control of the dilator muscles in the former is cholinergic (Young 1933b) (Fig. 12.2).

In the cod, *Gadus morhua*, the changes in pupillary diameter elicited by direct light or nerve stimulation are small compared to the large movements described in *Uranoscopus* and *Lophius*. There is no innervation of the iris by oculomotor fibres and iris dilator muscles appear to be absent. The iris sphincter contracts in response to light and a pupillary constriction of similar size can be produced by stimulating the ipsilateral sympathetic chain. Contrary to the situation in *Uranoscopus*, however, the fibres from the sympathetic chain in *Gadus morhua* are adrenergic and constrict the iris sphincter by an action via α-adrenoceptors (Nilsson 1980).

It would thus seem that the teleost pupil is constricted by a direct effect of light on the iris sphincter, at least in most species, and by the action of fibres from the cephalic sympathetic chains innervating the iris sphincter. These fibres may be cholinergic (*Uranoscopus*) or adrenergic (*Gadus*). An innervation of the dilator muscles of the iris by oculomotor fibres in the same way as in elasmobranchs, has been demonstrated in *Uranoscopus* but is absent in *Lophius* and *Gadus* (Fig. 12.2).

12.3 The Amphibian Iris

A direct effect of light on the amphibian iris sphincter is well documented (Brown-Séquard 1847, Steinach 1892, Guth 1901, Armstrong and Bell 1968, Morris 1976). The most detailed study of the autonomic nervous control of the amphibian pupil was made by Morris (1976) on a toad (*Bufo marinus*). In this species a *dilator pupillae* is absent and all effects of nerve stimulation are therefore on the iris sphincter muscle.

The iris sphincter of *Bufo* constricts in response to incident light, while dilation of the pupil can be achieved by stimulation of the ipsilateral sympathetic chain (Armstrong and Bell 1968, Morris 1976). The inhibitory fibres to the iris sphincter are solely adrenergic, acting via a β-adrenoceptors. The preganglionic outflow from the spinal cord takes place in the 2nd to 4th spinal roots (Morris 1976).

In the toad the direct pupilloconstrictor effect of light is thus antagonized by a β-adrenoceptor mediated inhibitory adrenergic innervation of the iris sphincter (Fig. 12.2).

12.4 The Reptilian Iris

The *sphincter pupillae* of reptiles is composed, at least mainly, of striated muscle and is controlled by excitatory cholinergic fibres running in the oculomotor nerve. Stimulation of the oculomotor fibres in reptiles thus produces narrowing of the pupil, which is contrary to the effects in fish, in which the oculomotor fibres cause pupillodilation by contracting of the dilator muscles.

In the alligator (*Alligator mississippiensis*), Iske (1929) showed a pupillo-constrictor effect of nicotine and pilocarpine applied directly to the eye. She could also show that atropine blocked the effects of stimulation of the oculomotor nerve, while curare, in doses that blocked the movements of the extraocular eye muscles, failed to block stimulation via the oculomotor nerve (Iske 1929). Iske (1929) also observed a pupillodilation after application of atropine in the turtle (*Emys blandingi*), further supporting the idea of a mainly muscarinic cholinoceptor mediated cholinergic effect on the sphincter. Koppányi and Sun (1926) make a passing comment on the effect of drugs on the iris of the alligator and conclude that the effects

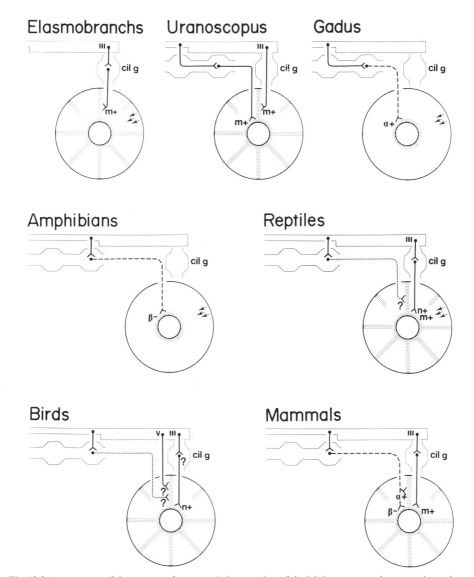

Fig. 12.2. A summary of the pattern of autonomic innervation of the iris in some vertebrate species and groups. The figure shows major effects of the innervation of the iris sphincter and dilator muscles. A direct effect of light on the iris sphincter is indicated by *arrows*. For explanation of symbols see Fig. 7.8

are similar to those observed in the pigeon iris, i.e. no effect of atropine and a pupillodilator effect of curare. No details about these experiments are, however, given, and the validity of the report is therefore uncertain.

There are no dilator muscles in the alligator iris, but in of turtles Gaskell and Gadow (1884) and Mills (1885) described a slight dilation of the pupil following stimulation of the sympathetic chains. It could thus be speculated that spinal autonomic fibres induce contraction of the dilator muscles (or possibly dilate the sphincter muscle) in the turtle, but the nature of these fibres is unknown.

The iris sphincter of reptiles is controlled by oculomotor fibres which cause pupillary constriction. Although the sphincter muscle is largely striated, pharmacological evidence from work on the alligator (Iske 1929) seems to indicate a mainly muscarinic cholinoceptor mechanism. Whether this result is due to aberrant striated muscle cholinoceptors in the reptilian iris, or to the presence of additional smooth muscle in the sphincter is not clear. Neither is it settled whether synapses in the oculomotor pathways are present in the ciliary ganglion. Painting the ganglion with 1% nicotine solution (Iske 1929) can hardly be regarded as a selective use of a pharmacological tool. Finally, the presence and nature of spinal autonomic fibres innervating the muscles of the reptilian iris needs clarification (Fig. 12.2).

12.5 The Avian Iris

The avian iris, like the reptilian, is composed mainly of striated muscle arranged into a sphincter and radial dilator muscles. The sphincter is controlled by constrictor fibres running in the oculomotor nerve. It seems clear that the population of large neurons in the ciliary ganglion represent the postganglionic neurons in the oculomotor pathways (Marwitt et al. 1971). The transmission in the ciliary ganglion is complex, involving both chemical and electrical junctions between pre- and postganglionic neurons (Marwitt et al. 1971, Gabella 1976).

It is doubtful if the dilator muscles of the avian iris are controlled by spinal autonomic fibres (see below), but in some early studies an innervation of the striated dilator muscles by motor fibres in the trigeminal nerve (V) is described (Zeglinski 1885, Jegorow 1890 cited in Koppányi and Sun 1926).

The effects on the avian iris of stimulation of the oculomotor nerve are blocked by curare, but are insensitive to atropine. The nerve endings in the muscle resemble more the motor end plate in skeletal muscle than the "en passant" autonomic nerve terminals (Iske 1929, Campbell and Smith 1962, Pilar and Vaughan 1969, Bennett 1974).

The anterior epithelial cell layer of the avian iris functions like radially arranged smooth dilator muscles. An innervation of this muscle layer by both cholinergic and adrenergic nerve terminals has been postulated (Ehinger 1967, Nishida and Sears 1970). The results regarding the control of the avian iris by fibres from the sympathetic chains are conflicting. No effect of sympathetic nerve stimulation was observed by Jegorow (1887), Langley (1904) or Koppányi and Sun (1926), but Gruenhagen (1887) described a pupillo-dilation following sympathetic chain stimulation.

The avian pupil is thus controlled mainly by oculomotor fibres innervating the striated muscle sphincter, and possibly by trigeminal motor fibres to the striated dilator muscles. The ganglionic transmission in the oculomotor pathways is peculiar. It is not known whether a peripheral ganglionic synapse exists in the postulated trigeminal pathways. As a matter of academic interest it should be noted that if a peripheral ganglionic synapse is absent, the nervous pathway does not belong to the autonomic nervous system as defined in Chap. 1 (Campbell 1970a). The possibility of an autonomic nervous control of the anterior epithelial smooth muscle and the presence of a spinal autonomic innervation of the avian iris need further attention (Fig. 12.2).

12.6 The Mammalian Iris

The iris sphincter and dilator muscles in mammals consist of smooth muscle, and both are innervated by cholinergic and adrenergic fibres (Ehinger and Falck 1965, 1966, Tranzer and Thoenen 1967, Ehinger et al. 1968, Ochi et al. 1968). Both α- and β-adrenoceptors have been described in the sphincter and dilator muscles of the cat (van Alphen et al. 1964, Ehinger et al. 1968) and rabbit (Persson and Sonmark 1971).

The sphincter pupillae of mammals is innervated by excitatory cholinergic fibres that run in the oculomotor and ciliary nerves (Fig. 12.1). Atropine applied to the eye produces a marked pupillary dilation (see Chap. 5.3.4). A weak dilatory effect on the sphincter by an adrenergic innervation acting via both α- and β-adrenoceptors is also evident. The dilator muscles are innervated by adrenergic fibres from the superior cervical ganglion, which cause contraction of the dilator muscles by an α-adrenoceptor mediated effect (Persson and Sonmark 1971).

Pupilloconstrictory NANC fibres in the iris sphincter, which act by releasing substance P, have also been postulated in the rabbit. It is possible that these fibres are sensory and that the NANC excitatory effect is due to antidromic release of substance P during experimental conditions (Leander et al. 1981 b).

The mammalian pupil is thus controlled by antagonistically acting cholinergic oculomotor fibres innervating the iris sphincter, and adrenergic pupillo-dilator fibres from the superior cervical ganglia which act by contracting the dilator muscles and relaxing the sphincter of the iris (Fig. 12.2).

Chapter 13
Chromatophores

Pigment-containing dermal and epidermal cells are a feature of practically all vertebrates. In fish, amphibians and reptiles, the pigment of these chromatophores (Greek: *chroma*, colour: *pherein*, to bear) can be aggregated or dispersed within the cell, so that the pigment of each chromatophore covers a smaller or larger area of the skin surface. Such intracellular re-arrangement of the pigment takes place as an adaptation of the animal to the background colour and is often referred to as "pysiological colour change", as opposed to "morphological colour change" which implies an alteration in the number of chromatophores or pigment content. The physiological colour change is relatively rapid, particularly in animals with a nervous control of the chromatophores (seconds to hours), while the morphological colour change generally requires days.

The movement of the intracellular pigment granules is achieved by microfilaments and microtubules, which cause aggregation or dispersion of the pigment (McGuire and Moellmann 1972, Bagnara and Hadley 1973). The control of the pigment movements in the vertebrate chromatophores is exerted by hormonal and nervous mechanisms, the relative importance of the two mechanisms depending on species. In amphibians and fish other than teleosts, the control is almost exclusively exerted by the polypeptide hormone melanocyte-stimulating hormone (MSH) from the pars intermedia of the hypophysis (pituitary gland). Circulating catecholamines may also be of importance in the chromatophore control in many vertebrates and melatonin (N-acetyl-5-methoxytryptamine) from the epiphysis (pineal organ) may also be involved. The role of melatonin in physiological colour change is, however, uncertain (Bagnara and Hadley 1973).

In many teleosts and in some reptilian species, notably chameleons, there is a direct autonomic control of the chromatophores which induces very rapid and spectacular colour changes (Brücke 1852). In some teleosts an autonomic control of the light-producing photophores has also been postulated (Nicol 1957).

Detailed treatment of chromatophore function and the physiology of colour change can be found in Parker (1948), Fingerman (1963) and Bagnara and Hadley (1973).

13.1 Teleost Chromatophores

Chromatophores of many colours occur in teleost fish but the physiological control of these cells has mostly been studied in the black melanophores. These cells are flattened with long branching processes into which the melanin granules move during pigment dispersion thus making the fish appear darker (Fig. 13.1). The de-

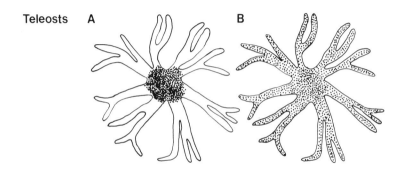
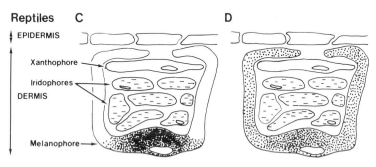

Fig. 13.1. A, B The teleost melanophore with the melanin pigment granules aggregated (**A**) and dispersed (**B**). **C, D** Diagrammatic figure of a section through the dermal chromatophore unit of a reptile. A yellow-green xanthophore is backed by light-reflecting iridophores which are partially enclosed by a melanophore with its pigment granules aggregated (**C**) and dispersed (**D**). When the melanin is aggregated the skin of the reptile will have the light colour of the xanthophore, whereas during melanin dispersion the xanthophore and the light reflecting iridophores will be covered by the dark melanin. (Based primarily on descriptions in Bagnara and Hadley 1973)

gree of aggregation and dispersion of the pigment is controlled by MSH from the pituitary gland as well as by autonomic nerve fibres from the sympathetic chains (von Frisch 1911). Visual stimuli induce the pigment movements within the melanophores and in some fish, notably flatfish, not only the general shade of the bottom (light or dark) but also the pattern (rocks, gravel, sand) is imitated. The rate at which the melanophore reaction takes place varies among species: in *Ameiurus* the change from one extreme to the other requires a full day (Parker 1934), while in other species the same process is completed in a matter of minutes (*Fundulus*, Parker 1948; *Pleuronectes*, Fernando and Grove 1974a, b; Fig. 13.2; *Phoxinus*, Grove 1969a) or even seconds (*Holocentrus*, Parker 1948).

The effect of MSH varies among species: in some teleosts the melanophore pigment aggregates after injection of MSH (e.g., *Phoxinus*, *Pleuronectes*), while in others pigment dispersion is induced (e.g., *Anguilla*, *Ameiurus*) (Parker 1934, Bagnara and Hadley 1973).

The nervous control of the melanophores is exerted by spinal autonomic pathways, in which the postganglionic fibres from the sympathetic chain ganglia re-en-

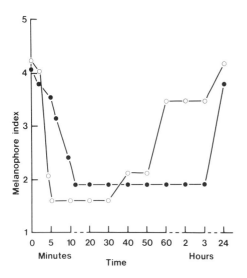

Fig. 13.2. Effect of tyramine (6×10^{-5} M; *open circles*) and noradrenaline (5×10^{-5} M; *closed circles*) on melanophores in plaice (*Pleuronectes platessa*) skin kept in 0.8% NaCl solution to induce pigment dispersion before the addition of the drugs. The melanophore index indicates the degree of melanin dispersion on a scale from *1* (fully aggregated pigment) to *5* (fully dispersed pigment). (Reproduced with permission as part of Fig. 2 from Fernando and Grove 1974b)

ter the spinal nerves and are distributed with these to the skin (von Frisch 1911, Grove 1969a). The fibres are generally adrenergic (with one exception, see below), and adrenergic nerve plexuses around melanophores have been demonstrated by Falck-Hillarp fluorescence histochemistry in *Salmo gairdneri* and *Tautogolabrus adspersus* (Jacobowitz and Laties 1968, Falck et al. 1969). The adrenergic fibres aggregate the pigment granules of the melanophores via an α-adrenoceptor mediated effect, which has been the subject of a remarkable number of physiological and pharmacological studies in different teleost species (Fujii 1961, Scheline 1963, Pye 1964a, b, Healey and Ross 1966, Grove 1969a, Fujii and Novales 1972, Fernando and Grove 1974a, b).

There is now evidence to suggest that adrenergic nerve fibres and catecholamines may also produce a pigment dispersion in teleost melanophores via a β-adrenoceptor effect (Miyashita and Fujii 1975), and that in some species an aggregating α-adrenoceptor-mediated effect may be absent (e.g., *Parasilurus asotus*: Enami 1939, Fujii and Miyashita 1976). It is difficult to understand how adrenergic fibres or circulating amines could produce pigment dispersion in species with a dominant α-adrenoceptor-mediated aggregating mechanism. The role of the β-adrenoceptor-mediated pigment dispersion is not clear.

One interesting exception to the rule of an adrenergic aggregating innervation has been demonstrated in the catfish (*Parasilurus asotus*). As mentioned above, the effect of catecholamines in this species is exclusively a pigment dispersion. Furthermore, although adrenaline/noradrenaline-induced pigment dispersion is prevented in *Parasilurus* by β-adrenoceptor antagonists, the pigment dispersion induced by dopamine can only be blocked by the dopamine receptor antagonists haloperidol or fluphenazine, demonstrating the presence of a dopamine receptor system (Miyashita and Fujii 1977). The spinal autonomic fibres innervating the melanophores of *Parasilurus* appear to act by release of acetylcholine, since the aggregating effect of nerve stimulation is blocked by atropine, enhanced by physostigmine and mimicked by cholinoceptor agonists (Fujii and Miyashita 1976).

Parker (1948) suggested the presence of cholinergic pigment dispersing fibres in the autonomic nervous supply ("chromatic tract") to the teleost melanophores, and thought that these fibres were of "parasympathetic" origin. No evidence for cholinergic-dispersing fibres was obtained by Grove (1969b) in a careful pharmacological study of the melanophore reactions in the minnow (*Phoxinus phoxinus*). The evidence for melanin-dispersing nerve fibres is, in fact, only indirect and the question of whether such fibres exist is not settled. It is possible that the supersensitivity to catecholamines caused by denervation of the skin, together with the documented presence of a β-adrenoceptor-mediated dispersion, could be sufficient to explain most of the results that have previously been interpreted in favour of dispersing fibres (Bagnara and Hadley 1973, Fernando and Grove 1974a). Effects of local, non-nervous factors such as histamine must also be considered (Fernando and Grove 1974b).

13.2 Reptilian Chromatophores

The "dermal chromatophore unit" in reptiles consists of three layers of pigment cells (Fig. 13.1). Uppermost, just under the basal lamina of the epidermis, is a xanthophore (a chromatophore with yellow pigment) and under this is a layer of iridophores (chromatophores with light-reflecting or light-scattering pigments). Under the iridophores, finally, is a melanophore and the dendritic processes of this envelop the iridophores and reach across the upper surface of the xanthophore. When the melanin is aggregated it is "tucked away" under the light-reflecting cells, causing the animal to be light yellow-green. During dispersion of the pigment in the melanophore, however, the melanin covers the xanthophore and light-reflecting layers so that the animal becomes dark brown (von Geldern 1921, Bagnara and Hadley 1973, Fig. 13.1).

A neural control of the rapidly changing chromatophores in African chameleons was first postulated by Brücke (1852). It is now well established that the pigment aggregation in species of *Chameleo* and *Phrynosoma* is controlled mainly, if not entirely, by adrenergic nerve fibres (Parker 1938, Fingerman 1963, Bagnara and Hadley 1973). The chromatophore unit of the American "chameleon" (*Anolis carolinensis*) is not innervated, but circulating catecholamines may exert some control over the melanin movements in this species (Kleinholz 1938a, b). It is clear that adrenoceptors are present in the chromatophores of *Anolis*, in which α-adrenoceptors mediate pigment aggregation and β-adrenoceptors mediate pigment dispersion similar to the situation in teleosts (Goldman and Hadley 1969, Bagnara and Hadley 1973).

There is no evidence for melanin-dispersing fibres in reptiles (Fingerman 1963, Bagnara and Hadley 1973).

Chapter 14

Concluding Remarks

The early research concerning function of the autonomic nervous system by Gaskell, Langley, Dale, Elliott, Loewi and others was focused mainly on two animal taxa: amphibians and, not surprisingly, mammals. Comparative physiological studies of autonomic nerve functions in elasmobranch and teleost fish was conducted in the early 1930's by JZ Young, who used the techniques and specific pharmacological tools adopted early in the century. Since then there has been a steady progress in our understanding of the function of the autonomic nervous systems in species representing all of the major vertebrate groups.

The intensitiy of the autonomic nervous research seems to be related to the availability of histochemical methods that can be used literally to *see* the neurons and transmitter substances involved. Thus the formaldehyde-induced fluorescence technique for the demonstration of monoamines in neurons has had an enormously stimulating influence on the research on adrenergic and other monoaminergic neurons in the autonomic nervous system and in the central nervous system. The development of immunohistochemical techniques for peptides and other constituents of the neurons also shows the same effect on research on central and autonomic neurons. Particularly the enteric nervous system, in which lack of appropriate methodology has long prevented major advances, is now the target of vigorous research activities.

For the comparative physiologist it is tempting to look for "phylogenetic trends" in the evolution of the autonomic nerve functions, i.e., an increasing degree of complexity from fish to mammals. The danger of this "phylogenetic philosophy", and any generalizations of the major vertebrate groups or classes, is that very few species in each group have been investigated. It is clear that great differences in the autonomic innervation patterns occur between related species, and that the behaviour and adjustment to environmental demands by the individual species may be of major importance in the evolution of the innervation patterns.

Bearing all these difficulties in mind, it may after all be possible to discern at least some patterns in the phylogenetic evoluton of certian systems. The adrenergic control of the heart and vasculature, for instance, appears to show a development from endogenous catecholamine storing cells (in the heart of cyclostomes and dipnoans), via an important control by catecholamines released from exogenous chromaffin cells to the direct and restricted control by adrenergic nerve fibres. In other systems, e.g., the autonomic innervation of the iris, phylogenetic trends are hard to discern because of an amazing variation in the patterns of autonomic nervous control of the iris sphincter and dilator muscles.

Two areas within the comparative physiology of the autonomic nervous system are particularly in need of further research. Firstly, there is a conspicuous lack of information about autonomic nervous reflexes in non-mammalian vertebrates, and the localization and function of sensory pathways of such reflex arcs. Secondly, there is a need for investigations that will clarify the *actual* function of the autonomic nervous system in the natural in vivo situation. For example telemetry systems, which can be used to transmit parameters of circulation (heart rate, blood flow, blood pressure etc.) from freely moving animals, will prove to be of great value for this type of research (cf. Chap. 7.8.2).

Our comprehension of the complexity of autonomic nerve function from the early dogmas of sympathetic = adrenergic and parasympathetic = cholinergic has increased rapidly over the past years. First of all it is now clear that autonomic neurons that are neither adrenergic nor cholinergic exist (NANC neurons). Furthermore there is now encouraging evidence in favour of some putative NANC transmitters – adenosine triphosphate, 5-hydroxytryptamine and several polypeptides. The diversity of neuron types in the autonomic nervous system, especially in the enteric nervous system, is thus much greater than previously believed. Elucidation of the mechanisms involved in NANC neurotransmission offers a fantastic challenge for physiologists, histochemists, anatomists, biochemists and pharmacologists. Furthermore, the demonstration of an ever-increasing number of presynaptic receptor systems that modulate the release of transmitter from the autonomic nerve terminals adds to the complexity of the autonomic control functions. Finally, the so-called "Dale principle" appears to have been invalidated by recent studies, in that there is now evidence from several systems of the co-existence of transmitter substances within the same neuron. The extent to which the "co-transmitters" are actually transmitters or just modulators of the action of a "main transmitter" at pre- and postsynaptic sites is likely to vary among the different systems.

Exciting evidence for the co-existence *and* co-function as neurotransmitters released from the same neuron comes from investigations on mammalian exocrine glands by Lundberg and co-workers (acetylcholine and VIP) and from work on the toad heart by Campbell and co-workers (acetylcholine and somatostatin). It is nice to see how the amphibian heart, once used by Loewi to demonstrate the chemical nature of neurotransmission, has again revealed new secrets about the function of a non-adrenergic, non-cholinergic transmitter (somatostatin) co-released with Loewi's "Vagusstoff" from autonomic nerve endings. Such discoveries strengthen the belief in the comparative physiological approach, and suggest that studies on non-mammalian vertebrates will provide important contributions to our general understanding of the autonomic nerve function in the vertebrates.

References

Abel JJ (1899) Über den blutdruckerregenden Bestandtheil der Nebenniere, das Epinephrin. Hoppe-Seyler's Z Physiol Chem 28:318–362

Abel JJ (1901) Further observations on epinephrine. Bull Johns Hopkins Hosp 12:80–84

Åberg G, Eränkö O (1967) Localization of noradrenaline and acetylcholinesterase in the taenia of the guinea-pig caecum. Acta Physiol Scand 69:383–384

Åberg G, Welin I (1967) Interactions of beta-receptor blocking agents with the nerve blocking effects of procaine in vitro. Life Sci 6:975–979

Åberg G, Dzedin T, Lundholm L, Olsson L, Svedmyr N (1969) A comparative study of some cardiovascular effects of sotalol (MJ 1999) and propranolol. Life Sci 8:353–365

Abrahamsson H (1973) Studies in the inhibitory nervous control of gastric motility. Acta Physiol Scand 88 Suppl 390:1–38

Abrahamsson T (1979a) Axonal transport of adrenaline, noradrenaline and phenylethanolamine-N-methyl transferase (PNMT) in sympathetic neurons of the cod, *Gadus morhua*. Acta Physiol Scand 105:316–325

Abrahamsson T (1979b) Phenylethanolamine-N-methyl transferase (PNMT) activity and catecholamine storage and release from chromaffin tissue of the spiny dogfish, *Squalus acanthias*. Comp Biochem Physiol 64C:169–172

Abrahamsson T (1980) The effect of SK&F 64139, an inhibitor of phenylethanolamine-N-methyl transferase (PNMT), on adrenaline and noradrenaline content in sympathetic neurons of the cod, *Gadus morhua*. Comp Biochem Physiol 67C:49–54

Abrahamsson T, Nilsson S (1975) Effects of nerve sectioning and drugs on the catecholamine content in the spleen of the cod, *Gadus morhua*. Comp Biochem Physiol 51C:231–233

Abrahamsson T, Nilsson S (1976) Phenylethanolamine-N-methyl transferase (PNMT) activity and catecholamine content in chromaffin tissue and sympathetic neurons in the cod, *Gadus morhua*. Acta Physiol Scand 96:94–99

Abrahamsson T, Holmgren S, Nilsson S, Pettersson K (1979a) On the chromaffin system of the African lungfish, *Protopterus aethiopicus*. Acta Physiol Scand 107:135–139

Abrahamsson T, Holmgren S, Nilsson S, Pettersson K (1979b) Adrenergic and cholinergic effects on the heart, the lung and the spleen of the African lungfish, *Protopterus aethiopicus*. Acta Physiol Scand 107:141–147

Abrahamsson T, Jönsson A-C, Nilsson S (1979c) Catecholamine synthesis in the chromaffin tissue of the African lungfish, *Protopterus aethiopicus*. Acta Physiol Scand 107:149–151

Abrahamsson T, Jönsson A-C, Nilsson S (1981) Activity of dopamine-β-hydroxylase (DBH) and phenylethanolamine-N-methyl transferase (PNMT) in heart, lung and chromaffine tissue from the Florida spotted gar, *Lepisosteus platyrhincus* (Holostei). Acta Physiol Scand 111:413–433

Ackerknecht EH (1974) The history of the discovery of the vegetative (autonomic) nervous system. Med Hist 18:1–8

Adams LA, Eddy S (1951) Comparative anatomy: an introduction to the vertebrates. John Wiley, New York

Adams WE (1942) Observations on the lacertilian sympathetic system. J Anat 77:6–11

Adler-Graschinsky E, Langer SZ (1975) Possible role of a β-adrenoceptor in the regulation of noradrenaline release by nerve stimulation through positive feed-back mechanism. Br J Pharmacol 53:43–50

Ahlman H, Lundberg JM, Dahlström A, Kewenter J (1976) A possible vagal adrenergic release of serotonin from enterochromaffin cells in the cat. Acta Physiol Scand 98:366–375

Ahlquist RP (1948) A study of the adrenotropic receptors. Am J Physiol 153:586–600

Aikawa T (1931) On the innervation of the frog's stomach. Jpn J Med Sci Biophys 2:91–129

Ajelis V, Björklund B, Falck B, Lindvall I, Lorén I, Walles B (1979) Application of the aluminium-formaldehyde (ALFA) histofluorescence method for demonstration of peripheral stores of catecholamines and indolamines in freeze-dried paraffine embedded tissue, cryostat sections and wholemounts. Histochemistry 65:1–15

Akers TK, Peiss CN (1963) Comparative study of effect of epinephrine on cardiovascular system of turtle, alligator, chicken and opossum. Proc Soc Exp Biol Med 112:396–399

Akester AR, Mann SP (1969) Ultrastructure and innervation of the tertiary bronchial unit in the lung of *Gallus domesticus*. J Anat 105:202–204

Akester AR, Akester B, Mann SP (1969) Catecholamines in the avian heart. J Anat 104:591

Aldrich TB (1901) A preliminary report on the active principle of the suprarenal gland. Am J Physiol 5:457–461

Allis EP Jr (1920) The branches of the branchial nerves of fishes, with special referenc to *Polyodon spathula*. J Comp Neurol 32:137–153

Alphen van GWHM, Robinette SL, Marci TJ (1964) The adrenergic receptors of the intraocular muscles of the cat. Int J Neuropharmacol 2:259–272

Alumets J, Håkanson R, Sundler F, Chang K-J (1978) Leu-enkephalin-like material in nerves and enterochromaffin cells in the gut. An immunohistochemical study. Histochemistry 56:187–196

Ambache N (1951) Unmasking, after cholinergic paralysis by botulinum toxin, of a reversed action of nicotine on the mammalian intestine, revealing the probable presence of local inhibitory ganglion cells in the enteric plexuses. Br J Pharmacol Chemother 6:51–67

Ambache N, Lessin AW (1955) Classification of intestinomotor drugs by means of type D botulinum toxin. J Physiol (London) 127:449–478

Ambache N, Zar MA (1970) Non-cholinergic transmission by post-ganglionic motor neurones in the mammalian bladder. J Physiol (London) 210:761–763

Andén NE, Henning M (1966) Adrenergic nerve function, noradrenaline level and noradrenaline uptake in cat nictitating membrane after reserpine treatment. Acta Physiol Scand 67:498–504

Andrew A (1974) Further evidence that enterochromaffin cells are not derived from the neural crest. J Embryol Exp Morphol 31:589–598

Angelakos ET, Glassman PM, Millards RW, King M (1965) Regional distribution and subcellular localization of catecholamines in the frog heart. Comp Biochem Physiol 15:313–324

Anton AH, Sayre DF (1962) A Study of the factors affecting the aluminium oxide-trihydroxyindole procedure for the analysis of catecholamines. J Pharmacol Exp Ther 138:360–375

Antonaccio MJ, Smith GB (1974) Effects of chronic pretreatment with small doses of reserpine upon adrenergic nerve function. J Pharmacol Exp Ther 188:654–667

Ariëns EJ (1954) Affinity and intrinsic activity in the theory of competitive inhibition. Arch Int Pharmacodyn Ther 99:32–49

Ariëns EJ (1966) Receptor theory and structure-action relationships. Adv Drug Res 3:235–285

Ariëns EJ, Rossum Van JM (1957) pD'_x, pA_x and pD_x values in the analysis of pharmacodynamics. Arch Int Pharmacodyn Ther 110:275–299

Ariëns EJ, Simonis AM (1961) Analysis of the action of drugs and drug combinations. In: Jonge de H (ed) Quantitative methods in pharmacology. North Holland, Amsterdam, pp 286–311

Ariëns EJ, Simonis AM (1964a) A molecular basis for drug action. J Pharm Pharmacol 16:137–157

Ariëns EJ, Simonis AM (1964b) A molecular basis for drug action. The interaction of one or more drugs with different receptors. J Pharm Pharmacol 16:289–312

Ariëns EJ, Simonis AM, Grot De WM (1955) Affinity and intrinsic-activity in the theory of competitive- and non-competitive inhibition and an analysis of some forms of dualism in action. Arch Int Pharmacodyn Ther 100:298–322

Ariëns EJ, Rossum Van JM, Simonis AM (1956) A theoretical basis of molecular pharmacology. Part II: Interactions of one or two compounds with two interdependent receptor systems. Arzneimittelforschung 6:611–621

Armstrong PB, Bell AL (1968) Pupillary responses in the toad as related to innervation of the iris. Am J Physiol 214:566–573

Ask JA, Stene-Larsen G, Helle KB (1980) Atrial beta$_2$-adrenoceptors in the trout. J Comp Physiol 139B:109–116

Augustinsson KB, Fänge R, Johnels A, Östlund E (1956) Histological, physiological and biochemical studies on the heart of two cyclostomes, hagfish (*Myxine*) and lamprey (*Lampetra*). J Physiol (London) 131:257–276

Axelrod J (1962) Purification and properties of phenylethanolamine-N-methyl transferase. J Biol Chem 237:1657–1660
Axelrod J (1966) Methylation reactions in the formation and metabolism of catecholamines and other biogenic amines. Pharm Rev 18:95–113
Axelrod J (1972) Dopamine-β-hydroxylase: Regulation of its synthesis and release from nerve terminals. Pharmacol Rev 24:233–243
Axelsson S, Björklund A, Falck B, Lindvall O, Svensson L-Å (1973) Glyoxylic acid condensation: A new fluorescence method for the histochemical demonstration of biogenic monoamines. Acta Physiol Scand 87:57–62
Azuma T, Binia A, Visscher M (1965) Adrenergic mechanisms in the bullfrog and turtle. Am J Physiol 209:1287–1299
Babkin BP, Bowie DJ, Nicholls JVV (1933) Circulatory reactions of the skate, *Raja*, to drugs. Contrib Can Biol Fish 8:209–219
Babkin BP, Friedman MHF, Mackay-Sawyer ME (1935) Vagal and sympathetic innervation of the stomach of the skate. J Biol Board Can 1:239–250
Bagnara JT, Hadley ME (1973) Chromatophores and colorchange. Prentice Hall, Englewood Cliffs
Balashov NV, Fänge R, Govyrin VA, Leont'eva GR, Nilsson S, Prozorovskaya MP (1981) On the adrenergic system of ganoid fish: the beluga, *Huso huso* (Chondrostei). Acta Physiol Scand 111:435–440
Balfour FM (1877) The development of elasmobranch fishes. J Anat 11:674–706
Bamford OS (1974) Oxygen reception in the rainbow trout (*Salmo gairdneri*). Comp Biochem Physiol 48A:69–76
Bänder A (1954) Über zwei chromaffine Zelltypen im Nebennierenmark und ihre Beziehung zum Vorkommen von Adrenalin und Arterenol. Arch Exp Pathol Pharmakol 223:140–147
Banister J, Mann SP (1966) An investigation of the adrenergic innervation of the heart and major blood vessels of the frog by Falck's method of fluorescence microscopy. J Physiol (London) 181:13–15P
Banister RJ, Portig PJ, Vogt M (1967) The content and localization of catecholamines in the carotid labyrinths and aortic arches of *Rana temporaria*. J Physiol (London) 192:529–535
Banks BEC, Brown C, Burgess GM, Burnstock G, Claret M, Cocks T, Jenkinson DH (1979) Apamine blocks certain neurotransmitter-induced increases in potassium permeability. Nature (London) 282:415
Barajas L, Wang P (1975) Demonstration of acetylcholinesterase in the adrenergic nerves of the renal glomerular arterioles. J Ultrastruct Res 53:244–253
Baumgarten HG (1967) Vorkommen und Verteilung adrenerger Nervenfasern im Darm der Schleie (*Tinca vulgaris*, Cuv.). Z Zellforsch Mikrosk Anat 76:248–259
Baumgarten HG, Holstein AF, Owman Ch (1970) Auerbach's plexus of mammals and man: electron microscopic identification of three different types of neuronal processes in myenteric ganglia of the large intestine from rhesus monkey, guinea-pig and man. Z Zellforsch 106:376–397
Baumgarten HG, Björklund A, Lachenmayer L, Nobin A, Rosengren E (1973) Evidence for existence of serotonin-, dopamine- and noradrenaline-containing neurons in the gut of *Lampetra fluviatilis*. Z Zellforsch 141:33–54
Bayliss WM, Starling EH (1899) The movements and the innervation of the small intestine. J Physiol (London) 24:99–143
Bayliss WM, Starling EH (1900) The movements and the innervation of the large intestine. J Physiol (London) 26:107–118
Belaud A, Peyraud-Waitzenegger M, Peyraud C (1971) Etude compaŕe des réactions vasomotrices des branchies perfusées de deux téléostéens: la Carpe et al Congre. C R Soc Biol 165:1114–1118
Bell C (1967) A histochemical study of the esterases in the bladder of the toad (*Bufo marinus*). Comp Biochem Physiol 21:91–98
Bell C (1969) Indirect cholinergic vasomotor control of intestinal blood flow in the domestic chicken. J Physiol (London) 205:317–327
Bell C (1982) Dopamine as a postganglionic autonomic transmitter. Neuroscience 7:1–8
Bell C, Burnstock G (1965) Cholinesterases in the bladder of the toad (*Bufo marinus*). Biochem Pharmacol 14:78–89
Benfey BG (1979a) The interconversion of adrenoceptors. In: Kalsner S (ed) Trends in autonomic pharmacology, vol I. Urban & Schwarzenberg, Baltimore München, pp 289–307

Benfey BG (1979b) Cardiac adrenoceptors at low temperature and the adrenoceptor interconversion hypothesis. Can J Physiol Pharmacol 57:771–777

Benfey BG, Grillo SA (1963) Antagonism of acetylcholine by adrenaline antagonists. Br J Pharmacol 20:528–533

Bennett MR (1966) Rebound excitation of the smooth muscle cells of the guinea-pig taenia coli after stimulation of intramural inhibitory nerves. J Physiol (London) 185:124–131

Bennett MR (1972) Autonomic neuromuscular transmission, 1st edn. Cambridge Univ Press, Cambridge

Bennett MR, Burnstock G (1968) Electrophysiology of the innervation of intestinal smooth muscle. In: Handbook of physiology, sect 6. Alimentary canal, vol IV. Motility. Am Physiol Soc, Washington DC, pp 1709–1732

Bennett T (1969a) Studies on the avian gizzard: Histochemical analysis of the extrinsic and intrinsic innervation. Z Zellforsch Mikrosk Anat 98:188–201

Bennett T (1969b) Nerve-mediated excitation and inhibition of the smooth muscle cells of the avian gizzard. J Physiol (London) 204:669–686

Bennett T (1969c) The effects of hyoscine and anticholinesterases on cholinergic transmission to the smooth muscle cells of the avian gizzard. Br J Pharmacol 37:585–594

Bennett T (1971) The adrenergic innervation of the pulmonary vasculature, the lung and the thoracic aorta, and on the presence of aortic bodies in the domestic fowl. (*Gallus gallus domesticus* L.). Z Zellforsch Mikrosk Anat 114:117–134

Bennett T (1974) Peripheral and autonomic nervous systems. In: Farner DS, King JS, Parkes KC (eds) Avian biology, vol IV. Academic Press, London New York, pp 1–77

Bennett T, Cobb JLS (1969a) Studies on the avian gizzard: Morphology and innervation of the smooth muscle. Z Zellforsch Mikrosk Anat 96:173–185

Bennett T, Cobb JLS (1969b) Studies on the avian gizzard: Auerbach's plexus. Z Zellforsch Mikrosk Anat 99:109–120

Bennett T, Malmfors T (1970) The adrenergic nervous system of the domestic fowl (*Gallus domesticus* L). Z Zellforsch 106:22–50

Bennett T, Malmfors T, Cobb JLS (1973) Fluorescence histochemical observations on catecholamine-containing cell bodies in Auerbach's plexus. Z Zellforsch 139:69–81

Bercovitz Z, Rogerts FT (1921) Contributions to the physiology of the stomach. IV The influence of the vagi on gastric tonus and motility in the turtle. Am J Physiol 55:323–338

Berger PJ (1971) The vagal and sympathetic innervation of the heart of the lizard *Tiliqua rugosa*. Aust J Exp Biol Med Sci 49:297–304

Berger PJ (1972) The vagal and sympathetic innervation of the isolated pulmonary artery of a lizard and a tortoise. Comp Gen Pharmacol 3:113–124

Berger PJ (1973) Autonomic innervation of the visceral and vascular smooth muscle of the lizard lung. Comp Gen Pharmacol 4:1–10

Berger PJ, Burnstock G (1979) Autonomic nervous system. In: Gans C (ed) Biology of the reptilia, vol X. Academic Press, London New York, pp 1–57

Berger PJ, Evans BK, Smith DG (1980) Localization of baroreceptors and gain of the baroreceptor-heart rate reflex in the lizard *Trachydosaurus rugosus*. J Exp Biol 86:197–209

Beringer T, Hadek R (1973) Ultrastructure of sinus venosus innervation in *Petromyzon marinus*. J Ultrastruct Res 42:312–323

Bernard C (1878) Leçons sur les phénomènes de la vie communs aux animaux et aux végétaux. Baillère, Paris

Berthelsen S, Pettinger WA (1977) A functional basis for classification of α-adrenergic receptors. Life Sci 21:595–606

Biber B (1974) Vasodilator mechanisms in the small intestine. Acta Physiol Scand 90 Suppl 401:1–31

Biber B, Lundgren O, Svanvik J (1971) Studies on the intestinal vasodilatation observed after mechanical stimulation of the mucosa of the gut. Acta Physiol Scand 82:177–190

Biber B, Fara J, Lundgren O (1973a) Intestinal vasodilatation in response to transmural electrical field stimulation. Acta Physiol Scand 87:277–282

Biber B, Fara J, Lundgren O (1973b) Intestinal vascular responses to 5-hydroxytryptamine. Acta Physiol Scand 87:526–534

Bidder F (1868) Die Endigungsweise der Herzzweige des N. vagus beim Frosch. Arch Anat Physiol Wiss Med 1–50

Birks RI, MacIntosh FC (1961) Acetylcholine metabolism of a sympathetic ganglion. Can J Biochem Physiol 39:787–827

Björklund A, Falck B, Owman Ch (1972) Fluorescence microscopic and microspectrofluorometric techniques for the cellular localization and characterization of biogenic amines. In: Berson SA (ed) Methods of investigative and diagnostic endocrinology, vol 1. The thyroid and biogenic amines. North Holland, Amsterdam, pp 318–368

Björklund A, Falck B, Lindvall O, Lorén I (1980) The aluminium-formaldehyde (ALFA) histofluorescence method for improved visualization of catecholamines and indolamines. 2. Model experiments. J Neurosci Meth 2:301–318

Blakely AGH (1968) The responses of the spleen to nerve stimulation in relation to the frequency of splenic nerve discharge. Proc R Soc London Ser B 171:201–211

Blaschko H (1939) The specific action of L-DOPA decarboxylase. J Physiol (London) 96:50–51P

Blaschko H, Comline RS, Schneider FH, Silver M, Smith AD (1967) Secretion of a chromaffin granule protein, chromogranin, from the adrenal gland after splanchnic stimulation. Nature (London) 215:58–59

Bloom G, Östlund E, Euler von US, Lishajko F, Ritzen M, Adams-Ray J (1961) Studies on catecholamine-containing granules of specific cells in cyclostome hearts. Acta Physiol Scand 53 Suppl 185:1–34

Bohr C (1894) The influence of section of the vagus nerve on disengagement of gases in the airbladder of fishes. J Physiol (London) 15:494–500

Bolis L, Rankin JC (1975) Adrenergic control of blood flow through fish gills: Environmental implications. In: Bolis L, Maddrell SHP, Schmidt-Nielsen K (eds) Comparative physiology. Elsevier/North Holland Biomedical Press, Amsterdam New York, pp 223–233

Bolton TB (1968a) Studies on the longitudinal muscle of the anterior mesenteric artery of the domestic fowl. J Physiol (London) 196:273–281

Bolton TB (1968b) Electrical and mechanical activity of the longitudinal muscle of the anterior mesenteric artery of the domestic fowl. J Physiol (London) 196:283–292

Bolton TB (1969) Spontaneous and evoked release of neurotransmitter substances in the longitudinal muscle of the anterior mesenteric artery of the domestic fowl. Br J Pharmacol 35:112–120

Bolton TB (1971a) The structure of the nervous system. In: Bell DJ, Freeman BM (eds) Physiology and biochemistry of the domestic fowl. Academic Press, London New York, pp 641–673

Bolton TB (1971b) The physiology of the nervous system. In: Bell DJ, Freeman BM (eds) Physiology and biochemistry of the domestic fowl. Academic Press, London New York, pp 675–703

Bolton TB (1976) Nervous system. In: Sturkie PD (ed) Avian physiology. Springer, Berlin Heidelberg New York, pp 1–28

Bolton TB, Raper C (1966) Innervation of domestic fowl and guinea pig ventricles. J Pharm Pharmacol 18:192–193

Bone Q (1963) Some observations upon the peripheral nervous system of the hagfish, *Myxine glutinosa*. J Mar Biol Assoc UK 43:31–37

Bonnet V (1929) De l'influence de l'hémorragie et de l'asphyxie sur le nombre des hématies dans le sang circulant des vertébrés inférieurs. J Physiol Pathol Gen 27:735–740

Born GVR (1970) 5-hydroxytryptamine receptors. In: Bülbring E, Brading AF, Jones AW, Tomita T (eds) Smooth muscle. Arnold, London, pp 418–450

Botazzi F (1902) Untersuchungen über das viscerale Nervensystem der Selachier. Z Biol (Munich) 43:372–442

Boura ALA, Green AF (1965) Adrenergic neurone blocking agents. Annu Rev Pharmacol 5:183–212

Boura ALA, Duncombe WG, McCoubrey A (1961) The distribution of some quaternary ammonium salts in the peripheral nervous system of cats in relation to the adrenergic blocking action of bretylium. Br J Pharmacol Chemother 17:92–100

Bourgeois P, Dupont W, Vaillant R (1978) Estimation du taux des catécholamines circulantes chez la Grenouille, à l'aide d'une technique de dosage radioenzymatique. C R Acad Sci 287:1149–1152

Boyd H, Burnstock G, Campbell G, Jowett A, O'Shea J, Wood M (1963) The cholinergic blocking action of adrenergic blocking agents in the pharmacological analysis of autonomic innervation. Br J Pharmacol 20:418–435

Boyd H, Burnstock G, Rogers D (1964) Innervation of the large intestine of the toad (*Bufo marinus*). Br J Pharmacol Chemother 23:151–163

Brandt HD, Offermeyer J (1977) An alternative model for non-competitive antagonism. Arch Int Pharmacodyn Ther 225:180–195

Brandt W (1922) Das Darmnervensystem von *Myxine glutinosa*. Z Anat Entwicklungsgesch 64:284–292

Brazeau P, Vale W, Burgus R, Ling N, Butcher M, Rivier J, Guillemin R (1973) Hypothalamic polypeptide that inhibits the secretion of immunoreactive pituitary growth hormone. Science 179:77–79

Brimijoin S (1975) Stop-flow: A new technique for measuring axonal transport, and its application to the transport of dopamine-β-hydroxylase. J Neurobiol 6:379–394

Brimijoin S (1979) Axonal transport and subcellular distribution of molecular forms of acetylcholinesterase in rabbit sciatic nerve. Mol Pharmacol 15:641–648

Brimijoin S, Wiermaa MJ (1977) Rapid axonal transport of tyrosine hydroxylase in rabbit sciatic nerves. Brain Res 120:77–96

Brimijoin S, Wiermaa MJ (1978) Rapid orthograde and retrograde axonal transport of acetylcholinesterase as characterized by the stop-flow technique. J Physiol (London) 285:120–142

Brimijoin S, Skau K, Wiermaa MJ (1978) On the origin and fate of external acetylcholinesterase in peripheral nerve. J Physiol (London) 285:143–158

Brimijoin S, Lundberg JM, Brodin E, Hökfelt T, Nilsson G (1980) Axonal transport of substance P in the vagus and sciatic nerves of the guinea pig. Brain Res 191:443–457

Brink van den FG (1969) Histamine and antihistamines. Janssen, NV Nijmegen

Brink van den FG (1973a) The model of functional interaction – I. Development and first check of a new model of functional synergism and antagonism. Eur J Pharmacol 22:270–278

Brink van den FG (1973b) The model of functional interaction – II. Experimental verification of a new model: the antagonism of β-adrenoceptor stimulants and other agonists. Eur J Pharmacol 22:279–286

Brodin E, Alumets J, Håkanson S, Leander S, Sundler F (1981) Immunoreactive substance P in the chicken gut: Distribution, development and possible functional significance. Cell Tissue Res 216:455–469

Brody JS, Klempfner G, Staum MM, Vidyasagar D, Kuhl DE, Waldhausen J (1972) Mucociliary clearance after lung denervation and bronchial transection. J Appl Physiol 32:160–164

Bronkhorst W, Dijkstra C (1940) Das neuromuskuläre System der Lunge. Anatomische und physiologische Untersuchungen über die Lungenmuskulatur und ihre Bedeutung für die Klinik der Tuberkulose. Beitr Klin Tuberk Spezifischen Tuberk Forsch 94:445–503

Brown CM, Burnstock G (1981) Evidence in support of the P_1/P_2 purinoceptor hypothesis in the guinea-pig taenia coli. Br J Pharmacol 73:617–624

Brown GL, Gillespie JS (1957) The output of sympathetic transmitter from the spleen of the cat. J Physiol (London) 138:81–102

Brown-Séquard M (1847) Recherches expérimentales sur l'action de la lumière et sur celle d'un changement de température sur l'iris, dans les cinq classes d'animaux vertébrés. CR Acad Sci 25:482–483

Brücke E (1852) Untersuchungen über den Farbwechsel des afrikanischen Chamäleons. Denkschr Akad Wiss Wien 4:179–210

Buchan AMJ, Polak JM, Pearse AGE (1980) Gut hormones in *Salamandra salamandra*. An immunocytochemical and electron microscopic investigation. Cell Tissue Res 211:331–343

Buchan AMJ, Polak JM, Bryant MG, Bloom SR, Pearse AGE (1981) Vasoactive intestinal polypeptide (VIP)-like immunoreactivity in anuran intestine. Cell Tissue Res 216:413–422

Bülbring E (1949) The methylation of noradrenaline by minced suprarenal tissue. Br J Pharmacol 4:234–244

Bülbring E, Burn JH (1935) The sympathetic dilator fibres in the muscles of the cat and dog. J Physiol (London) 83:483–501

Bülbring E, Crema A (1958) Observations concerning the action of 5-hydroxytryptamine on the peristaltic reflex. Br J Pharmacol 13:444–457

Bülbring E, Gershon MD (1967) 5-Hydroxytryptamine participation in the vagal inhibitory innervation of the stomach. J Physiol (London) 192:823–846

Bülbring E, Lin RCY (1958) The effect of intraluminal application of 5-hydroxytryptamine and 5-hydroxytryptophan on peristalsis, the local production of 5-hydroxytryptamine and its release in relation to intraluminal pressure and propulsive activity. J Physiol (London) 140:381–407

Burger JW, Bradley SE (1951) The general form of the circulation in the dogfish (*Squalus acanthias*). J Cell Comp Physiol 37:389–402

Burggren WW (1975) A quantitative analysis of ventilation tachycardia and its control in two chelonians, *Pseudemys scripta* and *Testudo graeca*. J Exp Biol 63:367–380
Burggren WW (1977) The pulmonary circulation of the chelonian reptile: morphology, haemodynamics and pharmacology. J Comp Physiol 116:303–323
Burggren WW (1978) Influence of intermittent breathing on ventricular depolarization patterns in chelonian reptiles. J Physiol (London) 278:349–364
Burn JH (1977a) Evidence that acetylcholine releases noradrenaline in the sympathetic fibre. J Pharm Pharmacol 6:325–329
Burn JH (1977b) The function of ACh released from sympathetic fibres. Clin Exp Pharmacol Physiol 4:59
Burn JH, Froede H (1963) The action of substances which block sympathetic postganglionic nervous transmission. Br J Pharmacol 20:378–387
Burn JH, Rand MJ (1960) Sympathetic postganglionic cholinergic fibres. Br J Pharmacol 15:56–66
Burn JH, Rand MJ (1965) Acetylcholine in adrenergic transmission. Annu Rev Pharmacol 5:163–182
Burnstock G (1958a) Reversible inactivation of nervous activity in a fish gut. J Physiol (London) 141:35–45
Burnstock G (1958b) The effect of drugs on spontaneous motility and on the response to stimulation of the extrinsic nerves of the gut of a teleostean fish. Br J Pharmacol Chemother 13:216–226
Burnstock G (1959) The innervation of the gut of the brown trout (*Salmo trutta*). Q J Microsc Sci 100:199–219
Burnstock G (1969) Evolution of the autonomic innervation of visceral and cardiovascular systems in vertebrates. Pharmacol Rev 21:247–324
Burnstock G (1970) Structure of smooth muscle and its innervation. In: Bülbring E, Brading AF, Jones AW, Tomita T (eds) Smooth muscle. Arnold, London, pp 1–69
Burnstock G (1972) Purinergic nerves. Pharmacol Rev 24:509–581
Burnstock G (1975) Comparative studies of purinergic nerves. J Exp Zool 194:103–133
Burnstock G (1976a) Purinergic receptors. J Theor Biol 62:491–503
Burnstock G (1976b) Do some nerves release more than one transmitter? Neuroscience 1:239–248
Burnstock G (1977) Cholinergic, adrenergic and purinergic neuromuscular transmission. Fed Proc 36:2434–2438
Burnstock G (1978a) A basis for distinguishing two types of purinergic receptor. In: Bolis L, Straub RW (eds) Cell membrane receptors for drugs and hormones: a multidisciplinary approach. Raven Press, New York, pp 107–118
Burnstock G (1978b) Do some sympathetic neurones release both noradrenaline and acetylcholine? Prog Neurobiol (Oxford) 11:205–222
Burnstock G (1979) The ultrastructure of autonomic cholinergic nerves and junctions. Prog Brain Res 49:3–21
Burnstock G (1980a) Purinergic modulation of cholinergic transmission. Gen Pharmacol 11:15–18
Burnstock G (1980b) Cholinergic and purinergic regulation of blood vessels. In: Geiger SR, Somlyo AD, Sparks HW (eds) Handbook of physiology, sect 2. The cardiovascular system. Waverly, Baltimore, pp 567–612
Burnstock G (1981) Review lecture: Neurotransmitters and trophic factors in the autonomic nervous system. J Physiol (London) 313:1–35
Burnstock G, Campbell G (1963) Comparative physiology of the vertebrate autonomic nervous system. II Innervation of the urinary bladder of the ringtail possum (*Pseudocheirus peregrinus*). J Exp Biol 40:421–437
Burnstock G, Costa M (1973) Inhibitory innervation of the gut. Gastroenterology 64:141–143
Burnstock G, Costa M (1975) Adrenergic neurons, 1st edn. Chapman & Hall, London
Burnstock G, Kirby S (1968) Absence of inhibitory effects of catecholamines on lower vertebrate arterial strip preparations. J Pharm Pharmacol 20:404–406
Burnstock G, Meghji P (1981) Distribution of P_1- and P_2-purinoceptors in the guinea-pig and frog heart. Br J Pharmacol 73:879–885
Burnstock G, Wood ME (1967a) Innervation of the urinary bladder of the sleepy lizard (*Trachysaurus rugosus*). II. Physiology and pharmacology. Comp Biochem Physiol 20:675–690
Burnstock G, Wood ME (1967b) Innervation of the lungs of the sleepy lizard (*Trachysaurus rugosus*). II. Physiology and pharmacology. Comp Biochem Physiol 22:815–831

Burnstock G, Campbell G, Bennett M, Holman M (1963a) Inhibition of the smooth muscle of the taenia coli. Nature (London) 200:581–582

Burnstock G, O'Shea J, Wood M (1963b) Comparative physiology of the vertebrate autonomic system. I. Innervation of the urinary bladder of the toad (*Bufo marinus*). J Exp Biol 40:403–420

Burnstock G, Campbell G, Bennett M, Holman M (1964) Innervation of the guinea-pig taenia coli: are there intrinsic inhibitory nerves which are distinct from sympathetic nerves? Int J Neuropharmacol 3:163–166

Burnstock G, Campbell G, Rand MJ (1966) The inhibitory innervation of the taenia of the guinea-pig caecum. J Physiol (London) 182:504–526

Burnstock G, Campbell G, Satchell DG, Smythe A (1970) Evidence that adenosine triphosphate or a related nucleotide is the transmitter substance released by non-adrenergic inhibitory nerves in the gut. Br J Pharmacol 40:668–688

Burnstock G, Evans B, Gannon BJ, Heath JW, James V (1971) A new method of destroying adrenergic nerves in adult animals using guanethidine. Br J Pharmacol 43:295–301

Burnstock G, Dumsday BH, Smythe A (1972) Atropine resistant excitation of the urinary bladder: the possibility of transmission via nerves releasing a purine nucleotide. Br J Pharmacol 44:451–461

Burnstock G, Cocks T, Paddle BM, Staszewska-Barczak J (1975) Evidence that prostaglandin is responsible for the 'rebound contraction' following stimulation of non-adrenergic, non-cholinergic ('purinergic') inhibitory nerves. Eur J Pharmacol 31:360–362

Burnstock G, Cocks T, Crowe R, Kasakov L (1978a) Purinergic innervation of the guinea-pig urinary bladder. Br J Pharmacol 63:125–138

Burnstock G, Cocks T, Kasakov L, Wong H (1978b) Direct evidence for ATP release from non-adrenergic, non-cholinergic ('purinergic') nerves in the guinea-pig taenia coli and bladder. Eur J Pharmacol 49:145–149

Butler PJ (1982) Respiratory and cardiovascular control during diving in birds and mammals. J Exp Biol 100:195–221

Butler PJ, Jones DR (1971) The effect of variations in heart rate and regional distribution of blood flow on the normal pressor response to diving in ducks. J Physiol (London) 214:457–479

Butler PJ, Jones DR (1982) Comparative physiology of diving in vertebrates. In: Lowenstein OE (ed) Advances in comparative physiology and biochemistry, vol VIII. Academic Press, London New York, in press

Butler PJ, Taylor EW (1971) Response of the dogfish (*Scyliorhinus canicula* L.) to slowly induced and rapidly induced hypoxia. Comp Biochem Physiol 39A:307–323

Butler PJ, Taylor EW, Short S (1977) The effects of sectioning cranial nerves V, VII, IX and X on the cardiac response of the dogfish *Scyliorhinus canicula* to environmental hypoxia. J Exp Biol 69:233–245

Butler PJ, Taylor EW, Capra MF, Davison W (1978) The effect of hypoxia on the levels of circulating catecholamines in the dogfish *Scyliorhinus canicula*, J Comp Physiol 127:325–330

Cabezas GA, Graf PD, Nadel JA (1971) Sympathetic versus parasympathetic nervous regulation of airways in dogs. J Appl Physiol 31:651–655

Cameron JS (1979) Autonomic nervous tone and regulation of heart rate in the goldfish, *Carassius auratus*. Comp Biochem Physiol 63C:341–349

Cameron JS, Brown SE (1981) Adrenergic and cholinergic responses of the isolated heart of the goldfish *Carassius auratus*. Comp Biochem Physiol 70C:109–116

Campbell G (1966a) The inhibitory nerve fibres in the vagal supply to the guinea-pig stomach. J Physiol (London) 185:600–612

Campbell G (1966b) Nerve mediated excitation of the taenia of the guinea-pig caecum. J Physiol (London) 185:148–159

Campbell G (1969) The autonomic innervation of the stomach of a toad (*Bufo marinus*). Comp Biochem Physiol 31:693–706

Campbell G (1970a) Autonomic nervous supply to effector tissues. In: Bülbring E, Brading A, Jones A, Tomita T (eds) Smooth muscle. Arnold, London, pp 451–495

Campbell G (1970b) Autonomic nervous systems. In: Hoar WS, Randall DJ (eds) Fish physiology, vol IV. Academic Press, London New York, pp 109–132

Campbell G (1971a) Autonomic innervation of the lung musculature of a toad (*Bufo marinus*). Comp Gen Pharmacol 2:281–286

Campbell G (1971 b) Autonomic innervation of the pulmonary vascular bed in a toad (*Bufo marinus*). Comp Gen Pharmacol 2:287–294

Campbell G (1975) Inhibitory vagal innervation of the stomach in fish. Comp Biochem Physiol 50C:169–170

Campbell G, Burnstock G (1968) Comparative physiology of gastrointestinal motility. In: Code CF (ed) Handbook of physiology. Am Physiol Soc, Washington DC, pp 2213–2266

Campbell G, Duxson MJ (1978) The sympathetic innervation of lung muscle in the toad *Bufo marinus*: a revision and an explanation. Comp Biochem Physiol 60C:65–73

Campbell G, Gannon BJ (1976) The splanchnic nerve supply to the stomach of the trout, *Salmo trutta* and *S. gairdneri*. Comp Biochem Physiol 55C:51–53

Campbell G, Gibbins IL (1979) Nonadrenergic, noncholinergic transmission in the autonomic nervous system: purinergic nerves. In: Kalsner S (ed) Trends in autonomic pharmacology, vol I. Urban & Schwarzenberg, Baltimore München, pp 103–144

Campbell G, Haller CJ, Rogers DC (1978) Fine structural and cytochemical study of the innervation of smooth muscle in an amphibian (*Bufo marinus*) lung before and after denervation. Cell Tissue Res 194:419–432

Campbell G, Gibbins IL, Morris JL, Furness JB, Costa M, Oliver AM, Beardsley AM, Murphy R (1982) Somatostatin is contained in and released from cholinergic nerves in the heart of the toad *Bufo marinus*. Neuroscience 7:2013–2023

Campbell HS, Smith JL (1962) The pharmacology of the pigeon pupil. Arch Ophthalmol 67:501–504

Cannon WB (1912) Peristalsis, segmentation and the myenteric reflex. Am J Physiol 30:114–128

Cannon WB (1929) Organization for physiological homeostasis. Physiol Rev 9:399–431

Cannon WB, Lieb CW (1911) The receptive relaxation of the stomach. Am J Physiol 29:267–273

Cannon WB, Uridil JE (1921) Studies on the conditions of activity in endocrine glands – VIII. Some effects on the denervated heart of stimulating the nerves of the liver. Am J Physiol 58:353–364

Capra MF, Satchell GH (1977a) The adrenergic responses of isolated saline-perfused prebranchial arteries and gills of the elasmobranch *Squalus acanthias*. Gen Pharmacol 8:67–71

Capra MF, Satchell GH (1977b) The differential haemodynamic responses of the elasmobranch, *Squalus acanthias*, to the naturally occurring catecholamines, adrenaline and noradrenaline. Comp Biochem Physiol 58C:41–47

Caravita S, Coscia L (1966) Les cellules chromaffines du coeur de la lamproie (*Lampetra zanandreai*). Étude au microscope éléctronique avant et après un traitement à la réserpine. Arch Biol 77:723–753

Carlson AJ (1904) Contributions to the physiology of the heart of the California hagfish (*Bdellostoma dombeyi*). Z Allg Physiol 4:259–288

Carlson AJ (1906) The presence of cardioregulative nerves in the lampreys. Am J Physiol 16:230–232

Carlson AJ, Luckhardt AB (1920a) Studies on the visceral sensory nervous system. I. Lung automatism and lung reflexes in the frog (*Rana pipiens* and *Rana catesbeiana*). Am J Physiol 54:55–95

Carlson AJ, Luckhardt AB (1920b) Studies on the visceral sensory nervous system. III. Lung automatism and lung reflexes in reptilia (Turtles: *Chrysemys elegans* and *Malacoclemmys lesueurii*. Snake: *Eutenia elegans*). Am J Physiol 54:261–306

Carlson AJ, Luckhardt AB (1921) Studies on the visceral sensory nervous system. X. The vagus control of the oesophagus. Am J Physiol 57:299–335

Carlsson A (1965) Drugs which block the storage of 5-hydroxytryptamine and related amines. In: Erspamer V (ed) Handbuch der experimentellen Pharmakologie, vol XIX. Springer, Berlin Heidelberg Göttingen, pp 529–592

Carlsson A (1966) Pharmacological depletion of catecholamine stores. Pharmacol Rev 18:541–549

Carlsson A, Falck B, Hillarp N-Å (1962) Cellular localization of brain monoamines. Acta Physiol Scand 56 Suppl 196:1–28

Carlsson E, Hedberg A (1976) Are the cardiac effects of noradrenaline and adrenaline mediated by different β-adrenoceptors? Acta Physiol Scand 98 Suppl 440:47

Carlsson E, Åblad B, Brändström A, Carlsson B (1972) Differentiated blockade of the chronotropic effects of various adrenergic stimuli in the cat heart. Life Sci 11:953–958

Carrier O (1972) Pharmacology of the periferal autonomic nervous system. Year Book Medical Publ, Chicago

Celander O (1954) The range of control exercised by the sympathico-adrenal system. Acta Physiol Scand 32 Suppl 116:1–132

Chan DKO, Chow NYS (1976) The effects of acetylcholine, biogenic amines and other vasoactive agents on the cardiovascular functions of the eel, *Anguilla japonica*. J Exp Biol 96:13–26

Chang MM, Leeman SE, Niall HD (1971) Amino acid sequence of substance P. Nature (London) 232:86–87

Chesher GB, Thorp RH (1965) The atropine resistance of the response to intrinsic nerve stimulation of the guinea-pig bladder. Br J Pharmacol Chemother 25:288–294

Chevrel R (1887–1890) Sur l'anatomie de système nerveux grand sympathique des elasmobranches et des poissons osseux. Arch Zool Exp Gen Suppl 5:1–196

Ciaranello RD (1978) Regulation of phenylthanolamine N-methyltransferase synthesis and degradation. Mol Pharmacol 14:478–489

Clark AJ (1926) The antagonism of acetylcholine by atropine. J Physiol (London) 61:547–556

Clark AJ (1937) General pharmacology. In: Heubner W, Schüller J (eds) Handbuch der experimentellen Pharmakologie, Bd IV, Springer, Berlin

Cobb JLS, Santer RM (1973) Electrophysiology of cardiac function in teleosts: cholinergically mediated inhibition and rebound excitation. J Physiol (London) 230:561–574

Cocks T, Burnstock G (1979) Effects of neuronal polypeptides on intestinal smooth muscle; a comparison with non-adrenergic, non-cholinergic nerve stimulation and ATP. Eur J Pharmacol 54:251–259

Cohen DH, Schnall AM, MacDonald RL, Pitts LH (1970) Medullary cells of origin of vagal cardioinhibitory fibres in the pigeon. J Comp Neurol 140:299–320

Cohen ML, Rosing E, Wiley KS, Slater IH (1978) Somatostatin inhibits adrenergic and cholinergic neurotransmission in smooth muscle. Life Sci 23:1659–1664 Coleman RA (1980) Purine antagonists in the identification of adenosinereceptors in guinea-pig trachea and the role of purines in non-adrenergic inhibitory neurotransmission. Br J Pharmacol 69:359–366

Comline RS (1946) Synthesis of acetylcholine by non-nervous tissue. J Physiol (London) 105:6P

Consolazione A, Milstein C, Wright B, Cuello AC (1981) Immunocytochemical detection of serotonin with monoclonal antibodies. J Histochem Cytochem 29:1425–1430

Conte del E (1977) Contiguity of the adrenaline-storing chromaffin cells with the interrenal tissue in the adrenal gland of a lizard. Gen Comp Endocrinol 32:1–6

Contejean CH (1892) Action des nerfs pneumogastrique et grand sympathique sur l'estomac chez les Batraciens. Arch Physiol Norm Pathol 4:640–650

Cook RD, Burnstock G (1976a) The ultrastructure of Auerbach's plexus in the guinea-pig. I. Neuronal elements. J Neurocytol 5:171–194

Cook RD, Burnstock G (1976b) The ultrastructure of Auerbach's plexus in the guinea-pig. II. Non-neuronal elements. J Neurocytol 5:195–206

Cook RD, King AS (1970) Observations on the ultrastructure of the smooth muscle and its innervation in the avian lung. J Anat 106:273–284

Cooke IRC, Campbell G (1980) The vascular anatomy of the gills of the smooth toadfish *Torquiginer glaber* (Teleostei: Tetraodontidae). Zoomorphologie 94:151–166

Corrodi H, Jonsson G (1967) The formaldehyde fluorescence method for the histochemical demonstration of biogenic monoamines. A review on the methodology. J Histochem Cytochem 15:65–78

Costa M, Furness JB (1976) The peristaltic reflex: An analysis of the nerve pathways and their pharmacology. Naunyn Schmiedebergs Arch Pharmakol 294:47–60

Costa M, Furness JB (1979a) The sites of action of 5-hydroxytryptamine in nerve-muscle preparations from guinea-pig small intestine and colon. Br J Pharmacol 65:237–248

Costa M, Furness JB (1979b) On the possibility that an indoleamine is a neurotransmitter in the gastrointestinal tract. Biochem Pharmacol 28:565–571

Costa M, Furness JB, Gabella G (1971) Catecholamine containing nerve cells in the mammalian myenteric plexus. Histochemie 25:103–106

Costa M, Furness JB, McLean JR (1976) The presence of aromatic L-amino acid decarboxylase in certain intestinal nerve cells. Histochemistry 48:129–143

Costa M, Cuello AC, Furness JB, Franco R (1980a) Distribution of enteric neurones showing immunoreactivity for substance P in the guinea-pig ileum. Neuroscience 5:323–332

Costa M, Furness JB, Buffa R, Said SI (1980b) Distribution of enteric nerve cell bodies and axons showing immunoreactivity for vasoactive intestinal polypeptide in the guinea-pig intestine. Neuroscience 5:587–596

Costa M, Furness JB, Llewellyn-Smith IJ, Cuello AC (1981) Projection of substance P-containing neurons within the guinea-pig small intestine. Neuroscience 6:411–424

Cotten de v M (1972) Regulation of catecholamine metabolism in the sympathetic nervous tissue. Pharmacol Rev 24:161–434

Coupland RE (1963) The innervation of the rat adrenal medulla. An electron microscopic study. J Anat 97:141–142

Coupland RE (1965) The natural history of the chromaffin cell. Longmans, London

Coupland RE (1971) Observations on the form and size distribution of chromaffin granules and on the identity of adrenaline- and noradrenaline-storing chromaffin cells in vertebrates and man. Mem Soc Endocrinol 19:611–635

Coupland RE (1972) The chromaffin system. In: Blaschko H, Muscholl E (eds) Handbook of experimental pharmacology, vol 33. Catecholamines, Springer, Berlin Heidelberg New York, pp 16–45

Coupland RE, Fujita T (1976) Chromaffin, enterochromaffin and related cells. Elsevier, Amsterdam

Coupland RE, Holmes RL (1957) The use of cholinesterase techniques for the demonstration of peripheral nervous structures. Q J Microsc Sci 98:327–330

Coupland RE, Holmes RL (1958) The distribution of cholinesterase in the adrenal glands of the rat, cat and rabbit. J Physiol (London) 141:97–106

Coupland RE, Hopwood D (1966) The mechanism of the differential staining reaction for adrenaline- and noradrenaline storage granules in tissues fixed in glutaraldehyde. J Anat 100:227–243

Coupland RE, Pyper AS, Hopwood D (1964) A method of differentiating between noradrenaline-storing cells in the light and electron microscope. Nature (London) 201:1240–1242

Couraud JY, Giamberardino L Di (1980) Axonal transport of the molecular forms of acetylcholinesterase in chick sciatic nerve. J Neurochem 35:1053–1066

Couteaux R, Laurent P (1957) Etude au microscope éléctronique du coeur de l'Anguille: observations sur la structure du tissue musculaire de l'oreillette et son innervation. CR Acad Sci 245:2097–2100

Couvreur E (1889) Influence de l'excitation du pneumogastrique sur la circulation pulmonaire de la grenouille. CR Acad Sci 109:823–825

Cowie AF, King AS (1969) Further observations on the bronchial muscle of birds. J Anat 104:177–178

Dahl E, Ehinger B, Falck B, Mecklenburg Cv, Myhrberg H, Rosengren E (1971) On the monoamine-storing cells in the heart of *Lampetra fluvialitis* and *L. planeri* (Cyclostomata). Gen Comp Endocrinol 17:241–246

Dahlén SE, Hedqvist P (1980) ATP, β-γ-methylene-ATP and adenosine inhibit non-cholinergic non-adrenergic transmission in rat urinary bladder. Acta Physiol Scand 109:137–142

Dahlöf C (1981) Studies on β-adrenoceptor mediated facilitation of sympathetic neurotransmission. Thesis, Univ Göteborg, Sweden

Dahlöf C, Åblad B, Borg KO, Ek L, Waldbeck B (1975) Prejunctional inhibition of adrenergic nervous vasomotor control due to β-receptor blockade. In: Almgren O, Carlsson A, Engel J (eds) Chemical tools in catecholamine research, vol II. North-Holland, Amsterdam, pp 201–210

Dahlöf C, Ljung B, Åblad B (1978a) Increased noradrenaline release in rat portal vein during sympathetic nerve stimulation due to activation of presynaptic β-adrenoceptors by noradrenaline and adrenaline. Eur J Pharmacol 50:75–78

Dahlöf C, Ljung B, Åblad B (1978b) Relative potency of β-adrenoceptor agonists on neuronal transmitter release in isolated rat portal vein. In: Szabadi E, Bradshaw CM, Bevan P (eds) Recent advances in the pharmacology of adrenoceptors. Elsevier/North-Holland Biomedical Press, Amsterdam New York, pp 355–356

Dahlström A (1965) Observations on the accumulation of noradrenaline in the proximal and distal parts of peripheral adrenergic nerves after compression. J Anat 99:677–689

Dahlström A (1971) Axoplasmic transport (with particular respect to adrenergic neurons). Philos Trans R Soc London Ser B 261:325–358

Dahlström A, Fuxe K (1964) A method for the demonstration of adrenergic nerve fibres in peripheral nerves. Z Zellforsch 62:602–607

Dahlström A, Häggendal J (1966) Studies on the transport and life-span of amine storage granules in a peripheral adrenergic neuron system. Acta Physiol Scand 67:278–288

Dahlström A, Häggendal J (1967) Studies on the transport and life-span of amine storage granules in the adrenergic neuron system of the rabbit sciatic nerve. Acta Physiol Scand 69:153–157

Dahlström AB, Zetterström BEM (1965) Noradrenaline stores in nerve terminals of the spleen. Changes during hemorrhargic shock. Science 147:1583–1585

Dahlström A, Fuxe K, Hillarp NÅ (1965) The site of action of reserpine. Acta Pharmacol Toxicol 22:277–292

Dahlström A, Häggendal J, Heilbronn E, Heiwall P-O, Saunders NR (1974) Proximodistal transport of acetylcholine in peripheral cholinergic neurons. In: Fuxe K, Olson L, Zotterman Y (eds) Dynamics of degeneration and growth in neurons. Pergamon Press, Oxford New York, pp 275–289

Dahlström A, Bööj S, Carlsson SS, Larsson P-A (1981) Rapid accumulation and axonal transport of "cholinergic vesicles" in rat sciatic nerve, studied by immunohistochemistry. Acta Physiol Scand 111: 217–219

Dale HH (1906) On some physiological actions of ergot. J Physiol (London) 34:163–206

Dale HH (1914) The action of certain esters and ethers of choline and their relation to muscarine. J Pharmacol Exp Ther 6:147–190

Dale HH (1933) Nomenclature of fibres in the autonomic system and their effects. J Physiol (London) 80:10–11P

Dale HH (1937a) The William Henry Welch lectures. Acetylcholine as a chemical transmitter of the effects of nerve impulses. I. History of ideas and evidence. Peripheral autonomic actions. Functional nomenclature of nerve fibres. J Mt Sinai Hosp NY 4:401–415

Dale HH (1937b) The William Henry Welch lectures. Acetylcholine as a chemical transmitter of the effects of nerve impulses. II. Chemical transmission at ganglionic synapses and voluntary motor nerve endings. Some general considerations. J Mt Sinai Hosp NY 4:416–429

Dale, HH (1954) The beginnings and prospects of neurohumoral transmission. Pharmacol Rev 6:7–13

Daly MJ, Levy GP (1979) The subclassification of β-adrenoceptors: Evidence in support of the dual β-adrenoceptor hypothesis. In: Kalsner S (ed) Trends in autonomic pharmacology. Urban & Schwarzenberg, Baltimore München, pp 347–385

Daly de Burgh M, Scott MJ (1961) The effects of acetylcholine on the volume and vascular resistance of the dog's spleen. J Physiol (London) 156:246–259

Davey MJ (1980) Relevant features of the pharmacology of prazosin. J Cardiovasc Pharmacol 2 Suppl 3: S287–S301

Davie PS (1981) Vascular responses of an eel tail preparation: alpha constriction and beta dilation. J Exp Biol 90:65–84

Davies BN, Withrington PG (1973) The actions of drugs on the smooth muscle of the capsule and blood vessels of the spleen. Pharm Rev 25:373–414

Davies BN, Gamble J, Withrington PG (1973) Frequency-dependent differences in the responses of the capsular and vascular smooth muscle of the spleen of the dog to sympathetic nerve stimulation. J Physiol (London) 228:13–25

Davies DT, Rankin JC (1973) Adrenergic receptors and vascular responses to catecholamines of perfused dogfish gills. Comp Gen Pharmacol 4:139–147

Daxboeck C, Holeton GF (1978) Oxygen receptors in the rainbow trout, *Salmo gairdneri*. Can J Zool 56:1254–1259

Day MD (1979) Autonomic Pharmacology (Experimental and clinical aspects). Churchill-Livingstone, Edinburgh London New York

Dean DM, Downie JW (1978) Interaction of prostaglandin and adenosine-5′-triphosphate in the non-cholinergic neurotransmission in rabbit detrusor. Prostaglandins 16:245–251

Deineka D (1905) Zur Frage über den Bau der Schwimmblase. Z Wiss Zool 78:149–164

Delaage MA, Puizillout JJ (1981) Radioimmunoassays for serotonin and 5-hydroxyindole acetic acid. J Physiol (Paris) 77:339–347

Delbro D (1981) Gastric excitatory motor responses conveyed by collaterals of thin afferent fibres in the vagal and splanchnic nerves. An experimental study in cat. Thesis, Univ Göteborg, Sweden

Denton EJ, Liddicoat JD, Taylor DW (1972) The permeability to gases of the swimbladder of the conger eel (*Conger conger*). J Mar Biol Assoc UK 52:727–746

Dittus P (1936) Interrenalsystem und chromaffine Zellen in Lebensablauf von *Ichtyophis glutinosus*. Z Wiss Zool 167:459–512

Dixon WE (1902) The innervation of the frog's stomach. J Physiol (London) 28:55–75

Dixon WE (1906) Vagus inhibition. Br Med J 2:180–181

Dixon WE (1907) On the mode of action of drugs. Med Mag (London) 16:454–457

Dockray GJ, Vaillant C, Dimaline R, Hutchison JB, Gregory RA (1979) Characterization of molecular forms of cholecystokinin, vasoactive intestinal polypeptide and bombesin like immunoreactivity in nerves and endocrine cells. In: Rosselin G, Fromageot P, Bonfils S (eds) Hormone receptors in digestion and nutrition. Elsevier-North Holland, Amsterdam, pp 501–511

Dorr LD, Brody MJ (1967) Hemodynamic mechanisms of erection in the canine penis. Am J Physiol 213:1526–1531

Douglas WW (1968) Stimulus secretion coupling: the concept and clues from chromaffin and other cells. Br J Pharmacol 34:451–474

Douglas WW (1970) Histamine and antihistamines: 5-hydroxytryptamine and antagonists. In: Goodman LS, Gilman A (eds) The pharmacological basis of therapeutics. Macmillan, London Toronto, pp 621–662

Douglas WW (1975) Secretomotor control of adrenal medullary secretion: synaptic, membrane and ionic events in stimulus secretion coupling. In: Blaschko H (ed) Handbook of physiology, sect 7, vol VI. Am Physiol Soc, Washington DC, pp 367–388

Downie JW (1981) The autonomic pharmacology of the urinary bladder and urethra: a neglected area. Trends Pharmacol Sci 2:163–165

Dreser H (1892) Notiz über eine Wirkung des Pilokarpins. Arch Exp Pathol Pharmakol 30:159–160

Dreyer NB (1949) The action of autonomic drugs on elasmobranch and teleost involuntary muscle. Arch Int Pharmacodyn Ther 78:63–66

Dreyfus CF, Bornstein MB, Gershon MD (1977) Synthesis of serotonin by neurons of the myenteric plexus in situ and in organotypic tissue culture. Brain Res 128:125–139

DuBois-Reymond EH (1877) Gesammelte Abhandlung der allgemeinen Muskel- und Nervenphysik. Reimer, Berlin

Dumsday B (1971) Atropine-resistance of the urinary bladder. J Pharm Pharmacol 23:222–225

Dunel S, Laurent P (1977) La vascularisation branchial chez l'Anguille: actione de l'acetylcholine et de l'adrénaline sur la répartition d'une résine polymérisable dans les différents compartiments vasculaires. CR Acad Sci Ser D 284:2011–2014

Eckert R, Randall D (1978) Animal physiology. Freeman, San Francisco

Edin R, Lundberg JM, Lidberg P, Dahlström A, Ahlman H (1980) Atropine sensitive contractile motor effects of substance P on the feline pylorus and stomach in vivo. Acta Physiol Scand 110:207–209

Edström A, Hanson M (1973) Temperature effects on fast axonal transport of protein in vitro in frog sciatic nerves. Brain Res 58:345–354

Edwards DJ (1972a) Electrical stimulation of isolated vagus nerve-muscle preparations of the stomach of the plaice *Pleuronectes platessa* L. Comp Gen Pharmacol 3:235–242

Edwards DJ (1972b) Reactions of the isolated plaice stomach to applied drugs. Comp Gen Pharmacol 3:235–242

Ehinger E (1967) Adrenergic nerves in the avian eye and ciliary ganglion. Z Zellforsch 82:577–588

Ehinger B, Falck B (1965) Noradrenaline and cholinesterase in concomitant nerve fibres in the rat iris. Life Sci 4:2097–2100

Ehinger B, Falck B (1966) Concomitant adrenergic and parasympathetic fibres in the rat iris. Acta Physiol Scand 67:201–207

Ehinger B, Falck B, Persson H, Rosengren A-M, Rosengren E (1966) Choline acetylase activity in the normal and denervated cat iris. Life Sci 5:481–483

Ehinger B, Falck B, Persson H (1968) Function of cholinergic nerve fibres in the cat iris dilator. Acta Physiol Scand 72:139–147

Ehinger B, Falck B, Sporrong B (1970) Possible axo-axonal synapses between peripheral adrenergic and cholinergic nerve terminals. Z Zellforsch Mikrosk Anat 107:508–521

Ekelund M, Ahrén B, Håkanson R, Lundquist I, Sundler F (1980) Quinacrine accumulates in certain peptide hormone-producing cells. Histochemistry 66:1–9

Ekström J, Elmer M (1977) Choline acetyltransferase activity in denervated urinary bladder of rat. Acta Physiol Scand 101:58–62

Elfvin L-G (1968) A new granule containing nerve cell in the inferior mesenteric ganglion of the rabbit. J Ultrastruct Res 22:37–44

Elfvin L-G, Hökfelt T, Goldstein M (1975) Fluorescence microscopical immunohistochemical and ultrastructural studies on sympathetic ganglia of the guinea-pig, with special reference to the SIF cells and their catecholamine content. J Ultrastruct Res 51:377–396

Elliott TR (1904) On the action of adrenaline. J Physiol (London) 31:20–21

Elliott TR (1905) The action of adrenaline. J Physiol (London) 32:401–467
Elliott TR (1907) The innervation of the bladder and the urethra. J Physiol (London) 35:367–445
Elliott TR (1913a) The innervation of the adrenal glands. J Physiol (London) 46:285–290
Elliott TR (1913b) Note on the quantitative estimation of adrenaline. J Physiol (London) 46:15–17P
Enami M (1939) Rôle de la sécrétion hypophysaire sur le changement de coloration chez un poisson-chat, *Parasilurus asotus* (L) CR Soc Biol 130:1498–1501
Enemar A, Falck B, Håkanson R (1965) Observations on the appearance of norepinephrine in the sympathetic nervous system of the chick embryo. Dev Biol 11:268–283
Engberg G, Svensson TH, Rosell S, Folkers K (1981) A synthetic peptide as an antagonist of substance P. Nature (London) 293:222–223
Epstein ML, Gershon MD (1978) Development of monoaminergic neurons in the enteric nervous system of the chick embryo. Neurosci Abstr 4:291
Eränkö O (1952) On the histochemistry of the adrenal medulla of the rat, with special reference to acid phosphatase. Acta Anat 16 Suppl 17:308
Eränkö O (1955) Distribution of adrenaline and noradrenaline in the adrenal medulla. Nature (London) 8:88–89
Eränkö O (1957) Distribution of adrenaline and noradrenaline in the hen adrenal gland. Nature (London) 179:417–418
Eränkö O (1976) SIF cells, chromaffin cells, granule-containing cells, and interneurons. In: Eränkö O (ed) SIF cells. Fogarty international center proceedings No 30, US Government Printing Office, Washington DC, pp 1–7
Eränkö O, Härkönen M (1963) Histochemical demonstration of fluorogenic amines in the cytoplasm of sympathetic ganglion cells of the rat. Acta Physiol Scand 58:285–286
Eränkö O, Härkönen M (1965) Monoamine-containing small cells in the superior cervical ganglion of the rat and an organ composed of them. Acta Physiol Scand 63:511–512
Eränkö O, Rechardt L, Eränkö L, Cunningham (1970) Light and electron microscopic histochemical observation on cholinesterase-containing sympathetic nerve fibres in the pineal body of the rat. Histochem J 2:479–489
Erlij D, Centrangolo R, Valadez R (1965) Adrenotropic receptors in the frog. J Pharmacol Exp Ther 149:65–70
Erspamer V (1954) Pharmacology of indolealkylamines. Pharmacol Rev 6:425–487
Erspamer V (1966) Occurence of indolealkylamines in nature. In: Erspamer V (ed) Handbook of experimental pharmacology, vol XIX. 5-Hydroxytryptamine and related indolealkylamines. Springer, Berlin Heidelberg New York, pp 132–181
Euler von US (1946) A specific sympathomimetic ergone in adrenergic nerve fibres (sympathin) and its relation to adrenaline and noradrenaline. Acta Physiol Scand 12:73–97
Euler von US (1956) Noradrenaline. Springfield, Illinois
Euler von US (1972) Synthesis, uptake and storage of catecholamines in adrenergic nerves. The effect of drugs In: Blaschko H, Muscholl E (eds) Catecholamines. Springer, Berlin Heidelberg New York, pp 186–230
Euler von US, Fänge R (1961) Catecholamines in nerves and organs of *Myxine glutinosa, Squalus acanthias* and *Gadus callarias*. Gen Comp Endocrinol 1:191–194
Euler von US, Gaddum JH (1931) An unidentified depressor substance in certain tissue extracts. J Physiol (London) 72:74–87
Euler von US, Lishajko F (1963) Effects of reserpine on the uptake of catecholamines in isolated nerve storage granules. Int J Neuropharmacol 2:127–134
Euler von US, Östlund E (1956) Occurence of a substance P-like polypeptide in fish intestine and brain. Br J Pharmacol 11:323–325
Euler von US, Östlund E (1957) Effects of certain biologically occurring substances on the isolated intestine of fish. Acta Physiol Scand 38:364–372
Evans B, Gannon BJ, Heath JW, Burnstock G (1972) Long-lasting damage to the internal male genital organs and their adrenergic innervation in rats following chronic treatment with the antihypertensive drug guanethidine. Fertil Steril 23:657–667
Evans MH (1972) Tetrodotoxin, saxitoxin, and related substances: their applications in neurobiology. Int Rev Neurobiol 15:83–166

Everett SD (1966) Pharmacological responses of the isolated oesophagus and crop of the chick. In: Horton-Smith C, Amoroso EC (eds) Physiology of the domestic fowl. Oliver & Boyd, Edinburgh, pp 261–273

Everett SD (1968) Pharmacological responses of the isolated innervated intestine and rectal caecum of the chick. Br J Pharmacol Chemother 33:342–356

Fahlén G (1967) Morphological aspects on the hydrostatic function of the gas bladder in *Clupea harengus*. Acta Univ Lund Sect 2 1:1–49

Fahlén G (1971) The functional morphology of the gas bladder of the genus *Salmo*. Acta Anat 78:161–184

Fahlén G, Falck B, Rosengren E (1965) Monoamines in the swimbladder of *Gadus callarias* and *Salmo irideus*. Acta Physiol Scand 64:119–126

Fahrenkrug J (1979) Vasoactive intestinal polypeptide: Measurement, distribution and putative neurotransmitter function. Digestion 19:149–169

Fahrenkrug J, Galbo H, Holst JJ, Schaffalitzky de Muckadell OB (1978a) Influence of the autonomic nervous system on the release of vasoactive intestinal polypeptide from the porcine gastrointestinal tract. J Physiol (London) 280:405–422

Fahrenkrug J, Haglund U, Jodal M, Lundgren O, Olbe L, Schaffalitzky de Muckadell OB (1978b) Nervous release of vasoactive intestinal polypeptide in the gastrointestinal tract of cats: Possible physiological implications. J Physiol (London) 284:291–305

Falck B (1962) Observations on the possibilities of the cellular localization of monoamines by a fluorescence method. Acta Physiol Scand 56 Suppl 197:1–25

Falck B, Owman Ch (1965) A detailed methodological description of the fluorescence method for the cellular demonstration of biogenic monoamines. Acta Univ Lund Sect 2 7:1–23

Falck B, Hillarp NÅ, Thieme G, Torp A (1962) Fluorescence of catecholamines and related compounds condensed with formaldehyde. J Histochem Cytochem 10:348–354

Falck B, Häggendal J, Owman Ch (1963) The localization of adrenaline and adrenergic nerves in the frog. QJ Exp Physiol 48:253–257

Falck B, Mecklenburg Cv, Myhrberg H, Persson H (1966) Studies on adrenergic and cholinergic receptors in the isolated hearts of *Lampetra fluviatilis* (Cyclostomata) and *Pleuronectes platessa* (Teleostei). Acta Physiol Scand 68:64–71

Falck B, Müntzing J, Rosengren AM (1969) Adrenergic nerves to the dermal melanophores of the rainbow trout, *Salmo gairdneri*. Z Zellforsch Mikrosk Anat 99:430–434

Fänge R (1948) Effects of drugs on the intestine of a vertebrate without sympathetic nervous system. Ark Zool 40A:1–9

Fänge R (1953) The mechanism of gas transport in the euphysoclist swimbladder. Acta Physiol Scand 30 Suppl 110:1–133

Fänge R (1966) Physiology of the swimbladder. Physiol Rev 46:299–322

Fänge R (1973) The physiology of the swimbladder. In: Bolis L, Schmidt-Nielsen K, Maddrell SHP (eds) Comparative physiology. North Holland, Amsterdam, pp 135–159

Fänge R (1976) Gas exchange in the swimbladder. In: Hughes GM (ed) Respiration in amphibious vertebrates. Academic Press, London New York, pp 189–211

Fänge R, Grove D (1979) Digestion. In: Hoar WS, Randall DJ (eds) Fish physiology. Academic Press, London New York, pp 161–260

Fänge R, Hanson A (1973) Comparative pharmacology of catecholamines. In: Michelson MJ (ed) Int Encycl Pharmacol Ther Sect 85. Pergamon Press, Oxford New York, pp 391–517

Fänge R, Holmgren S (1982) Choline acetyltransferase activity in the fish swimbladder. J Comp Physiol 146:57–61

Fänge R, Johnels AG (1958) An autonomic nerve plexus control of the gall bladder in *Myxine*. Acta Zool (Stockholm) 39:1–8

Fänge R, Östlund E (1954) The effects of adrenaline, noradrenaline, tyramine and other drugs on the isolated heart from marine vertebrates and a cephalopod (*Eledone cirrosa*). Acta Zool (Stockholm) 35:289–305

Fänge R, Sundell G (1969) Lymphomyeloid tissues, blood cells and plasma proteins in *Chimaera monstrosa* (Pisces, Holocephali). Acta Zool (Stockholm) 50:155–168

Fänge R, Johnels AG, Enger PS (1963) The autonomic nervous system. In: Brodal A, Fänge R (eds) The biology of myxine. Univ Forlaget, Oslo, pp 124–136

Fänge R, Holmgren S, Nilsson S (1976) Autonomic nerve control of the swimbladder of the goldsinny wrasse, *Ctenolabrus rupestris*. Acta Physiol Scand 97:292–303

Fara JW, Rubinstein EH, Sonnenschein RR (1972) Intestinal hormones in mesenteric vasodilatation after intraduodenal agents. Am J Physiol 223:1058–1067

Fasth S, Hultén L (1973) The effect of bradykinin on the consecutive vascular sections of the small and large intestine. Acta Chir Scand 139:707–715

Fasth S, Hultén L, Lundgren O, Nordgren S (1977) Vascular responses to mechanical stimulation of the mucosa of the cat colon. Acta Physiol Scand 101:98–104

Feldberg W, Toh CC (1953) Distribution of 5-hydroxytryptamine (serotonin, enteramine) in the wall of the digestive tract. J Physiol (London) 119:352–362

Fernando MM, Grove DJ (1974)a) Melanophore aggregation in the plaice (*Pleuronectes platessa* L.) – I. Changes in in vivo sensitivity to sympathomimetic amines. Comp Biochem Physiol 48A:711–721

Fernando MM, Grove DJ (1974b) Melanophore aggregation in the plaice (*Pleuronectes platessa* L.). – II. In vitro effects of adrenergic drugs. Comp Biochem Physiol 48A:723–732

Ferry CB (1966) Cholinergic link hypothesis in adrenergic neuroeffector transmission. Physiol Rev 46:420–456

Fillenz M (1970) Innervation of the cat spleen. Proc Soc London Ser B 174:459–468

Finch L, Haeusler G, Kuhn H, Thoenen H (1973) Rapid recovery of vascular adrenergic nerves in the rat after chemical sympathectomy with 6-hydroxydopamine. Br J Pharmacol 48:59–72

Fingerman M (1963) The control of chromatophores. Pergamon Press, Oxford New York

Flavahan NA, McGrath JC (1981) α_1 adrenoceptors can mediate chronotropic responses in the rat heart. Br J Pharmacol 73:586–588

Folkers K, Hörig J, Rosell S, Björkroth U (1981) Chemical design of antagonists of substance P. Acta Physiol Scand 111:505–506

Folkow B (1952) Impulse frequency in sympathetic vasomotor fibres correlated to the release and elimination of the transmitters. Acta Physiol Scand 25:49–76

Folkow B, Neil E (1971) Circulation. Oxford Univ Press, London Toronto

Folkow B, Uvnäs B (1948) The chemical transmission of vasoconstrictor impulses to the hind limbs and splanchnic region of the cat. Acta Physiol Scand 15:365–388

Folkow B, Yonce LR (1967) The negative inotropic effect of vagal stimulation on the heart ventricles of the duck. Acta Physiol Scand 71:77–84

Folkow B, Haeger K, Uvnäs B (1948) Cholinergic vasodilator nerves in the sympathetic outflow to the muscles of the hind limb of the cat. Acta Physiol Scand 15:401–411

Folkow B, Fuxe K, Sonnenschein RR (1966) Responses of skeletal musculature and its vasculature during "diving" in the duck. Peculiarities of the adrenergic vasoconstrictor innervation. Acta Physiol Scand 67:327–342

Folkow B, Nilsson NJ, Yonce LR (1967) Effects of "diving" on cardiac output in ducks. Acta Physiol Scand 70:347–361

Folkow B, Häggendal J, Lisander B (1968) Extent of release and elimination of noradrenaline at peripheral adrenergic nerve terminals. Acta Physiol Scand 72 suppl 307

Fontaine-Perus J, Chanconie M, Polak JM, LeDouarin NM (1981) Origin and development of VIP and substance P containing neurons in the embryonic avian gut. Histochemistry 71:313–324

Forster ME (1976a) Effects of catecholamines on the heart and on branchial and peripheral resistances of the eel, *Anguilla anguilla* (L.) Comp Biochem Physiol 55C:27–32

Forster ME (1976b) Effects of adrenergic blocking drugs on the cardiovascular system of the eel, *Anguilla anguilla* (L.) Comp Biochem Physiol 55C:33–36

Fouchereau-Peron M, Laburthe M, Besson J, Rosselin G, Le Gal Y (1980) Characterization of the vasoactive intestinal polypeptide (VIP) in the gut of fishes. Comp Biochem Physiol 65A:489–492

Fozard J (1979) Cholinergic mechanisms in adrenergic function. In: Kalsner S (ed) Trends in autonomic pharmacology. Urban & Schwarzenberg, Baltimore München, pp 145–194

Fozard JR, Spedding M, Palfreyman MG, Wagner J, Möhring J, Koch-Weser J (1980) Depression of sympathetic nervous function by DL-α-monofluoromethyldopa, an enzymeactivated, irreversible inhibitor of L-aromatic amino acid decarboxylase. J Cardiovasc Pharmacol 2:229–245

Francis ETB (1934) The anatomy of the salamander. Clarendon, Oxford

Franco R, Costa M, Furness JB (1979a) Evidence for the release of endogenous substances P from intestinal nerves. Naunyn Schmiedebergs Arch Pharmacol 306:185–201

Franco R, Costa M, Furness JB (1979b) Evidence that axons containing substance P in the guinea-pig ileum are of intrinsic origin. Naunyn Schmiedebergs Arch Pharmacol 307:57–63

Fredholm BB, Hedqvist P (1980) Modulation of neurotransmission by purine nucleotides and nucleosides. Biochem Pharmacol 29:1635–1643

Friedman MH (1935) A study of the innervation of the stomach of *Necturus* by means of drugs. Trans R Soc Can 29:175–185

Frisch von K (1911) Beiträge zur Physiologie der Pigmentzellen in der Fischhaut. Pflügers Arch Gesamte Physiol Menschen Tiere 138:319–387

Fujii R (1961) Demonstration of the adrenergic nature of transmission at the junction between melanophore-concentrating nerve and melanophore in bony fish. J Fac Sci Imp Univ Tokyo Sect IV, 9:171–196

Fujii R, Miyashita Y (1976) Receptor mechanisms in fish chromatophores. – III. Neurally controlled melanosome aggregation in a siluroid (*Parasilurus asotus*) is strangely mediated by cholinoceptors. Comp Biochem Physiol 55:43–49

Fujii R, Novales RR (1972) Nervous control of melanosome movements in vertebrate melanophores. In: Riley V (ed) Pigmentation: its genesis and biologic control. Appleton-Century-Crofts, New York, pp 315–326

Furchgott RF (1967) The pharmacological differentiation of adrenergic receptors. Ann NY Acad Sci 139:553–570

Furchgott RF, Jurkiewicz A, Jurkiewicz NH (1973) Antagonism of propranolol to isoproterenol in guinea-pig trachea: some cautionary findings. In: Usdin E, Snyder SH (eds) Frontiers in catecholamine research. Pergamon Press, Oxford New York, pp 295–299

Furness JB, Costa M (1971) Morphology and distribution of intrinsic adrenergic neurones in the proximal colon of the guinea-pig. Z Zellforsch Mikrosk Anat 120:346–363

Furness JB, Costa M (1973) The nervous release and the action of substances which affect intestinal muscle through neither adrenoceptors nor cholinoceptors. Philos Trans R Soc London Ser B 265:123–133

Furness JB, Costa M (1974) The adrenergic innervation of the gastrointestinal tract. Ergeb Physiol Biol Chem Exp Pharmacol 69:1–51

Furness JB, Costa M (1975) The use of glyoxylic acid for the fluorescence histochemical demonstration of peripheral stores of noradrenaline and 5-hydroxytryptamine in whole mounts. Histochemistry 41:335–352

Furness JB, Costa M (1978) Distribution of intrinsic nerve cell bodies and axons which take up aromatic amines and their percursors in the small intestine of the guinea-pig. Cell Tissue Res 188:527–543

Furness JB, Costa M (1979a) Projections of intestinal neurons showing immunoreactivity for vasoactive intestinal polypeptide are consistent with these neurons being the inhibitory neurons. Neurosci Lett 15:199–204

Furness JB, Costa M (1979b) Actions of somatostatin on excitatory and inhibitory nerves in the intestine. Eur J Pharmacol 56:69–74

Furness JB, Costa M (1980) Types of nerves in the enteric nervous system. Neuroscience 5:1–20

Furness JB, Moore J (1970) The adrenergic innervation of the cardio-vascular system of the lizard *Trachysaurus rugosus*. Z Zellforsch 108:150–176

Furness JB, Costa M, Freeman CG (1979) Absence of tyrosine hydroxylase activity and dopamine β-hydroxylase immunoreactivity in intrinsic nerves of guinea-pig ileum. Neuroscience 4:305–310

Furness JB, Costa M, Franco R, Llewellyn-Smith IJ (1980a) Neuronal peptides in the intestine: distribution and possible functions. Adv Biochem Psychopharmacol 22:601–618

Furness JB, Eskay RL, Brownstein MJ, Costa M (1980b) Characterization of somatostatin-like immunoreactivity in intestinal nerves by high pressure liquid chromatography and radioimmunoassay. Neuropeptides 1:97–103

Furshpan EJ, MacLeish PR, O'Lague PH, Potter DD (1976) Chemical transmission between rat sympathetic neurons and cardiac myocytes developing in microcultures: Evidence for cholinergic, adrenergic and dual-function neurons. Proc Natl Acad Sci USA 73:4225–4229

Gabe M (1970) The adrenal. In: Gans C, Parsons TS (eds) Biology of the reptilia. Morphology C, vol III. Academic Press, London New York, pp 263–318

Gabella G (1972) Fine structure of the myenteric plexus in the guinea-pig ileum. J Anat 111:69–97

Gabella G (1976) Structure of the autonomic nervous system. Chapman & Hall, London

Gabella G (1979) Innervation of the gastrointestinal tract. Int Rev Cytol 59:129–193

Gabella G (1981) Ultrastructure of the nerve plexuses of the mammalian intestine: The enteric glial cells. Neuroscience 6:425–436

Gaddum JH, Picarelli ZP (1957) Two kinds of tryptamine receptor. Br J Pharmacol 12:323–328

Gannon BJ (1971) A study of the dual innervation of teleost heart by a field stimulation technique. Comp Gen Pharmacol 2:175–183

Gannon BJ, Burnstock G (1969) Excitatory adrenergic innervation of the fish heart. Comp Biochem Physiol 29:765–773

Gannon BJ, Campbell GD, Satchell GH (1972) Monoamine storage in relation to cardiac regulation in the Port Jackson shark *Heterodontus portusjacksoni*. Z Zellforsch 131:437–450

Gaskell JF (1912) The distribution and physiological action of the suprarenal medullary tissue in *Petromyzon fluviatilis*. J Physiol (London) 44:59–67

Gaskell WH (1883) On the innervation of the heart, which especial reference to the heart of the tortoise. J Physiol (London) 4:43–127

Gaskell WH (1884) On the augmentor (accelerator) nerves of the heart of coldblooded animals. J Physiol (London) 5:46–48

Gaskell WH (1886) On the structure, distribution and function of the nerves which innervate the visceral and vascular systems. J Physiol (London) 7:1–80

Gaskell WH, Gadow H (1884) On the anatomy of the cardiac nerves in certain coldblooded vertebrates. J Physiol (London) 5:362–372

Geffen LB, Livett BG (1971) Synaptic vesicles in sympathetic neurons. Physiol Rev 51:98–157

Geffen LB, Ostberg A (1969) Distribution of granular vesicles in normal and constricted sympathetic neurons. J Physiol (London) 204:583–592

Geldern von CE (1921) Color changes and structures of the skin of *Anolis carolinensis*. Proc Calif Acad Sci 10:77–117

Gershon MD (1967) Effects of tetrodotoxin on innervated smooth muscle preparations. Br J Pharmacol Chemother 29:259–279

Gershon MD (1970) The identification of neurotransmitters to smooth muscle. In: Bülbring E, Brading A, Jones A, Tomita T (eds) Smooth muscle. Arnold, London, pp 496–524

Gershon MD (1981) The enteric nervous system. Annu Rev Neurosci 4:227–272

Gershon MD, Altman RF (1971) An analysis of the uptake of 5-hydroxytryptamine by the myenteric plexus of the small intestine of the guinea pig. J Pharmacol Exp Ther 179:29–41

Gershon MD, Bursztajn S (1978) Properties of the enteric nervous system: Limitation of access of intravascular macromolecules to the myenteric plexus and muscularis externa. J Comp Neurol 180:467–488

Gershon MD, Jonakait GM (1979) Uptake and release of 5-hydroxytryptamine by enteric serotonergic neurons: Effects of fluoxetine (Lilly 110140) and chlorimipramine. Br J Pharmacol 66:7–9

Gershon MD, Dreyfus CF, Pickel VM, Joh TH, Reis DJ (1977) Serotonergic neurons in the peripheral nervous system: Identification in gut by immunohistochemical localization of tryptophan hydroxylase. Proc Natl Acad Sci USA 74:3086–3089

Gershon MD, Dreyfus CF, Rothman TP (1979) The mammalian enteric nervous system: A third autonomic division. In: Kalsner S (ed) Trends in autonomic pharmacology, vol I Urban & Schwarzenberg, Baltimore München, pp 59–101

Gershon MD, Sherman D, Dreyfus CF (1980) Effects of indolic neurotoxins on enteric serotonergic neurons. J Comp Neurol 190:581–596

Giacomini E (1902a) Contributo alla conoscenza delle capsule surrenali dei Ciclostomi. Sulle capsule surrenali dei Petromizonti. Monit Zool Ital 13:143–162

Giacomini E (1902b) Contributo alla conoscenza delle capsule surrenali dei Ciclostomi. Sulle capsule dei Missinoidi. R Accad Sci Bologna 7:135–140

Giacomini E (1902c) Sulla esistenza della sostanza midollare nelle capsule surrenale del Teleostei. Monit Zool Ital 13:183–189

Giacomini E (1904) Contributo alla conoscenza delle capsule surrenali dei ganoidi e particolarmente sull'esistenza della loro sostanza midollare. Monit Zool Ital 15:19–32

Giacomini E (1906) Sulle capsule surrenali e sul simpatico dei Dipnoi ricerche in *Protopterus annectens*. RC Acad Lincei 15:394–398

Gibbins IL (1982) Lack of correlation between ultrastructural and pharmacological types of non-adrenergic autonomic nerves. Cell Tissue Res 221:551–582

Gillespie JS, Kirpekar SM (1966) The uptake and release of radioactive noradrenaline by the splenic nerves of cats. J Physiol (London) 187:51–68

Glossman H, Hornung R, Presek P (1980) The use of ligand binding for the characterization of α-adrenoceptors. J Cardiovasc Pharmacol 2 Suppl 3:S303–S324

Gold WM, Kessler GF, Yu DYD (1972) Role of vagus nerves in experimental asthma in allergic dogs. J Appl Physiol 33:719–725

Goldberg LI (1972) Cardiovascular and renal actions of dopamine: potential clinical applications. Pharm Rev 24:1–29

Goldman JM, Hadley ME (1969) The beta adrenergic receptor and cyclic 3′,5′-adenosine monophosphate: possible roles in the regulation of melanophore responses of the spadefoot toad, *Scaphiopus couchi*. Gen Comp Endocrinol 13:151–163

Goldstein M (1966) Inhibition of norepinephrine biosynthesis at the dopamine-β-hydroxylase stage. Pharmacol Rev 18:77–82

Gonella J (1978) La motricité digestive et sa régulation nerveuse. J Physiol (Paris) 74:131–140

Gonella J (1981) The physiological role of peripheral serotoninergic neurones. A review. J Physiol (Paris) 77:515–519

Goodall McC (1951) Studies of adrenaline and noradrenaline in mammalian heart and suprarenals. Acta Physiol Scand 24, Suppl 85:1–51

Goodman LS, Gilman A (1970) The pharmacological basis of therapeutics, 4th edn. Macmillan, London

Goodrich JT, Bernd P, Sherman DL, Gershon MD (1980) Phylogeny of enteric serotonergic neurons. J Comp Neurol 190:15–28

Govyrin VA (1977) Development of vasomotor adrenergic innervation in onto- and phylogenesis. J Evol Biochem Physiol 13:614–620

Govyrin VA, Leont'eva GR (1965) Distribution of catecholamines in vertebrate myocardium. J Evol Biochem Physiol 1:38–44

Govyrin VA, Popova DI (1969) On the nature of nervous cells in the intestine of the lamprey *Lampetra fluviatilis*. Biochem Physiol 5:592–593

Greeff K, Kasperat H, Oswald W (1962) Paradoxe Wirkungen der elektrischen Vagusreizung am isolierten Magen und Herzvorhofpräparat des Meerschweinchens sowie deren Beeinflussung durch Ganglienblocker, Sympathicolytica, Reserpin und Cocain. Naunyn Schmiedebergs Arch Exp Pathol Pharmakol 243:528–545

Greene CW (1902) Contributions to the physiology of the California hagfish, *Polistotrema stouti*. II. The absence of regulative nerves for the systemic heart. Am J Physiol 6:318–324

Grillo MA (1966) Electron microscopy of sympathetic tissues. Pharmacol Rev 18:387–399

Grillo MA, Jacobs L, Comroe JH (1974) A combined flourescence histochemical and electron microscopic method for studying special monoamine-containing cells (SIF). J Comp Neurol 153:1–14

Groth H-P (1972) Licht- und fluoreszenzemikroskopische Untersuchungen zur Innervation des Luftsachsystems der Vögel. Z Zellforsch 127:87–115

Grove DJ (1969a) The effects of adrenergic drugs on melanophores of the minnow, *Phoxinus phoxinus* (L.). Comp Biochem Physiol 28:37–54

Grove DJ (1969b) Melanophore dispersion in the minnow, *Phoxinus phoxinus* (L.). Comp Biochem Physiol 28:55–65

Grove DJ, Campbell G (1979a) The role of extrinsic and intrinsic nerves in the co-ordination of gut motility in the stomachless flatfish *Rhombosolea tapirina* and *Ammotretis rostrata* Guenther. Comp Biochem Physiol 63C:143–159

Grove DJ, Campbell G (1979b) Effects of extrinsic nerve stimulation on the stomach of the flathead *Platycephalus bassensis* Cuvier and Valenciennes. Comp Biochem Physiol 63C:373–380

Grove DJ, Starr CR, Allard DR, Davies W (1972) Adrenaline storage in the pronephros of the plaice, *Pleuronectes platessa* L. Comp Gen Pharmacol 3:205–212

Grove DJ, O'Neill JG, Spillett PB (1974) The action of 5-hydroxytryptamine on longitudinal gastric smooth muscle of the plaice, *Pleuronectes platessa*. Comp Gen Pharmacol 5:229–238

Gruenhagen A (1887) Über den Einfluß des Sympathicus auf die Vogelpupille. Pflüger's Arch Gesamte Physiol Menschen Tiere 40:65–67

Grynfeltt E (1904) Notes histologiques sur la capsule surrénale des amphibiens. J Anat Paris 40:180–220

Guillemin R (1976) Somatostatin inhibits the release of acetylcholine induced electrically in the myenteric plexus. Endocrinology 99:1653–1654

Guimarães S (1969) Reversal by pronethalol of dibenamine blockade: a study on the seminal vesicle of guinea-pig. Br J Pharmacol 36:594–601

Gulati OD, Parikh HM, Ragunath PR (1973) Further studies on the alpha-adrenoceptor blocking action of beta-adrenoceptor blocking agents. Eur J Pharmacol 22:196–205

Gunn M (1951) A study of the enteric plexuses in some amphibians. QJ Microsc Sci 92:55–78

Gustafsson LE (1980) Studies on modulation of transmitter release and effector responsiveness in autonomic cholinergic neurotransmission. Acta Physiol Scand 110 Suppl 489:1–28

Guth E (1901) Untersuchungen über die direkte motorische Wirkung des Lichtes auf den sphincter Pupillae des Aal- und Froschauges. Pflügers Arch 85:119–142

Gyermek L (1966) Drugs which antagonize 5-hydroxytryptamine and related indolealkylamines. In: Ersparmer V (ed) Handbook of experimental pharmacology. 5-Hydroxytryptamine and related indolealkylamines, vol XIX. Springer, Berlin Heidelberg New York, pp 471–528

Hagen P (1962) Observations on the substrate specificity of dopa decarboxylase from ox adrenal medulla, human phaechromocytoma and human argentaffinoma. Br J Pharmacol 18:175–182

Häggendal J, Saunders NR, Dahlström A (1971) Rapid accumulation of acetylcholine in nerve above a crush. J Pharm Pharmacol 223:552–555

Hahn HL, Wilson AG, Graf PD, Fischer SP, Nadel JA (1978) Interaction between serotonin and efferent vagus nerves in dog lungs. J Appl Physiol 44:144–149

Hallbäck D-A (1973) Acetylcholinesterase-containing structures in the intestine of *Myxine glutinosa* L. Acta Regiae Soc Sci Litt Gothob Zool 8:24–25

Han GL Van, Emaus TL, Meester WD (1973) Adrenergic receptors in turtle ventricle myocardium. Eur J Pharmacol 24:145–150

Harden Jones FR, Marshall NB (1953) The structure and functions of the teleostean swimbladder. Biol Rev 28:16–83

Harris WS, Morton MJ (1968) A cardiac intrinsic mechanism that relates heart rate to filling pressure. Circulation 38 Suppl 6:95

Hato H (1958a) Effects of the electric stimulation of the peripheral cut end of the rami communicantes and of the splanchnic nerve upon the movements of the frog's stomach. J Physiol Soc Jpn 20:388–396

Hato H (1958b) Effects of the electric stimulation of the peripheral cut end of the vagosympathetic nerves on the movements of the frog's stomach. J Physiol Soc Jpn 20:401–406

Healey EG, Ross DM (1966) The effects of drugs on the background response of the minnow *Phoxinus phoxinus* L. Comp Biochem Physiol 19:545–580

Heath JW, Burnstock G (1977) Selectivity of neuronal degeneration produced by chronic guanethidine treatment. J Neurocytol 6:397–405

Heath JW, Evans BK, Burnstock G (1973) Axon retraction following guanethidine treatment. Studies of sympathetic neurons in vivo. Z Zellforsch 146:439–451

Hebb CO, Ratković D (1962) Choline acetylase in the placenta of man and other species. J Physiol (London) 163:307–313

Hedberg A (1980) Studies on beta-1- and beta-2-adrenoceptor mechanisms in the mammalian heart with special reference to the action of prenalterol. Thesis, Univ Göteborg, Sweden

Hedberg A, Nilsson S (1975) Vago-sympathetic innervation of the heart of the puff-adder, *Bitis arietans*. Comp Biochem Physiol 53C:3–8

Hedberg A, Nilsson S (1976) Inotropic responses to drugs of isolated heart strips from a snake. Acta Physiol Scand 96:35A

Hedberg A, Minneman KP, Molinoff PB (1980) Differential distribution of beta-1 and beta-2 adrenergic receptors in cat and guinea pig heart. J Pharmacol Exp Ther 212:503–508

Hedqvist P (1977) Basic mechanisms of prostaglandin actions on autonomic neurotransmission. Annu Rev Pharmacol 17:259–279

Hedqvist P, Gustafsson L, Hjemdahl P, Svanborg K (1980) Aspects of prostaglandin action on autonomic neuroeffector transmission. Adv Prostaglandin Thromboxane Res 8:1245–1248

Heidenhain R (1872) Über die Wirkung einiger Gifte auf die Nerven der Glandula Submaxillaris. Pflüger's Arch Gesamte Physiol Menschen Tiere 5:309–318

Helgason SS, Nilsson S (1973) Drug effects on pre- and post-branchial blood pressure and heart rate in a free-swimming marine teleost, *Gadus morhua*. Acta Physiol Scand 88:533–540

Henderson VE, Roepke MH (1934) The role of acetylcholine in bladder contractile mechanisms and in parasympathetic ganglia. J Pharmacol Exp Ther 51:97–111

Henle J (1865) Über das Gewebe der Nebenniere und der Hypophysis. Z Rat Mat 24:143–152

Henning M, Johansson P (1981) Central action of L-Dopa in blood pressure regulation and effect on catecholamine content in the frog, *Xenopus laevis*. Comp Biochem Physiol 70C:117–121

Hiatt EP (1943) The action of adrenaline, acetylcholine and potassium in relation to the innervation of the isolated auricle of the spiny dogfish (*Squalus acanthias*). Am J Physiol 139:45–48

Hill CE, Hendry IA (1977) Development of neurons synthesizing noradrenaline and acetylcholine in the superior cervical ganglion of the rat in vivo and in vitro. Neuroscience 2:741–749

Hill CE, Hendry IA, MacLennan IS (1980) Development of cholinergic neurones in cultures of rat superior cervical ganglia. Role of calcium and macromolecules. Neuroscience 5:1027–1032

Hillarp N-Å (1946) Structure of the synapse and the peripheral innervation apparatus of the autonomic nervous system. Acta Anat Suppl 4:1–153

Hirsch EF, Jellinek M, Cooper T (1964) Innervation of the hagfish heart. Circ Res 14:212–217

Hirst GDS, Neild TO (1980) Evidence for two populations of excitatory receptors for noradrenaline on arteriolar smooth muscle. Nature (London) 283:767–768

Hirst GDS, Holman M, McKirdy HC (1975) Two descending nerve pathways activated by distension of guinea-pig small intestine. J Physiol (London) 244:113–127

Hirt A (1921) Der Grenzstrang des Sympathicus bei einigen Sauriern. Z Anat Entwicklungsgesch 62:536–551

Hohnke LA (1975) Regulation of arterial blood pressure in the common green iguana. Am J Physiol 228:386–391

Hökfelt T, Kellert J-O, Nilsson G, Pernow B (1975) Substance P: localization in the central nervous system and in some primary sensory neurons. Science 190:889–890

Hökfelt T, Elfvin LG, Elde R, Schultzberg M, Goldstein M, Luft R (1977a) Occurrence of somatostatin-like immunoreactivity in some peripheral sympathetic noradrenergic neurons. Proc Natl Acad Sci USA 74:3587–3591

Hökfelt T, Johansson O, Kellerth J-O, Ljungdahl A, Nilsson G, Nygårds A, Pernow B (1977b) Immunohistochemical distribution of substance P. In: Euler von US, Pernow B (eds) Substance P. Nobel symposium 37. Raven Press, New York, pp 117–145

Hökfelt T, Johansson O, Ljungdahl Å, Lundberg JM, Schultzberg M (1980a) Peptidergic neurons. Nature (London) 284:515–521

Hökfelt T, Lundberg JM, Schultzberg M, Johansson O, Skirboll L, Änggård A, Fredholm B, Hamberger B, Pernow B, Rehfeld J, Goldstein M (1980b) Cellular localization of peptides in neural structures. Proc R Soc London Ser B 210:63–77

Hökfelt T, Lundberg JM, Schultzberg M, Fahrenkrug J (1981) Immunohistochemical evidence for a local VIP-ergic neuron system in the adrenal gland of the rat. Acta Physiol Scand 113:575–576

Holmes W (1950) The adrenal homologues in the lungfish *Protopterus*. Proc R Soc London Ser B 137:549–565

Holmgren S (1977) Regulation of the heart of a teleost, *Gadus morhua*, by autonomic nerves and circulating catecholamines. Acta Physiol Scand 99:62–74

Holmgren S (1978) Sympathetic innervation of the coeliac artery from a teleost, *Gadus morhua*. Comp Biochem Physiol 60C:27–32

Holmgren S (1981) Choline acetyltransferase activity in the heart from two teleosts, *Gadus morhua* and *Salmo gairdneri*, Comp Biochem Physiol 69C:403–405

Holmgren S (1982) The effects of VIP, substance P and ATP on isolated muscle strips from the stomach of the rainbow trout, *Salmo gairdneri*. Acta Physiol Scand 114:39A

Holmgren S, Campbell G (1978) Adrenoceptors in the lung of the toad, *Bufo marinus*: regional differences in response to amines and to sympathetic nerve stimulation. Comp Biochem Physiol 60C:11–18

Holmgren S, Fänge R (1981) Effects of cholinergic drugs on the intestine and gallbladder of the hagfish, *Myxine glutinosa* L., with a report on the inconsistent effects of catecholamines. Mar Biol Lett 2:265–277

Holmgren S, Nilsson S (1974) Drug effects on isolated artery strips from two teleosts, *Gadus morhua* and *Salmo gairdneri*. Acta Physiol Scand 90:431–437

Holmgren S, Nilsson S (1975) Effects of some adrenergic and cholinergic drugs on isolated spleen strips from the cod, *Gadus morhua*. Eur J Pharmacol 32:163–169

Holmgren S, Nilsson S (1976) Effects of denervation, 6-hydroxydopamine and reserpine on the cholinergic and adrenergic responses of the spleen of the cod, *Gadus morhua*. Eur J Pharmacol 39:53–59

Holmgren S, Nilsson S (1980) Studies on the non-adrenergic, non-cholinergic innervation of the stomach of the rainbow trout, *Salmo gairdneri*, using a "Loewi"-preparation. Acta Physiol Scand 109:19A

Holmgren S, Nilsson S (1981) On the non-adrenergic, non-cholinergic innervation of the rainbow trout stomach. Comp Biochem Physiol 70C:65–69

Holmgren S, Nilsson S (1982) Neuropharmacology of adrenergic neurons in teleost fish. Comp Biochem Physiol 72C:289–302

Holmgren S, Vaillant C, Dimaline R (1982) VIP-, substance P-, gastrin/CCK-, bombesin-, somatostatin-, and glucagon-like immunoreactivities in the gut of the rainbow trout, *Salmo gairdneri*. Cell Tissue Res 223:141–153

Holmstedt B (1957a) A modification of the thiocholine method for the determination of cholinesterases. I. Biochemical evaluation of selective inhibitors. Acta Physiol Scand 40:322–330

Holmstedt B (1957b) A modification of the thiocholine method for the determination of cholinesterases. II. Histochemical application. Acta Physiol Scand 40:331–337

Holtz P (1939) Dopadecarboxylase. Naturwissenschaften 43:724–725

Holtz P, Heise R, Ludtke K (1938) Fermentativer Abbau von L-Dioxyphenylalanin (Dopa) durch Niere. Arch Exp Pathol Pharmakol 191:87–118

Holtzbauer M, Sharman DF (1972) The distribution of catecholamines in vertebrates. In: Blaschko H, Muscholl E (eds) Catecholamines. Handbuch der experimentellen Pharmakologie, vol XXXIII. Springer, Berlin Heidelberg New York, pp 110–185

Holzer P, Emson PC, Iversen LL, Sharman DF (1981) Regional differences in the response to substance P of the longitudinal muscle and the concentration of substance P in the digestive tract of the guinea-pig. Neuroscience 6:1433–1441

Hooper M, Spedding M, Sweetman AJ, Weetman DF (1974) 2-2'-pyridylisatogen tosylate: an antagonist of the inhibitory effects of ATP on smooth muscle. Br J Pharmacol 50:458–459

Hopf H (1911) Über den hemmenden und erregenden Einfluß des Vagus auf den Magen des Frosches. Z Biol 55:409–459

Houssay BA, Wassermann GF, Tramezzani JH (1962) Formation et sécrétion differentielles d'adrénaline et de noradrénaline surrénales. Arch Int Pharmacodyn Ther 140:84–91

Hughes J (1975) Isolation of an endogeneous compound from the brain with pharmacological properties similar to morphine. Brain Res 88:295–308

Huković S, Rand MJ, Vanov S (1965) Observation on the isolated, innervated preparation of rat urinary bladder. Br J Pharmacol Chemother 24:178–188

Hulme EC, Berrie CP, Birdsall NJM, Burgen ASV (1981) Two populations of binding sites for muscarinic antagonists in the rat heart. Eur J Pharmacol 73:137–142

Ignarro LJ, Titus E (1968) The presence of antagonistically acting alpha- and beta-adrenergic receptors in the mouse spleen. J Pharmacol Exp Ther 160:72–80

Innes IR, Nickerson M (1970) Drugs inhibiting the action of acetylcholine on structures innervated by postganglionic parasympathetic nerves (antimuscarine or atropinic drugs). In: Goodman LS, Gilman A (eds) The pharmacological basis of therapeutics. Macmillan, London Toronto, pp 524–548

Ishii K, Ishii K (1978) A reflexogenic area controlling the blood pressure in toad (*Bufo vulgaris formosa*). Jpn J Physiol 28:423–431

Ishii K, Honda K, Ishii K (1966) The function of the carotid labyrinth in the toad. Tohoku J Exp Med 88:103–116

Iske MS (1929) A study of the iris mechanism of the alligator. Anat Rec 44:57–77

Ito Y, Kuriyama H (1971) Nervous control of the motility of the alimentary canal of the silver carp. J Exp Biol 55:469–487

Iversen LL (1967) The uptake and storage of noradrenaline in sympathetic nerves. Cambridge Univ Press, Cambridge

Iversen LL (1974) Uptake mechanism for neurotransmitter amines. Biochem Pharmacol 23:1927–1935

Iversen LL (1979) Criteria for establishing a neurotransmitter. In: Burnstock G, Gershon MD, Hökfelt T, Iversen LL, Kosterlitz HW, Szurszewski JH (eds) Neurosciences research program bulletin, vol 17. Non-adrenergic, non-cholinergic autonomic neurotransmission mechanisms. MIT Press, Cambridge, pp 388–391

Izquierdo JJ (1930) On the influence of the extracardiac nerves upon sino-auricular conduction in the heart of *Scyllium*. J Physiol (London) 69:29–47

Jackson DE, Pelz MD (1918) A contribution to the physiology and pharmacology of Chelonian lungs. J Lab Clin Med 3:344–347
Jacobowitz DM, Laties AM (1968) Direct adrenergic innervation of a teleost melanophore. Anat Rec 163:501–504
Jacobowitz DM, Cooper T, Barner HB (1967) Histochemical and chemical studies of adrenergic and cholinergic nerves in normal and denervated cat hearts. Circ Res 20:289–298
Jager LP, Schevers JAM (1980) A comparison of effects evoked in guinea-pig taenia caecum by purine nucleotides and by "purinergic" nerve stimulation. J Physiol (London) 299:75–83
Jansson G (1969) Extrinsic nervous control of gastric motility. Acta Physiol Scand 75 Suppl 326:1–42
Jarvik ME (1970) Drugs used in the treatment of psychiatric disorders. In: Goodman LS, Gilman A (eds) The pharmacological basis of therapeutics. Macmillan, London Toronto, pp 151–203
Jegorow J (1887) Ueber den Einfluss des Sympathicus auf die Vogelpupille. Pflüger's Arch Gesamte Physiol Menschen Tiere 41:326–348
Jenkin PM (1928) Note on the nervous system of *Lepidosiren paradoxa*. Proc R Soc Edinburgh 48:55–69
Jenkinson DH (1981) Peripheral actions of apamin. Trends Pharmacol Sci 2:318–320
Jenkinson DH, Morton JKM (1967) The role of α- and β-adrenergic receptors in some actions of catecholamines on intestinal smooth muscle. J Physiol (London) 188:387–402
Jensen D (1961) Cardioregulation in an aneural heart. Comp Biochem Physiol 2:181–201
Jensen D (1965) The aneuronal heart of the hagfish. Ann N Y Acad Sci 127:443–458
Jensen D (1969) Intrinsic cardiac rate regulation in the sea lamprey, *Petromyzon marinus* and rainbow trout, *Salmo gairdneri*. Comp Biochem Physiol 30:685–690
Jensen D (1970) Intrinsic cardiac rate regulation in elasmobranchs: the horned shark, *Heterodontus francisci*, and thornback ray, *Platyrhinodis triseriata*. Comp Biochem Physiol 34:289–296
Jessen KR, Saffrey MJ, Van Noorden S, Bloom SR, Polak JM, Burnstock G (1980) Immunohistochemical studies of the enteric nervous system in tissue culture and in situ: Localization of vasoactive intestinal polypeptide. Neuroscience 5:1717–1736
Joh TH, Goldstein M (1973) Isolation and characterization of multiple forms of phenylethanolamine N-methyltransferase. Mol Pharmacol 9:117–129
Johansen K (1960) Circulation in the hagfish, *Myxine glutinosa* L. Biol Bull 118:289–295
Johansen K (1963) Cardiovascular dynamics in the amphibian *Amphiuma tridactylum* Cuvier. Norwegian monographs on medical science, Universitetsforl, Oslo Acta Med Scand Suppl 402
Johansen K (1964) Regional distribution of circulating blood during submersion asphyxia in the duck. Acta Physiol Scand 62:1–9
Johansen K (1971) Comparative physiology: gas exchange and circulation in fishes. Annu Rev Physiol 33:569–612
Johansen K (1982) Blood circulation and the rise of air-breathing (passes and bypasses) In: Taylor CR, Johansen K, Bolis L (eds) A companion to animal physiology. Cambridge University Press, pp 91–105
Johansen K, Aakhus T (1963) Central cardiovascular responses to submersion asphyxia in the duck. Am J Physiol 205:1167–1171
Johansen K, Burggren W (1980) Cardiovascular function in the lower vertebrates. In: Bourne GH (ed) Hearts and heart-like organs, vol I. Academic Press, London New York, pp 61–117
Johansen K, Hanson D (1968) Functional anatomy of the hearts of lungfishes and amphibians. Am Zool 8:191–210
Johansen K, Millard RW (1974) Cold-induced neurogenic vasodilation in skin of the giant fulmar, *Macronectes giganteus*. Am J Physiol 227:1232–1235
Johansen K, Reite OB (1964) Cardiovascular responses to vagal stimulation and cardioacceleration blockade in birds, duck (*Anas boscas*) and seagull (*Larus argentatus*). Comp Biochem Physiol 12:479–488
Johansen K, Reite OB (1967) Effects of acetylcholine and biogenic amines on pulmonary smooth muscle in the African lungfish, *Protopterus aethiopicus*. Acta Physiol Scand 71:248–252
Johansen KK, Reite OB (1968) Influence of acetylcholine and biogenic amines on branchial, pulmonary and systemic vascular resistance in the African lungfish, *Protopterus aethiopicus*. Acta Physiol Scand 74:465–471
Johansen K, Krog J, Reite OB (1964) Autonomic nervous influence on the heart of the hypothermic hibernator. Finn Acad Sci A IV:245–255

Johansen K, Franklin DL, VanClitters RL (1966) Aortic blood flow in free swimming elasmobranchs. Comp Biochem Physiol 19:151–166

Johansen K, Lenfant C, Hanson D (1968) Cardiovascular dynamics in the lungfishes. Z Vergl Physiol 59:157–186

Johansen K, Lenfant C, Hanson D (1970) Phylogenetic development of pulmonary circulation. Fed Proc 29:1135–1140

Johansson O, Lundberg JM (1981) Ultrastructural localization of VIP-like immunoreactivity in large dense-core vesicles of "cholinergic-type" nerve terminals in cat exocrine glands. Neuroscience 6:847–862

Johansson P (1979) Antagonistic effects of synthetic alpha adrenoceptor agonists on isolated artery strips from the cod, *Gadus morhua*. Comp Biochem Physiol 63C:267–268

Johansson P, Henning M (1981) The effect of L-Dopa on blood pressure and catecholamine content in the brain and peripheral organs in the Atlantic cod,, *Gadus morhua*. Comp Biochem Physiol 70C:249–253

Johnels AG (1956) On the peripheral autonomic nervous system of the trunk region of *Lampetra planeri*. Acta Zool (Stockholm) 37:251–286

Johnels AG (1957) On the cardiac plexus of the vagus intestinalis nerves in *Myxine*. Acta Zool (Stockholm) 38:283–288

Johnels AG, Östlund E (1958) Anatomical and physiological studies on the enteron of *Lampetra fluviatilis* (L). Acta Zool (Stockholm) 39:9–12

Jonakait GM, Tamir H, Rapport MM, Gershon MD (1977) Detection of a soluble serotonin binding protein in the mammalian myenteric plexus and other peripheral sites of serotonin storage. J Neurochem 28:277–284

Jonakait GM, Tamir H, Gintzler AR, Gershon MD (1979) Release of [^{3}H]-serotonin and its binding protein from enteric neurons. Brain Res 174:55–69

Jones AC (1926) Innervation and nerve terminations of the reptilian lung. J Comp Neurol 40:371–388

Jones DR (1973) Systemic arterial baroreceptors in ducks and the consequence of their denervation on some cardiovascular responses to diving. J Physiol (London) 234:499–518

Jones DR, Randall DJ (1978) The respiratory and circulatory systems during exercise. In: Hoar WS, Randall DJ (eds) Fish physiology, vol VII. Locomotion. Academic Press, London New York, pp 425–501

Jönsson A-C (1982) Dopamine-β-hydroxylase activity in the axillary bodies, the heart and the splanchnic nerve in two elasmobranchs, *Squalus acanthias* and *Etmopterus spinax*. Comp Biochem Physiol 71C:191–194

Jönsson A-C, Nilsson S (1979) Effects of pH, temperature and Cu^{2+} on the activity of dopamine-β-hydroxylase from the chromaffin tissue of the cod, *Gadus morhua*. Comp Biochem Physiol 62C:5–8

Jönsson A-C, Nilsson S (1981) Axonal transport and subcellular distribution of dopamine-β-hydroxylase in the cod, *Gadus morhua*. Acta Physiol Scand 111:441–445

Jönsson A-C, Nilsson S (1982) Activity of tyrosine hydroxylase in chromaffin tissue of the Atlantic cod, *Gadus morhua*. Acta Physiol Scand 111:45A

Jullien A, Ripplinger J (1954) De l'existence de cellules ganglionnaires sur le trajet du pneumogastrique de la grenouille. Ann Sci Univ Besançon 1:79–83

Jullien A, Ripplinger J (1957) Physiologie du coeur des poissons et de son innervation extrinsique. Ann Sci Univ Besançon 9:35–92

Jung MJ, Palfreyman MG, Wagner J, Bey P, Ribereau-Gayon G, Zraïka M, Koch-Weser J (1980) Inhibition of monoamine synthesis by irreversible blockade of aromatic amino acid decarboxylase with α-monofluoromethyldopa. Life Sci 24:1037–1042

Kalsner S (1979) Trends in autonomic pharmacology. Urban & Schwarzenberg, Baltimore-München

KarenKan K-S, Chao LP (1981) Localization of choline acetyltransferase at neuromuscular junctions. Muscle Nerve 4:91–93

Karnovsky MJ, Roots L (1964) A "directcoloring" thiocholine method for cholinesterase. J Histochem Cytochem 12:219–221

Kása P, Mann SP, Hebb C (1970) Localization of choline acetyltransferase. Nature (London) 226:812–814

Katayama Y, North RA (1978) Does substance P mediate slow synaptic excitation within the myenteric plexus? Nature (London) 274:387–388

Katayama Y, North A, Williams JT (1979) The action of substance P on neurones of the myenteric plexus of the guinea-pig small intestine. Proc R Soc London Ser B 206:191–208

Katz B (1966) Nerve, muscle and synapse. McGraw-Hill Book Co, New York

Kaufman S, Friedman S (1965) Dopamine-β-hydroxylase. Pharmacol Rev 17:71–100

Kemp KW, Santer RM, Lever JD, Lu K-S, Presley R (1977) A numerical relationship between chromaffin-positive and small intensely fluorescent cells in sympathetic ganglia. J Anat 124:269–274

Keys A, Bateman JB (1932) Branchial responses to adrenaline and pitressin in the eel. Biol Bull Mar Biol Lab 63:327–338

Khalil F, Malek SPA (1952) Studies on the nervous control of the heart of *Uromastyx aegyptica*. Physiol Comp Oecol 2:386–390

Kilbinger H, Wessler I (1979) Increase by α-adrenolytic drugs of acetylcholine release evoked by field stimulation of the guinea-pig ileum. Naunyn Schmiedebergs Arch Pharmacol 309:255–257

Kilbinger H, Wessler I (1980) Inhibition by acetylcholine of the stimulation-evoked release of [^{3}H]-acetylcholine from guinea-pig myenteric plexus. Neuroscience 5:1331–1340

King AS, Cowie AF (1969) The functional anatomy of the bronchial muscle of the bird. J Anat 105:323–336

Kirby S, Burnstock G (1969a) Comparative pharmacological studies of isolated spiral strips of large arteries from lower vertebrates. Comp Biochem Physiol 28:307–320

Kirby S, Burnstock G (1969b) Pharmacological studies of the cardiovascular system in the anaesthetised sleepy lizard (*Tiliqua rugosa*) and toad (*Bufo marinus*). Comp Biochem Physiol 28:321–332

Kirschner LB (1969) Ventral aortic pressure and sodium fluxes in perfused eel gills. Am J Physiol 217:596–604

Kirshner N, Goodall McC (1957) The formation of adrenaline from noradrenaline. Biochim Biophys Acta 24:658–659

Kirtisinghe P (1940) The myenteric nerve-plexus in some lower chordates. Q J Microsc Sci 81:521–539

Kisch B (1950) Reflex cardiac inhibition in the ganoid *Acipenser sturio*. Am J Physiol 160:552–555

Kissling G, Reutter K, Sieber G, Nyuyen-Duong H, Jacob R (1972) Negative Inotropie von endogenem Acetylcholin beim Katzen- und Hühnerventrikelmyokard. Pflüger's Arch 333:35–50

Kitabgi P, Vincent JP (1981) Neurotensin is a potent inhibitor of guinea-pig colon contractile activity. Eur J Pharmacol 74:311–318

Klaverkamp JF, Dyer DC (1974) Autonomic receptors in isolated rainbow trout. Eur J Pharmacol 28:25–34

Kleinholtz LH (1938a) Studies in reptilian colour changes – II. The pituitary and adrenal glands in the regulation of the melanophores of *Anolis carolinensis*. J Exp Biol 15:474–491

Kleinholtz LH (1938b) Studies in reptilian colour changes – III. Control of the light phase and behaviour of isolated skin. J Exp Biol 15:492–499

Klinge E, Sjöstrand NO (1974) Contraction and relaxation of the retractor penis muscle and the penile artery of the bull. Acta Physiol Scand 93 Suppl 420:1–88

Knowlton FP (1942) An investigation of inhibition by direct stimulation of the turtle's heart. Am J Physiol 135:446–451

Kobayashi S, Serizawa Y, Fujita T, Coupland RE (1978) SGC (small granule chromaffin) cells in the mouse adrenal medulla: Light and electron microscopic identification using semi-thin and ultrathin sections. Endocrinol Jpn 25:467–476

Kobinger W, Pichler L (1981) α_1- and α_2-adrenoceptor subtypes: selectivity of various agonists and relative distribution of receptors as determined in rats. Eur J Pharmacol 73:313–322

Koelle GB (1950) The histochemical differentiation of types of cholinesterase and their localization in tissues of the cat. J Pharmacol Exp Ther 100:158–179

Koelle GB (1970a) Neurohumoral transmission and the autonomic nervous system. In: Goodman LS, Gilman A (eds) The pharmacological basis of therapeutics. Macmillan, London Toronto, pp 402–441

Koelle GB (1970b) Anticholinesterase agents. In: Goodman LS, Gilman A (eds) The pharmacological basis of therapeutics. Macmillan, London Toronto, pp 442–465

Koelle GB (1970c) Parasympathomimetic agents. In: Goodman LS, Gilman A (eds) The pharmacological basis of therapeutics. Macmillan, London Toronto, pp 466–477

Kohn A (1902) Das chromaffin Gewebe. Ergeb Anat Entwicklungsgesch 12:253–348

Kojima H, Ishibashi Y, Anraku S (1976) The fluorescence histochemical studies on cellular localization of catecholamines in bullfrog adrenals. Kurume Med J 23:181–182

Komori S, Ohashi H, Okada T, Takewaki T (1979) Evidence that adrenaline is released from adrenergic neurones in the rectum of the fowl. Br J Pharmacol 65:261–269

Komuro T, Baluk P, Burnstock G (1982) An ultrastructural study of nerve profiles in the myenteric plexus of the rabbit colon. Neuroscience 7:295–305

Kopin IJ (1964) Storage and metabolism of catecholamines: role of monoamine oxidase. Pharmacol Rev 16:171–199

Kopin IJ (1968) False adrenergic transmitters. Annu Rev Pharmacol 8:377–394

Kopin IJ, Axelrod J, Gordon E (1961) The metabolic fate of H^3-epinephrine and C^{14}-metanephrine in the rat. J Biol Chem 236:2109–2113

Kopin IJ, Breese GR, Krauss KR, Weise VK (1968) Selective release of newly synthesized norepinephrine from the cat spleen during nerve stimulation. J Pharmacol Exp Ther 161:271–278

Koppányi T, Sun KH (1926) Comparative studies on pupillary reactions in tetrapods. III. The reactions of the avian iris. Am J Physiol 78:364–367

Kose W (1907) Die Paraganglien bei den Vögeln. Arch Mikr Anat 69:563–790

Kosterlitz HW, Lees GM (1964) Pharmacological analysis of intrinsic intestinal reflexes. Pharmacol Rev 16:301–339

Kosterlitz HW, Thompson JW, Wallis DI (1964) The compound action potential in the nerve supplying the medial smooth muscle of the nictitating membrane of the cat. J Physiol (London) 171:426–433

Kottegoda SR (1970) Peristalsis of the small intestine. In: Bülbring E, Brading A, Jones A, Tomita T (eds) Smooth muscle. Arnold, London, pp 525–541

Krawkow NP (1913) Über die Wirkung von Giften auf die Gefäße isolierter Fischkiemen. Pflüger's Arch Gesamte Physiol Menschen Tiere 151:583–603

Kunos G, Nickerson M (1976) Temperature induced interconversion of alpha- and beta-adrenoceptors in the frog heart. J Physiol (London) 256:23–40

Kunos G, Nickerson M (1977) Effects of sympathetic innervation and temperature on the properties of rat heart adrenoceptors. Br J Pharmacol 59:603–614

Kunos G, Szentivanyi M (1968) Evidence favouring the existence of a single adrenergic receptor. Nature (London) 217:1077–1078

Kunos G, Yong MS, Nickerson M (1973) Transformation of adrenergic receptors in the myocardium. Nature (London) 241:119–120

LaBrosse EH, Axelrod J, Kopin IJ, Kety SS (1961) Metabolism of 7-^{3}H-epinephrine-D-bitartrate in normal young men. J Clin Invest 40:253–260

Laffont J, Labat R (1966) Action de l'adrénaline sur la fréquence cardiaque de le Carpe commune. Effect de la température du milieu sur l'intensité de la réaction. J Physiol (Paris) 58:351–355

Lande de la IS, Tyler MJ, Pridmore BR (1962) Pharmacology of the heart of *Tiliqua (Trachysaurus) rugosa*. Aust J Exp Biol 40:129–138

Lands AM, Arnold A, McAuliff JP, Luduena FP, Brown TG (1967a) Differentation of receptor systems activated by sympathomimetic amines. Nature (London) 214:597–598

Lands AM, Luduena FP, Buzzo HH (1967b) Differentiation of receptors responsive to isoproterenol. Life Sci 6:2241–2249

Lands AM, Luduena FP, Buzzo HJ (1969) Adrenotrophic β-receptors in the frog and chicken. Life Sci 8:373–382

Langer M, Noorden Van S, Polak JM, Pearse AGE (1979) Peptide hormone-like immunoreactivity in the gastrointestinal tract and endocrine pancreas of eleven teleost species. Cell Tissue Res 199:493–508

Langer SZ (1970) The metabolism of (^{3}H)noradrenaline released by electrical stimulation from the isolated nictitating membrane of the cat and from the vas deferens of the rat. J Physiol (London) 208:515–546

Langley JN (1898) On the union of cranial autonomic (visceral) fibres with the nerve cells of the superior cervical ganglion. J Physiol (London) 23:240–270

Langley JN (1901) Observations on the physiological action of extracts of the suprarenal bodies. J Physiol (London) 27:237–256

Langley JN (1904) On the sympathetic system of birds, and on the muscles which move the feathers. J Physiol (London) 30:221–252

Langley JN (1905) On the reaction of cells and of nerve endings to certain poisons, chiefly as regards the reaction of striated muscle to nicotine and to curari. J Physiol (London) 33:374–413

Langley JN (1911) The origin and course of the vaso-motor fibres of the frog's foot. J Physiol (London) 41:483–498

Langley JN (1921) The autonomic nervous system, part I. Heffer, Cambridge

Langley JN, Anderson HK (1895) The innervation of the pelvic and adjoining viscera. J Physiol (London) 19:71–139

Langley JN, Orbeli LA (1911) Observations on the sympathetic and sacral autonomic system of the frog. J Physiol (London) 41:450–482

Larsson I (1981) Studies on the extrinsic neural control of serotonin release from the small intestine. Acta Physiol Scand 113 Suppl 499:1–43

Larsson L-I (1977) Ultrastructural localization of a new neuronal peptide (VIP). Histochemistry 54:173–176

Larsson L-I (1980) Gastrointestinal cells producing endocrine, neurocrine and paracrine messengers. Clin Gastroenterol 9:485–516

Laurent P (1962) Contribution à l'étude morphologique et physiologique de l'innervation du coeur des Téléostéens. Arch Anat Microsc Morphol Exp 51:339–458

Laurent P (1967) Neurophysiologie – la pseudobranchie des téléostéens: preuves électrophysiologiques de ses fonctions chémoréceptrice et baroréceptrice. C R Acad Sci 264:1879–1882

Laurent P, Dunel S (1976) Functional organization of the teleost gill. – I. Blood pathways. Acta Zool (Stockholm) 57:189–209

Laurent P, Dunel S (1980) Morphology of gill ephitelia in fish. Am J Physiol 238:R147–R149

Leander S, Håkanson R, Sundler F (1981 a) Nerves containing substance P, vasoactive intestinal polypeptide, enkephalin or somatostatin in the guinea pig taenia coli. Distribution, ultrastructure and possible functions. Cell Tissue Res 215:21–40

Leander S, Håkanson R, Rosell S, Folkers K, Sundler F, Tornqvist K (1981 b) A specific substance P antagonist blocks smooth muscle contractions induced by non-cholinergic, non-adrenergic nerve stimulation. Nature (London) 294:467–469

Leban J, Rockur G, Yamaguchi I, Folkers K, Björkroth U, Rosell S, Yanaihara N, Yanaihara C (1979) Synthesis of substance P analogs and agonistic and antagonistic activities. Acta Chem Scand Ser B 33:664–668

Lee FL (1967) The relation between norepinephrine content and response to sympathetic nerve stimulation of various organs of cats pretreated with reserpine. J Pharmacol Exp Ther 156:137–141

Lefkowitz RJ (1978) Identification and regulation of alpha- and beta-adrenergic receptors. Fed Proc 37:123–129

Leivestad H, Andersen H, Scholander PF (1957) Physiological response to air exposure in codfish. Science 126:505

Leknes IL (1980) Ultrastructure of atrial endocardium and myocardium in three species of Gadidae (Teleostei). Cell Tissue Res 210:1–10

Leont'eva GR (1966) Distribution of catecholamines in blood vessel walls of cyclostomes, fishes, amphibians and reptiles. J Evol Biochem Physiol 2:31–36

Leont'eva GR (1978) Vasomotor innervation of arteries and veins in the frog Rana temporaria. J Evol Biochem Physiol 14:492–496

Lever JD, Santer RM, Lu K-S, Presley R (1976) Chromaffin-positive and small intensely fluorescent cells in normal and amine-depleted sympathetic ganglia. In: Coupland RE, Fujita T (eds) Chromaffin, enterochromaffin and related cells. Elsevier, Amsterdam New York, pp 83–93

Lever JD, Santer RM, Lu K-S, Presley R (1977) Electron probe x-ray microanalysis of small granulated cells in rat sympathetic ganglia after sequential aldehyde and dichromate treatment. J Histochem Cytochem 25:275–279

Levin EY, Levenberg B, Kaufman S (1960) Enzymatic conversion of 3,4-dihydroxyphenylethylamine to norepinephrine. J Biol Chem 235:2080–2086

Leydig F (1853) Anatomische-histologische Untersuchungen über Fische und Reptilien. Reimer, Berlin

Libet B (1970) Generation of slow inhibitory and exitatory postsynaptic potentials. Fed Proc 29:1945–1956

Libet B (1976) The SIF cell as a functional dopamine-releasing interneuron in the rabbit superior cervical ganglion. In: Eränkö O (ed) SIF cells. Fogarty international center proceedings No 30, US Government Printing Office, Washington DC, pp 163–177

Lillywhite HB, Seymour RS (1978) Regulation of arterial blood pressure in Australian tiger snakes. J Exp Biol 75:65–79

Lin Y-C, Sturkie PD (1968) Effect of environmental temperature on the catecholamines of chickens. Am J Physiol 214:237–240

Lindmar R, Wolf U (1975) The phenylethanolamine-N-methyl transferase (PNMT) activity in various organs of the chicken. Arch Pharmacol 287:R2

Lindström T (1949) On the cranial nerves of the cyclostomes with special reference to n trigeminus. Acta Zool (Stockholm) 30:315–458

Linnoila RI, Diaugustine RP, Hervonen A, Miller RJ (1980) Distribution of [Met^5]- and [Leu^5]-enkephalin-, vasoactive intestinal polypeptide and substance P-like immunoreactivities in human adrenal glands. Neuroscience 5:2247–2259

Loewi O (1921) Über humorale Übertragbarkeit der Herznervenwirkung. I. Mitteilung. Pflüger's Arch Gesamte Physiol Menschen Tiere 189:239–242

Loewi O, Navratil E (1926a) Über humorale Übertragbarkeit der Herznervenwirkung. X. Mitteilung. Über das Schicksal des Vagusstoffs. Pflüger's Arch Gesamte Physiol Menschen Tiere 214:678–688

Loewi O, Navratil E (1926b) Über humorale Übertragbarkeit der Herznervenwirkung. XI. Mitteilung. Über den Mechanismus der Vaguswirkung von Physostigmin und Ergotamin. Pflüger's Arch Gesamte Physiol Menschen Tiere 214:689–696

Lokhandwala MF (1979) Inhibition of cardiac sympathetic neurotransmission by adenosine. Eur J Pharmacol 60:353–357

Lorén I, Björklund A, Falck B, Lindvall O (1980) The aluminium-formaldehyde (ALFA) histofluorescence method for improved visualization of catecholamines and indoleamines. 1. A detailed account of the methodology for central nervous tissue using paraffin, cryostat or vibratome sections. J Neurosci Meth 2:277–300

Lovenberg W, Weissbach H, Udenfriend S (1962) Aromatic L-amino acid decarboxylase. J Biol Chem 237:89–93

Lu K-S, Lever JD, Santer RM, Presley R (1976) Small granulated cell types in rat superior cervical and coeliac-mesenteric ganglia. Cell Tissue Res 172:331–343

Lubinska L, Niemierko S (1971) Velocity and intensity of bidirectional migration of acetylcholinesterase in transected nerves. Brain Res 27:329–342

Luckhardt AB, Carlsson AJ (1920) Studies on the visceral sensory nervous system. II. Lung automatism and lung reflexes in the salamanders (*Necturus, Axolotl*). Am J Physiol 54:122–137

Luckhardt AB, Carlsson AJ (1921a) Studies on the visceral sensory nervous system. IV. The action of certain drugs on the lung motor mechanisms of the reptilia (Turtles: *Chrysemys elegans* and *Malacoclemmys lesueurii*. Snake: *Eutenia elegans*). Am J Physiol 55:13–30

Luckhardt AB, Carlsson AJ (1921b) Studies on the visceral sensory nervous system. VI. Lung automatism and lung reflexes in *Cryptobranchus*, with further notes on the physiology of the lung of *Necturus*. Am J Physiol 55:212–222

Luckhardt AB, Carlsson AJ (1921c) Studies on the visceral sensory nervous system. VIII. On the presence of vasomotor fibres in the vagus nerve to the pulmonary vessels of the amphibian and the reptilian lung. Am J Physiol 56:72–112

Lukomskaya NJ, Michelson MJ (1972) Pharmacology of the isolated heart of the lamprey, *Lampetra fluviatilis*. Comp Gen Pharmacol 3:213–225

Lund DD, Schmid PG, Roskoski R (1979) Choline acetyltransferase activity in rat and guinea pig heart following vagotomy. Am J Physiol 236:4620–4623

Lundberg JM (1981) Evidence for coexistence of vasoactive intestinal polypeptide (VIP) and acetylcholine in neurons of cat exocrine glands. Morphological, biochemical and functional studies. Acta Physiol Scand 112 Suppl 496:1–57

Lundberg JM, Hökfelt T, Schultzberg M, Uvnäs-Wallensten K, Köhler C, Said SI (1979) Occurrence of vasoactive intestinal polypeptide (VIP)-like immunoreactivity in certain cholinergic neurons of the cat: evidence from combined immunohistochemistry and acetylcholinesterase staining. Neuroscience 4:1539–1559

Lundberg JM, Änggård A, Fahrenkrug J, Hökfelt T, Mutt V (1980a) Vasoactive intestinal polypeptide in cholinergic neurons of exocrine glands: Functional significance of coexisting transmitters for vasodilation and secretion. Proc Natl Acad Sci USA 77:1651–1655

Lundberg JM, Hökfelt T, Änggård A, Uvnäs-Wallensten K, Brimijoin S, Brodin E, Fahrenkrug J (1980 b) Peripheral peptide neurons: distribution, axonal transport, and some aspects on possible function. In: Costa E, Trabucchi M (eds) Neural peptides and neuronal communication. Raven Press, New York, pp 25–36

Lundberg JM, Fahrenkrug J, Brimijoin S (1981 a) Characteristics of the axonal transport of vasoactive intestinal polypeptide (VIP) in nerves of the cat. Acta Physiol Scand 112:427–436

Lundberg JM, Änggård A, Fahrenkrug J (1981 b) Complementary role of vasoactive intestinal polypeptide (VIP) and acetylcholine for cat submandibular gland blood flow and secretion. – I. VIP release. Acta Physiol Scand 113:317–327

Lundberg JM, Änggård A, Fahrenkrug J (1981 c) Complementary role of vasoactive intestinal polypeptide (VIP) and acetylcholine for cat submandibular gland blood flow and secretion – II. Effects of cholinergic antagonists and VIP antiserum. Acta Physiol Scand 113:329–336

Lutz BR (1930 a) The effect of adrenaline on the auricle of elasmobranch fishes. Am J Physiol 94:135–139

Lutz BR (1930 b) Reflex cardiac and respiratory inhibition in the elasmobranch *Scyllium canicula*. Biol Bull Mar Biol Lab 59:170–178

Lutz BR (1930 c) The innervation of the heart of the elasmobranch, *Scyllium canicula*. Biol Bull Mar Biol Lab 59:211–216

Lutz BR (1931) The innervation of the stomach and rectum and the action of adrenaline in elasmobranch fishes. Biol Bull Mar Biol Lab 61:93–100

Lutz BR, Wyman LC (1927) The chromaphil tissue and interrenal bodies of elasmobranchs and the occurence of adrenine. J Exp Zool 47:295–307

Lutz BR, Wyman LC (1932) The effect of adrenaline on the blood pressure of the elasmobranch, *Squalus acanthias*. Biol Bull Mar Biol Lab 62:17–22

MacIntosh FC, Birks RI, Sastry PB (1956) Pharmacological inhibition of acetylcholine synthesis. Nature (London) 178:1181

MacKenzie I, Burnstock G (1980) Evidence against vasoactive intestinal polypeptide being the non-adrenergic, non-cholinergic inhibitory transmitter released from nerves supplying the smooth muscle of the guinea-pig taenia coli. Eur J Pharmacol 67:255–264

Macmillan WH (1959) A hypothesis concerning the effect of cocaine on the action of sympathomimetic amines. Br J Pharmacol Chemother 14:385–391

Maguire MH, Satchell DG (1979) The contribution of adenosine to the inhibitory actions of adenine nucleotides on the guinea-pig taenia coli: Studies with phosphate-modified adenine nucleotide analogs and dipyridamole. J Pharmacol Exp Ther 211:626–631

Malmfors G, Leander S, Brodin E, Håkanson R, Holmin T, Sundler F (1981) Peptide-containing neurons intrinsic to the gut wall. An experimental study in the pig. Cell Tissue Res 214:225–238

Malmfors T, Sachs Ch (1968) Degeneration of adrenergic nerves produced by 6-hydroxydopamine. Eur J Pharmacol 3:89–92

Marchbanks RM (1968) The uptake of ^{14}C-choline into synaptosomes in vitro. Biochem J 110:533–541

Marcus H (1910) Über den Sympathicus. S B Ges Morphol Physiol München 25:119–131

Märki F, Axelrod J, Witkop B (1962) Catecholamines and methyltransferases in the South American toad (*Bufo marinus*). Biochim Biophys Acta 58:367–369

Marques M, Serrano L (1960) Adrenalina e noradrenalina na suprarenal de tartarugas normais e hipofisoprivas. Rev Bras Biol 20:251–256

Marshall PB (1955) Some chemical and physical properties associated with histamine antagonism. Br J Pharmacol 10:270–278

Martinson J (1964) The effect of graded stimulation of efferent vagal nerve fibres on gastric motility. Acta Physiol Scand 62:256–262

Martinson J (1965) Vagal relaxation of the stomach. Experimental re-investigation of the concept of the transmission mechanism. Acta Physiol Scand 64:453–462

Martinson J, Muren A (1963) Excitatory and inhibitory effects of vagus stimulation on gastric motility in the cat. Acta Physiol Scand 57:309–316

Marwitt R, Pilar G, Weakly JN (1971) Characterization of two ganglion cell populations in avian ciliary ganglia. Brain Res 25:317–334

Mastrolia L, Gallo V, Manelli H (1976) Cytological and histochemical observations on chromaffin cells of the suprarenal gland of *Triturus cristatus* (urodele amphibian). Boll Zool 43:27–36

Mazeaud MM (1971) Recherches sur la biosynthèse, la sécrétion et le catabolisme de l'adrénaline et de la noradrénaline chez quelques espèces de cyclostomes et de poissons. Ph D thesis, Univ Paris

Mazeaud MM (1972) Epinephrine biosynthesis in *Petromyzon marinus* (cyclostoma) and *Salmo gairdneri* (teleost). Comp Gen Pharmacol 32:457–468

Mazeaud MM, Mazeaud F (1973) Excretion and catabolism of catecholamines in fish. Part 2. Catabolites. Comp Gen Pharmacol 4:209–217

McGuire J, Moellmann G (1972) Cytochalasin B: Effects on microfilaments and movement of melanin granules within melanocytes. Science 175:642–644

McLean JR, Burnstock G (1966) Histochemical localization of catecholamines in the urinary bladder of the toad (*Bufo marinus*). J Histochem Cytochem 14:538–548

McLean JR, Burnstock G (1967a) Innervation of the urinary bladder of the sleepy lizard (*Trachysaurus rugosus*). I. Fluorescent histochemical localization of catecholamines. Comp Biochem Physiol 20:667–673

McLean JR, Burnstock G (1967b) Innervation of the lung of the toad (*Bufo marinus*). II. Fluorescent histochemical localization of catecholamines. Comp Biochem Physiol 22:767–773

McLean JR, Burnstock G (1967c) Innervation of the lungs of the sleepy lizard (*Trachysaurus rugosus*). I. Fluorescent histochemical localization of catecholamines. Comp Biochem Physiol 22:809–813

McLean JR, Nilsson S (1981) A histochemical study of the gas gland innervation in the Atlantic cod, *Gadus morhua*. Acta Zool (Stockholm) 62:187–194

McLean JR, Bell C, Burnstock G (1967) Histochemical and pharmacological studies of the innervation of the urinary bladder of the frog (*Rana temporaria*). Comp Biochem Physiol 21:383–392

McLelland J (1969) Observations with the light microscope on the ganglia and nerve plexuses of the intrapulmonary bronchi of the bird. J Anat 105:202

McNeill JH (1979) Cyclic AMP and myocardial contraction. In: Kalsner S (ed) Trends in autonomic pharmacology. Urban & Schwarzenberg, Baltimore München, pp 421–441

McWilliam JA (1885) Cardiac inhibition in the newt. J Physiol (London) 6:16P

Meyer F (1927) Versuche über Blutdruckzügler beim Frosch. Pflüger's Arch Gesamte Physiol 215:545–552

Michaelis L, Menten ML (1913) The kinetics of invertase action. Biochem Z 49:333

Michelson MJ, Zeimal EV (1973) Acetylcholine; an approach to the molecular mechanism of action. Pergamon Press, Oxford New York

Middendorf WJ, Russel JA (1978) Innervation of tracheal smooth muscle in baboons. Fed Proc 37:1782

Midtgård U (1980) Blood vessels in the hind limb of the mallard (*Anas platyrhynchos*): Anatomical evidence for a sphincteric action of shunt vessels in connection with the arterio-venous heat exchange system. Acta Zool (Stockholm) 61:39–49

Midtgård U (1981) The *rete tibiotarsale* and arterio-venous association in the hind limb of birds: A comparative morphological study on counter-current heat exchange systems. Acta Zool (Stockholm) 62:67–87

Midtgård U, Bech C (1981) Responses to catecholamines and nerve stimulation of the perfused *rete tibiotarsale* and associated blood vessels in the hind limb of the mallard (*Anas platyrhynchos*). Acta Physiol Scand 112:77–81

Miles FA (1969) Excitable cells. William Heinemann Medical Books, London

Millard RW, Johansen K, Milsom WK (1973) Radiotelemetry of cardiovascular responses to exercise and diving in penguins. Comp Biochem Physiol 46A:227–240

Miller JW (1967) Adrenergic receptors in the myometrium. Ann NY Acad Sci 139:788–798

Mills TW (1884) Some observations on the influence of the vagus and accelerators on the heart etc. of the turtle. J Physiol (London) 5:359–361

Mills TW (1885) The innervation of the heart of the slider terrapin (*Pseudemys rugosa*). J Physiol (London) 6:248–286

Milochin AA (1960) Über synaptische Verbindungen im Darmplexus der Cyclostomen. Z Mikrosk Anat Forsch 66:42–52

Milsom WK, Langille BL, Jones DR (1977) Vagal control of pulmonary vascular resistance in the turtle, *Chrysemys scripta*. Can J Zool 55:359–367

Miyashita Y, Fujii R (1975) Receptor mechanisms in fish chromatophores – II. Evidence for beta adrenoceptors mediating melanosome dispersion in guppy melanophores. Comp Biochem Physiol 51C:179–187

Miyashita Y, Fujii R (1977) Evidence for dopamine receptors mediating pigment dispersion in the melanophores of the siluroid *Parasilurus asotus* – a preliminary note. J Sapporo Med Coll 18:67–70

Mohsen T, Lattouf H, Jadoun G (1974) Variations des effects de l'adrénaline sur le coeur isolé de *Protopterus annectens* (Poisson Dipneuste) selon la dose et selon la phase du cycle biologique de l'animal. C R Soc Biol 168:915–919

Moore A, Hiatt RB (1967) Action of epinephrine on gastrointestinal motility in the spiny dogfish. Bull Mt Desert Isl Biol Lab 7:32–33

Moran NC, Perkins ME (1958) Adrenergic blockade of the mammalian heart by a dichloro analogue of isoproterenol. J Pharmacol Exp Ther 124:223–237

Morris JL (1976) Motor innervation of the toad iris (*Bufo marinus*). Am J Physiol 231:1272–1278

Morris JL (1982) Seasonal variation in responses of the toad renal vasculature to adrenaline. Naunyn Schmiedebergs Arch Pharmacol 320:246–254

Morris JL, Gibbins IL, Clevers J (1981) Resistance of adrenergic neurotransmission in the toad heart to adrenoceptor blockade. Naunyn Schmiedebergs Arch Pharmacol 317:331–338

Mott JC (1951) Some factors affecting the blood circulation in the common eel (*Anguilla anguilla*). J Physiol (London) 114:387–398

Müller E, Liljestrand G (1918) Anatomische und experimentelle Untersuchungen über das autonome Nervensystem der Elasmobranchier nebst Bemerkungen über die Darmnerven bei den Amphibien und Saugtieren. Arch Anat 1918:137–172

Muscholl E (1961) Effect of cocaine and related drugs on the uptake of noradrenaline by heart and spleen. Br J Pharmacol Chemother 16:352–359

Muscholl E (1972) Adrenergic false transmitters. In: Blaschko H, Muscholl E (eds) Handbuch der experimentellen Pharmakologie, vol XXXIII. Catecholamines. Springer, Berlin Heidelberg New York, pp 618–660

Muscholl E (1979) Presynaptic muscarine receptors and inhibition of release. In: Paton DM (ed) The release of catecholamines from adrenergic neurones. Pergamon Press, Oxford New York, pp 87–110

Mutt V, Said SI (1974) Structure of the porcine vasoactive intestinal octacosapeptide: The amino-acid sequence. Use of kallikrein in its determination. Eur J Biochem 42:581–589

Nachmansohn D, Machado AL (1943) The formation of acetylcholine. A new enzyme: choline acetylase. J Neurophysiol 6:397–403

Nagatsu I, Kondo Y (1975) Studies on axonal flow of dopamine β-hydroxylase and tyrosine hydroxylase in sciatic nerves by immunofluorescent and biochemical methods. Acta Histochem Cytochem 8:279–287

Nagatsu I, Kondo Y, Inagaki S, Karasawa N, Kato T, Nagatsu T (1977) Immunofluorescent studies on tyrosine hydroxylase: application for its axoplasmic transport. Acta Histochem Cytochem 10:494–499

Nagatsu I, Kondo Y, Inagaki S, Kojima H, Nagatsu T (1979) Immunofluorescent and biochemical studies on tyrosine hydroxylase and dopamine-β-hydroxylase of the bullfrog sciatic nerves. Histochemistry 61:103–109

Nagatsu T, Levitt M, Udenfriend S (1964) Tyrosine hydroxylase. The initial step in norepinephrine biosynthesis. J Biol Chem 239:2910–2917

Nakano T, Tomlinson N (1967) Catecholamine and carbohydrate concentrations in rainbow trout (*Salmo gairdneri*) in relation to physical disturbance. J Fish Res Board Can 24:1701–1715

Nakazato Y, Sato H, Ohga A (1970) Evidence for a neurogenic rebound contraction of the smooth muscle of the chicken proventriculus. Experientia 26:50–51

Nandi J (1961) New arrangement of interrenal and chromaffin tissues of teleost fishes. Science 134:389–390

Nandi J (1964) The structure of the interrenal gland in teleost fishes. Univ Calif Publ Zool 65:129–195

Narahashi T, Moore JW, Scott WR (1964) Tetrodotoxin blockage of sodium conductance increase in lobster giant axons. J Gen Physiol 47:965–974

Nayler WG, Howells JE (1965) Phosphorylase a/b ratio in the lamprey heart. Nature (London) 207:81

Nicholls JVV (1934) Reaction of the smooth muscle of the gastro-intestinal tract of the skate to stimulation of autonomic nerves in isolated nerve-muscle preparations. J Physiol (London) 83:56–67

Nickerson M (1970) Drugs inhibiting adrenergic nerves and structures innervated by them. In: Goodman LS, Gilman A (eds) The pharmacological basis of therapeutics. Macmillan, London Toronto, pp 549–584

Nicol JAC (1952) Autonomic nervous systems in lower chordates. Biol Rev Cambridge Philos Soc 27:1–49
Nicol JAC (1957) Observations on photophores and luminescence in the *Porichthys*. Q J Microsc Sci 98:179–188
Nielsen KC, Owman Ch (1968) Difference in cardiac adrenergic innervation between hibernators and non-hibernating mammals. Acta Physiol Scand 74 Suppl 316:1–30
Nielsen KC, Owman Ch, Santini M (1969) Anastomosing adrenergic nerves from the sympathetic trunk to the vagus at the cervical level in the cat. Brain Res 12:1–9
Nilsson A (1975) Structure of the vasoactive intestinal octacosapeptide from chicken intestine. The amino acid sequence. FEBS Lett 60:322–326
Nilsson S (1970) Excitatory and inhibitory innervation of the urinary bladder and gonads of a teleost (*Gadus morhua*). Comp Gen Pharmacol 1:23–28
Nilsson S (1971) Adrenergic innervation and drug responses of the oval sphincter in the swimbladder of the cod (*Gadus morhua*). Acta Physiol Scand 83:446–453
Nilsson S (1972) Autonomic vasomotor innervation in the gas gland of the swimbladder of a teleost, *Gadus morhua*. Comp Gen Pharmacol 3:371–375
Nilsson S (1973) Fluorescent histochemistry of the catecholamines in the urinary bladder of a teleost, *Gadus morhua*. Comp Gen Pharmacol 4:17–21
Nilsson S (1976) Fluorescent histochemistry and cholinesterase staining in sympathetic ganglia of a teleost, *Gadus morhua*. Acta Zool (Stockholm) 57:69–77
Nilsson S (1978) Sympathetic innervation of the spleen of the cane toad, *Bufo marinus*. Comp Biochem Physiol 61C:133–149
Nilsson S (1980) Sympathetic nervous control of the iris sphincter of the Atlantic cod, *Gadus morhua*. J Comp Physiol 138:149–155
Nilsson S (1981) On the adrenergic system of ganoid fish: the Florida spotted gar, *Lepisosteus platyrhincus* (Holostei). Acta Physiol Scand 111:447–454
Nilsson S, Fänge R (1967) Adrenergic receptors in the swimbladder and gut of a teleost (*Anguilla anguilla*). Comp Biochem Physiol 23:661–664
Nilsson S, Fänge R (1969) Adrenergic and cholinergic vagal effects on the stomach of a teleost (*Gadus morhua*). Comp Biochem Physiol 30:691–694
Nilsson S, Grove DJ (1974) Adrenergic and cholinergic innervation of the spleen of the cod, *Gadus morhua*. Eur J Pharmacol 28:135–143
Nilsson S, Holmgren S (1976) Uptake and release of catecholamines in sympathetic nerve fibres in the spleen of the cod, *Gadus morhua*. Eur J Pharmacol 39:41–51
Nilsson S, Pettersson K (1981) Sympathetic nervous control of blood flow in the gills of the Atlantic cod, *Gadus morhua*. J Comp Physiol 144:157–163
Nilsson S, Holmgren S, Grove DJ (1975) Effects of drugs and nerve stimulation on the spleen and arteries of two species of dogfish, *Scyliorhinus canicula* and *Squalus acanthias*. Acta Physiol Scand 95:219–230
Nilsson S, Abrahamsson T, Grove DJ (1976) Sympathetic nervous control of adrenaline release from the head kidney of the cod, *Gadus morhua*. Comp Biochem Physiol 55C:123–127
Nishida S, Sears M (1970) Fine structure of the anterior epithelial cell layer of the iris of the hen. Exp Eye Res 9:241–245
Nolf P (1934) Les nerfs extrinsèques de l'intestin de l'oiseau. III. Le nerf de Remak. Arch Int Physiol 39:227–256
Noon L (1900) Some observations on the nerve cell connections of the efferent vagus fibres in the tortoise. J Physiol (London) 26:5P–7P
Noorden Van S, Patent GJ (1980) Vasoactive intestinal polypeptide-like immunoreactivity in nerves of the pancreatic islet of the teleost fish *Gillichthys mirabilis*. Cell Tissue Res 212:139–146
Norberg K-Å, Hamberger B (1964) The sympathetic adrenergic neurone, Some characteristics revealed by histochemical studies on the intraneuronal distribution of the transmitter. Acta Physiol Scand 63 Suppl 238: 1–42
Norberg K-Å, Sjöqvist F (1966) New possibilities for adrenergic modulation of ganglionic transmission. Pharmacol Rev 18:743–751
North RA (1973) The calcium-dependent slow after-hyperpolarization in myenteric plexus neurones with tetrodotoxin-resistant action potentials. Br J Pharmacol 49:709–711

North RA, Katayama Y, Williams JT (1980) Actions of peptides on enteric neurons. Adv Biochem Psycopharm 22:83–92

Ochi J, Konishi M, Yoshikawa H, Sano Y (1968) Fluorescence and electron microscopic evidence for the dual innervation of the iris sphincter muscle of the rabbit. Z Zellforsch 91:90–95

Oesch F, Otten U, Thoenen H (1973) Relationship between the rate of axoplasmic transport and subcellular distribution of enzymes involved in the synthesis of norepinephrine. J Neurochem. 20:1691–1706

Oguri M (1960) Studies on the adrenal glands of teleost: On the distribution of chromaffin cells and interrenal cells in the head kidneys of fishes. Bull Jpn Soc Sci Fish 26:443–447

Ohnesorge FK, Rehberg M (1963) Der Darm der Schleie (*Tinca vulgaris*) als pharmakologisches Versuchsobjekt. Naunyn Schmiedebergs Arch Exp Pathol Pharmakol 246:81–82

Oliver G, Schäfer EA (1895) The physiological effects of extracts of the suprarenal capsules. J Physiol (London) 18:230–276

Olson L, Malmfors T (1970) Growth characteristics of adrenergic nerves in the adult rat. Acta Physiol Scand 80 Suppl 348:1–112

Olson L, Ålund M, Norberg K-Å (1976) Fluorescence microscopical demonstration of a population of gastrointestinal nerve fibres with a selective affinity for quinacrine. Cell Tissue Res 171:407–423

Opdyke DF, Opdyke NE (1971) Splenic responses to stimulation in *Squalus acanthias*. Am J Physiol 221:623–625

Opdyke DF, McGreehan JR, Messing S, Opdyke NE (1972) Cardiovascular responses to spinal cord stimulation and autonomically active drugs in *Squalus acanthias*. Comp Biochem Physiol 42:611–620

Osborne NN (1979) Is Dale's principle valid? Trends Neurosci 2:73–75

Östlund E (1954) The distribution of catecholamines in lower animals and their effect on the heart. Acta Physiol Scand 31 Suppl 112:1–67

Östlund E, Fänge R (1962) Vasodilation by adrenaline and noradrenaline and the effects of some other substances on perfused fish gills. Comp Biochem Physiol 5:307–309

Otorii T (1953) Pharmacology of the heart of *Entosphenus japonicus*. Acta Med Biol 1:51–59

Otsuka N, Chihara J, Sakurada H, Kanda S (1977) Catecholamine-storing cells in the cyclostome heart. Arch Histol Jpn 40 (Suppl) 241–244

Owman Ch, Björklund A (1978) Current research on the histochemistry and function of biogenic amines. A tribute to Bengt Falck. Acta Physiol Scand 102 Suppl 452:5–8

Parker GH (1934) Colour changes in the catfish *Ameiurus* in relation to neurohumours. J Exp Zool 69:199–234

Parker GH (1938) The colour changes in lizards, particularly in *Phrynosoma*. J Exp Biol 15:48–73

Parker GH (1948) Animal colour changes and their neurohumours. Cambridge Univ Press, Cambridge

Parker WN (1892) On the anatomy and physiology of *Protopterus annectens*. Trans R Irish Acad (Dublin) 30:109–230

Paton DN (1912) On the extrinsic nerves of the heart of the bird. J Physiol (London) 45:106–114

Paton WDM (1961) A theory of drug action based on the rate of drug receptor combination. Proc R Soc London Ser B 154:21–69

Paton WDM (1964) The mechanism of action of acetylcholine. Pharmacology of smooth muscle. 2nd Int Pharmacol Meet, vol VI, Pergamon Press, Oxford, pp 71–79

Paton WDM (1979) The release of catecholamines from adrenergic neurons. Pergamon Press, Oxford

Paton WDM, Vizi ES (1969) The inhibitory action of noradrenaline and adrenaline on acetylcholine output by guinea-pig ileum longitudinal muscle strip. Br J Pharmacol 35:10–28

Paton WDM, Zaimis E (1949) The pharmacological actions of polymethylase bistrimethylammonium salts. Br J Pharmacol 4:381–400

Paton WDM, Zar MA (1968) The origin of acetylcholine released from guinea-pig intestine and longitudinal muscle strips. J Physiol (London) 194:13–33

Patterson PH, Reichardt LF, Chun LLY (1978) Biochemical studies on the development of primary sympathetic neurons in cell culture. Symp Quant Biol 40:389–397

Patterson TL (1928) The influence of the vagi on the motility of the empty stomach in *Necturus*. Am J Physiol 84:631–640

Patterson TL, Fair E (1933) The action of the vagus on the stomach-intestine of the hagfish. Comparative studies. VIII. J Cell Comp Physiol 3:113–119

Payan P, Girard JP (1977) Adrenergic receptors regulating patterns of blood flow through the gills of trout. Am J Physiol 232:H18–H23

Pearse AGE (1968) Common cytochemical and ultrastructural characteristics of cells producing polypeptide hormones (the APUD series) and their relevance to thyroid and ultimobranchial C cells and calcitonin. Proc R Soc London Ser B 170:71–80

Pearse AGE (1969) The cytochemistry and ultrastructure of polypeptide hormone-producing cells of the APUD series, and the embryologic, physiologic and pathologic implications of the concept. J Histochem Cytochem 17:303–313

Pearse AGE (1976) Neurotransmitters and the APUD concept. In: Coupland RE, Fujita T (eds) Chromaffin, enterochromaffin and related cells. Elsevier, Amsterdam New York, pp 147–154

Pearse AGE, Polak JM (1971) Neural crest origin of the endocrine polypeptide (APUD) cells in the gastrointestinal tract and pancreas. Gut 12:783–788

Pederson R, O'Dorisio T, Howe B, McIntosh C, Mueller M, Brown J, Cataland S (1981) Vagal release of IR-VIP and IR-gastrin from the isolated perfused rat stomach. Mol Cell Endocrinol 23:225–231

Pendleton RG, Kaiser C, Gessner G (1976) Studies on adrenal phenylethanolamine N-methyltransferase (PNMT) with SK&F 64139, a selective inhibitor. J Pharmacol Exp Ther 197:623–632

Pendleton RG, Gessner G, Sawyer J (1980) Comparison of the effects of SK&F 29661 and 64139 upon adrenal and cardiac catecholamines. Eur J Pharmacol 68:117–127

Peng JH, Kimura H, McGeer PL, McGeer EG (1981) Anti-choline acetyltransferase fragments antigen binding (Fab) for immunohistochemistry. Neurosci Lett 21:281–285

Persson H, Sonmark B (1971) Adrenoceptors and cholinoceptors in the rabbit iris. Eur J Pharmacol 15:240–244

Pettersson G (1979) The neural control of the serotonin content in mammalian enterochromaffin cells. Acta Physiol Scand 107 Suppl 470:1–30

Pettersson K, Johansen K (1982) Hypoxic vasoconstriction and the effects of adrenaline on gas exchange efficiency in fish gills. J Exp Biol 97:263–272

Pettersson K, Nilsson S (1979a) Nervous control of the branchial vascular resistance of the Atlantic cod, *Gadus morhua*. J Comp Physiol 129:179–183

Pettersson K, Nilsson S (1979b) Catecholamine stores in the holocephalan fish *Chimaera monstrosa L.* Mar Biol Lett 1:41–46

Pettersson K, Nilsson S (1980) Drug induced changes in cardio-vascular parameters in the Atlantic cod, *Gadus morhua*. J Comp Physiol 137:131–138

Peyrin L, Cier JF, Peres G (1969) La méthylation de la noradrénaline et les interrelations corticomédullaires chez la Roussette. Ann Endocrinol (Paris) 30:1–38

Peyrin L, Peres G, Cier JF (1971) Recherches sur la transméthylation enzymatique de la noradrénaline dans les corps axillaires de la petite Roussette (*Scyliorhinus canicula*). I. Etude du donneur de methyle. C R Soc Biol Paris 165:2145–2148

Peyrin L, Peres G, Cier JF (1972) Recherches sur la transméthylation enzymatique de la noradrénaline dans les corps axillaires de la petite Roussette (*Schyliorhinus canicula*). II. Propriétés physicochimiques du système de méthylation. C R Soc Biol Paris 166:99–103

Phipps RJ, Richardson PS (1976) The nervous and pharmacological control of tracheal mucus secretion in the goose. J Physiol (London) 258:116–117

Pick J (1970) The autonomic nervous system. Morphological, comparative, clinical and surgical aspects. Lippincott, Philadelphia Toronto

Piezzi RS (1966) Two types of synapses in the chromaffin tissue of the toad's adrenal. Acta Physiol Lat Am 16:282–285

Pilar G, Vaughan PC (1969) Electrophysiological investigations of the pigeon iris neuromuscular junctions. Comp Biochem Physiol 29:51–72

Pisano JJ, Creveling CR, Udenfriend S (1960) Enzymic conversion of p-tyramine to p-hydroxyphenylethanolamine (norsynephrin). Biochem Biophys Acta 43:566–568

Porlier GA, Nadeau RA, de Champlain J, Bichet DG (1977) Increased circulating plasma catecholamines and plasma renin activity in dogs after chemical sympathectomy with 6-hydroxydopamine. Can J Physiol Pharmacol 55:724–733

Potter GE (1927) Respiratory function of the swimbladder in *Lepidosteus*. J Exp Zool 49:45–76

Potter LT (1967) Uptake of propranolol by isolated guinea-pig atria. J Pharmacol Exp Ther 155:91–100

Potter LT (1970) Synthesis, storage and release of [^{14}C]acetylcholine in isolated rat diaphragm muscles. J Physiol (London) 206:145–166

Potter de WP (1973) Release of amines from sympathetic nerves. In: Usdin E, Snyder S (eds) Frontiers in catecholamine research. Pergamon Press, Oxford

Powell CE, Slater IH (1958) Blocking of inhibitory adrenergic receptors by a dichloro analog of isoproterenol. J Pharmacol Exp Ther 122:480–488

Priede IG (1974) The effect of swimming activity and section of the vagus nerves on heart rate in rainbow trout. J Exp Biol 60:305–319

Priola DV, O'Brien WJ, Dail WG, Simpson WW (1981) Cardiac catecholamine stores after cardiac sympathectomy, 6-OH-DA, and cardiac denervation. Am J Physiol 240:H889–H895

Probert L, De Mey J, Polak JM (1981) Distinct subpopulations of enteric p-type neurones contain substance P and vasoactive intestinal polypeptide. Nature (London) 294:470–471

Purves MJ (1975) The control of the avian cardiovascular system. Symp Zool Soc Lond 35:13–32

Pye JD (1964a) Nervous control of chromatophores in teleost fishes – 1. Electrical stimulations in the minnow [*Phoxinus phoxinus* (L)]. J Exp Biol 41:525–534

Pye JD (1964b) Nervous control of chromatophores in teleost fishes – 2. The influence of certain drugs in the minnow [*Phoxinus phoxinus* (L)]. J Exp Biol 41:535–541

Randall DJ (1966) The nervous control of cardiac activity in the tench (*Tinca tinca*) and the goldfish (*Carassius auratus*). Physiol Zool 39:185–192

Randall DJ (1968) Functional morphology of the heart in fishes. Am Zool 8:179–189

Randall DJ, Jones DR (1973) The effect of deafferentiation of the pseudobranch on the respiratory response in hypoxia of the trout (*Salmo gairdneri*). Respir Physiol 17:291–301

Randall DJ, Stevens ED (1967) The role of adrenergic receptors in cardiovascular changes associated with exercise in salmon. Comp Biochem Physiol 21:415–424

Rankin JC, Maetz J (1971) A perfused gill preparation: Vascular actions of neurohypophyseal hormones and catecholamines. J Endocrinol 51:621–635

Rapport MM, Green AA, Page IH (1948) Serum vasoconstrictor (serotonin). IV. Isolation and characterization. J Biol Chem 176:1243–1251

Ray DL (1950) The peripheral nervous system of *Lampanyctus leucopsarus*. J Morphol 87:61–178

Read JB, Burnstock G (1968) Comparative histochemical studies of adrenergic nerves in the enteric plexuses of vertebrate large intestine. Comp Biochem Physiol 27:505–517

Read JB, Burnstock G (1969a) Adrenergic innervation of the gut musculature in vertebrates. Histochemie 17:263–272

Read JB, Burnstock G (1969b) A method for the localization of adrenergic nerves during early development. Histochemie 20:197–200

Reite OB (1969) The evolution of vascular smooth muscle responses to histamine and 5-hydroxytryptamine. – I. Occurrence of stimulatory actions in fish. Acta Physiol Scand 75:221–239

Reite OB (1970) The evolution of vascular smooth muscle responses to histamine and 5-hydroxytryptamine. – III. Manifestation of dual actions of either amine in reptiles. Acta Physiol Scand 78:213–231

Reite OB, Millard RW, Johansen K (1977) Effects of low tissue temperature on peripheral vascular control mechanisms. Acta Physiol Scand 101:247–253

Rennick BR, Pryor MZ, Basch BG (1965) Urinary metabolites of epinephrine and norepinephrine in the chicken. J Pharmacol Exp Ther 148:270–276

Richards BD, Fromm PO (1969) Patterns of blood flow through filaments and lamellae of isolated-perfused rainbow trout (*Salmo gairdneri*) gills. Comp Biochem Physiol 29:1063–1070

Richards JG, Lorez HP, Tranzer JP (1973) Indolealkylamine nerve terminals in cerebral ventricles: identification by electron microscopy and fluorescence histochemistry. Brain Res 57:277–288

Richardson JB (1979) Nerve supply to the lungs. Am Rev Respir Dis 119:785–802

Richardson JB, Bouchard T (1973) Demonstration of a nonadrenergic inhibitory nervous system in the trachea of the guinea-pig. J Allergy Clin Immunol 56:473

Richardson KC (1964) Fine structure of the albino rabbit iris with special reference to the identification of adrenergic and cholinergic nerves and nerve endings in its intrinsic muscles. Am J Anat 114:173–205

Ristori MT (1970) Réflexe de barosensibilité chez un poisson téléostéen (*Cyprinus carpio* L). C R Soc Biol Paris 164:1512–1516

Robb JS (1953) Specialized (conducting) tissue in the turtle heart. Am J Physiol 172:7–13

Roberts JL (1968) Activity level and cardioregulation of the northern sea robin, *Prionotus carolinus* (L). Biol Bull Woods Hole 135:433–434

Robison GA, Butcher RW, Sutherland EW (1971) Cyclic AMP. Academic Press, London New York

Rockur G, Yamaguchi I, Leban JJ, Björkroth U, Rosell S, Folkers K (1979) Synthesis of peptides related to substance P and their activities as agonists and antagonists. Acta Chem Scand B33:375–378

Romer AS (1962) The vertebrate body, 3rd edn. Saunders, Philadelphia London

Roskoski R Jr, McDonald RI, Roskoski LM, Marvin WJ, Hermsmeyer K (1977) Choline acetyltransferase activity in heart: evidence for neuronal and not myocardial origin. Am J Physiol 233:H642–H646

Ross LG (1978) The innervation of the resorptive structures in the swimbladder of a physoclist fish, *Pollachius virens* (L). Comp Biochem Physiol 61C:385–388

Rossum van JM (1963) Cumulative dose-response curves. – II. Technique for the making of dose-response curves in isolated organs and the evaluation of drug parameters. Arch Int Pharmacodyn Ther 143:299–330

Rossum van JM, Brink van den FG (1963) Cumulative dose response curves. – I. Introduction to the technique. Arch Int Pharmacodyn Ther 143:238–246

Rothman TP, Ross LL, Gershon MD (1976) Separately developing axonal uptake of 5-hydroxytryptamine and norepinephrine in the fetal ileum of the rabbit. Brain Res 115:437–456

Rovainen CM (1979) Neurobiology of lampreys. Physiol Rev 59:1007–1077

Rubin W, Gershon MD, Ross LL (1971) Electron microscopic radioautographic identification of serotonin-synthesizing cells in the mouse gastric mucosa. J Cell Biol 50:399–415

Saavedra JM, Coyle JT, Axelrod J (1973) The distribution and properties of the nonspecific N-methyltransferase in brain. J Neurochem 20:743–752

Saetersdal TS, Justesen N-P, Krohnstad AW (1974) Ultrastructure and innervation of the teleostean atrium. J Mol Cell Cardiol 6:415–437

Saetersdal TS, Sorensen E, Myklebust R, Helle KB (1975) Granule containing cells and fibres in the sinus venosus of elasmobranchs. Cell Tissue Res 163:471–490

Saffrey MJ, Polak JM, Burnstock G (1982) Distribution of vasoactive intestinal polypeptide-, substance P-, enkephalin- and neurotensin-like immunoreactive nerves in the chicken gut during development. Neuroscience 7:279–293

Said SI, Mutt V (1970) Polypeptide with broad biological activity. Isolation from small intestine. Science 169:1217–1218

Said SI, Mutt V (1972) Isolation from porcine intestine of a vasoactive octacosapeptide related to secretin and glucagon. Eur J Biochem 28:199–204

Said SI, Giachetti A, Nicosia S (1980) VIP: Possible functions as a neural peptide. In: Costa E, Trabucchi M (eds) Neural peptides and neuronal communication. Raven Press, New York, pp 75–82

Saint-Aubain de ML, Wingstrand KG (1979) A sphincter in the pulmonary artery of the frog *Rana temporaria* and its influence on blood flow in skin and lungs. Acta Zool (Stockholm) 60:163–172

Saito T (1973) Nervous control of intestinal motility in goldfish. Jpn J Smooth Musc Res 9:79–86

Saito T, Tenma K (1976) Effects of left and right vagal stimulation on excitation and conduction of the carp heart (*Cyprinus carpio*). J Comp Physiol 111:39–53

Sakharov DA, Salimova NB (1980) Serotonin neurons in the peripheral nervous system of the larval lamprey, *Lampetra planeri*. A histochemical, microspectro-fluorimetric and ultrastructural study. Zool Jahrb Physiol 84:231–239

Santer RM (1972) Ultrastructural and histochemical studies on the innervation of the heart of a teleost (*Pleuronectes platessa* L.). Z Zellforsch 131:519–528

Santer RM (1977) Monoaminergic nerves in the central and peripheral nervous systems of fishes. Gen Pharmacol 8:157–172

Santer RM, Cobb JLS (1972) The fine structure of the heart of the teleost *Pleuronectes platessa* L. Z Zellforsch 131:1–14

Santer RM, Lu K-S, Lever JD, Presley R (1975) A study of the distribution of chromaffin-positive (CH+) and small intensely fluorescent (SIF) cells in sympathetic ganglia of the rat of various ages. J Anat 119:589–599

Santis de VP, Längsfeld W, Lindmar R, Löffelholz K (1975) Evidence for noradrenaline and adrenaline as sympathetic transmitters in the chicken. Br J Pharmacol 55:343–350

Satchell DG (1981) Nucleotide pyrophosphate antagonizes responses to adenosine 5-triphosphate and non-adrenergic, non-cholinergic inhibitory nerve stimulation in the guinea-pig isolated taenia coli. Br J Pharmacol 74:319–321

Satchell DG, Lynch A, Bourke PM, Burnstock G (1972) Potentiation of the effects of exogenously applied ATP and purinergic nerve stimulation on the guinea-pig taenia coli by dipyridamole and hexobendine. Eur J Pharmacol 19:343–350

Satchell DG, Burnstock G, Dann P (1973) Antagonism of the effects of purinergic nerve stimulation and exogenously applied ATP on the guinea-pig taenia coli by 2-substituted imidazolines and related compounds. Eur J Pharmacol 23:264–269

Satchell GH (1961) The response of the dogfish to anoxia. J Exp Biol 38:531–543

Satchell GH (1970) A functional appraisal of the fish heart. Fed Proc 29:1120–1123

Satchell GH (1971) Sites of cholinergic vasoconstriction in trout (*Salmo gairdneri*). Can J Zool 56:1678–1683

Satchell GH, Hanson D, Johansen K (1970) Differential blood flow through the afferent branchial arteries of the skate, *Raja rhina*. J Exp Biol 52:721–726

Sato H (1969) The role of the autonomic nerves on the motility of the stomach of the domestic fowl. Jpn J Vet Res 17:88–89

Sato H, Ohga A, Nakazato Y (1970) The excitatory and inhibitory innervation of the stomachs of the domestic fowl. Jpn J Pharmacol 20:382–397

Sauerbier I (1977) Seasonal variation in tissue catecholamine level and its turnover in the frog (*Rana temporaria*). Gen Comp Endocrinol 31:183–188

Schäfer EA, Moore B (1896) On the contractility and innervation of the spleen. J Physiol (London) 20:1–50

Scheline RR (1963) Adrenergic mechanisms in fish: Chromatophore pigment concentration in the cuckoo wrasse *Labrus ossifragus* L. Comp Biochem Physiol 9:215–227

Scheuermann DW (1979) Untersuchungen hinsichtlich der Innervation des Sinus venosus und des Aurikels von *Protopterus annectens*. Acta Morphol Neerl Scand 17:231–232

Scheuermann DW, Stilman C, Reinhold Ch, DeGroodt-Lasseel MHA (1981) Microspectrofluorometric study of monoamines in the auricle of the heart of *Protopterus aethiopicus*. Cell Tissue Res 217:443–449

Schild HO (1947) pA, a new scale for the measurement of drug antagonism. Br J Pharmacol 2:189–206

Schmidt RF (1978) Fundamentals of neurophysiology, 2nd edn. Springer, Berlin Heidelberg New York

Schofield GC (1968) Anatomy of muscular and neural tissues in the alimentary canal. In: Code CF (ed) Handbook of physiology, sect 6: Alimentary canal. Am Physiol Soc, Washington DC, pp 1579–1627

Schoot van der JB, Creveling CR (1964) Substrates and inhibitors of dopamine-β-hydroxylase (DBH). Adv Drug Res 2:47–88

Schueler FW (1960) The mechanism of action of the hemicholiniums. Int Rev Neurobiol 2:77–97

Schultz R, Wüster M, Simantov R, Snyder S, Hertz A (1977) Electrically stimulated release of opiate-like material from the myenteric plexus of the guinea-pig ileum. Eur J Pharmacol 41:347–348

Schultzberg M, Hökfelt T, Nilsson G, Terenius L, Rehfeld JF, Brown M, Elde R, Goldstein M, Said S (1980) Distribution of peptide- and catecholamine-containing neurons in the gastro-intestinal tract of rat and guinea-pig: Immunohistochemical studies with antisera to substance P, vasoactive intestinal polypeptide, enkephalins, somatostatin, gastrin/cholecystokinin, neurotensin and dopamine-β-hydroxylase. Neuroscience 5:689–744

Scott JL, Neudeck LD (1972) Catabolism of epinephrine-^{14}C by American chameleons (*Anolis carolinensis*). Gen Comp Endocrinol 18:37–42

Semba T, Hiraoka T (1957) Motor responses of the stomach and small intestine caused by stimulation of the peripheral end of the splanchnic nerve, thoracic sympathetic trunk and spinal roots. Jpn J Physiol 7:64–71

Setler PE, Pendleton RG, Rinlay E (1975) The cardiovascular actions of dopamine and the effects of central and peripheral catecholaminergic receptor blocking drugs. J Pharmacol Exp Ther 192:702–712

Shelton G (1970) The effect of lung ventilation on blood flow to the lungs and body of the amphibian, *Xenopus laevis*. Respir Physiol 9:183–196

Shelton G, Burggren W (1976) Cardiovascular dynamics of the chelonia during apnoea and lung ventilation. J Exp Biol 64:323–343

Shepherd DM, West GB (1951) Noradrenaline and the suprarenal medulla. Br J Pharmacol 6:665–674

Shepherd DM, West GB (1953) Hydroxytyramine and the adrenal medulla. J Physiol (London) 120:15–19

Shepherd DM, West GB, Erspamer V (1953) Chromaffin bodies of various species of dogfish. Nature (London) 172:509

Shibata Y, Yamamoto T (1976) Fine structure and cytochemistry of specific granules in the lamprey atrium. Cell Tissue Res 172:487–502

Shimadu K, Kobayasi S (1966) Neural control of the pulmonary smooth muscle in the toad. Acta Med Biol (Niigata) 13:297–303

Shiozu K (1965) A typical beta adrenergic receptor in toad rectum. Bull Osaka Med Sch 11:61–72

Short S, Butler PJ, Taylor EW (1977) The relative importance of nervous, humoral and intrinsic mechanisms in the regulation of heart rate and stroke volume in the dogfish *Scyliorhinus canicula*. J Exp Biol 70:77–92

Sippel TO (1955) Properties and development of cholinesterase in the hearts of certain vertebrates. J Exp Zool 128:165–184

Sjöstrand NO (1965) The adrenergic innervation of the vas deferens and the accessory male genital glands. Acta Physiol Scand 65 Suppl 257:1–82

Skadhauge E (1981) Osmoregulation in birds. Springer, Berlin Heidelberg New York

Smith AD, Winkler H (1972) Fundamental mechanisms in the release of catecholamines. In: Blaschko H, Muscholl E (eds) Catecholamines. Springer, Berlin Heidelberg New York, pp 538–617

Smith DG (1977) Sites of cholinergic vasoconstriction in trout gills. Am J Physiol 233:R222–R229

Smith DG (1978a) Neural regulation of blood pressure in rainbow trout (*Salmo gairdneri*). Can J Zool 56:1678–1683

Smith DG (1978b) Evidence for pulmonary vasoconstriction during hypercapnia in the toad, *Bufo marinus*. Can J Zool 56:1530–1534

Smith DG, Campbell G (1976) The anatomy of the pulmonary vascular bed in the toad, *Bufo marinus*. Cell Tissue Res 165:199–213

Smith DG, Macintyre DH (1979) Autonomic innervation of the visceral and vascular smooth muscle of a snake lung (Ophidia: colubridae). Comp Biochem Physiol 62C:187–191

Smith DG, Berger PJ, Evans BK (1981) Baroreceptor control of heart rate in the conscious toad, *Bufo marinus*. Am J Physiol 241:R307–R311

Smith FM, Jones DR (1978) Localization of receptors causing hypoxic bradycardia in trout (*Salmo gairdneri*). Can J Zool 56:1260–1265

Sneddon JD, Smythe A, Satchell D, Burnstock G (1973) Investigation of the identity of the transmitter substance released by non-adrenergic, non-cholinergic excitatory nerves in the gut of lower vertebrates. Comp Gen Pharmacol 4:53–60

Sosa RP, McKnight AT, Hughes J, Kosterlitz HW (1977) Incorporation of labelled amino acids into the enkephalins. FEBS Lett 84:195–198

Spector S, Sjoerdsma A, Udenfriend S (1965) Blockade of endogenous norepinephrine synthesis with α-methyltyrosine, an inhibitor of tyrosinehydroxylase. J Pharmacol Exp Ther 147:86–95

Spector S, Gordon R, Sjoerdsma A, Udenfriend S (1967) Endproduct inhibition of tyrosine hydroxylase as a possible mechanism for regulation of norepinephrine synthesis. Mol Pharmacol 3:549–555

Stabrovskii EM (1967) The distribution of adrenaline and noradrenaline in the organs of the baltic lamprey *Lampetra fluviatilis* at rest and during various functional stresses. J Evol Biochem Physiol 3:216–221

Stabrovskii EM (1968) Adrenaline and noradrenaline in the organs of carp, *Cyprinus carpio*, at rest and under functional stresses. J Evol Biochem Physiol 4:337–341

Stannius H (1849) Das peripherische Nervensystem der Fische. Stiller, Rostock

Starke K (1977) Regulation of noradrenaline release by presynaptic receptor systems. In: Adrian RG, Holzer H, Kramer K, Linden RJ, Miescher PA, Rasmussen H, Trendelenburg U, Vogt W (eds) Reviews of physiology, biochemistry and pharmacology. Springer, Berlin Heidelberg New York, pp 1–124

Starke K (1979) Presynaptic regulation of release in the central nervous system. In: Paton DM (ed) The release of catecholamines from adrenergic neurons. Pergamon Press, Oxford, pp 143–183

Starke K, Docherty JR (1980) Recent developments in α-adrenoceptor research. J Cardiovasc Pharmacol 2:S269–S286

Starke K, Montel H, Gayk W, Merker R (1974) Comparison of the effects of clonidine on pre- and postsynaptic adrenoceptors in the rabbit pulmonary artery. Naunyn Schmiedebergs Arch Pharmacol 285:133–150

Starke K, Endo T, Taube HD (1975a) Relative pre- and postsynaptic potencies of α-adrenoceptor agonists in the rabbit pulmonary artery. Naunyn Schmiedebergs Arch Pharmacol 291:55–78

Starke K, Endo T, Taube HD (1975b) Pre- and postsynaptic components in effect of drugs with α-adrenoceptor affinity. Nature (London) 254:440–441

Starkey RR, Brimijoin S (1979) Stop-flow analysis of the axonal transport of dopa decarboxylase (EC 4.1.1.26) in rabbit sciatic nerves. J Neurochem 32:437–441

Steen JB (1963) The physiology of the swimbladder of the eel *Anguilla vulgaris.* – III. The mechanism of gas secretion. Acta physiol Scand 59:221–241

Steen JB (1970) The swimbladder as a hydrostatic organ. In: Hoar WS, Randal DJ (eds) Fish physiology, vol IV. Academic Press, London New York, pp 413–443

Steinach E (1890) Untersuchungen zur vergleichenden Physiologie der Iris. Erste Mitteilung. Pflügers Arch Gesamte Physiol Menschen Tiere 46:289–340

Steinach E (1892) Untersuchungen zur vergleichenden Physiologie der Iris. Pflügers Arch Gesamte Physiol Menschen Tiere 52:495–525

Steinach E, Weiner H (1895) Motorische Funktionen hinterer Spinalnervenwurzeln. Arch Ges Physiol 60:593–622

Stene-Larsen G (1981) Comparative aspects of cardiac adrenoceptors: characterization of the β_2-adrenoceptor as a common "adrenaline"-receptor in vertebrate hearts. Comp Biochem Physiol 70C:1–12

Stene-Larsen G, Helle KB (1978a) Cardiac β_2-adrenoceptor in the frog. Comp Biochem Physiol 60C:165–173

Stene-Larsen G, Helle KB (1978b) Evidence against a transformation of the β_2-adrenoceptor in the frog by changes in temperature or metabolic state. Life Sci 23:2681–2688

Stephenson RP (1956) A modification of receptor theory. Br J Pharmacol 11:379–393

Stevens ED, Randall DJ (1967) Changes in blood pressure, heart rate and breathing rate during moderate swimming activity in rainbow trout. J Exp Biol 46:307–315

Stevens ED, Bennion GR, Randall DJ, Shelton G (1972) Factors affecting arterial pressure and blood flow from the heart in intact, unrestrained lingcod, *Ophiodon elongatus.* Comp Biochem Physiol 43A:681–695

Stevenson SV, Grove DJ (1977) The extrinsic innervation of the stomach of the plaice *Pleuronectes platessa* L. – I. The vagal nerve supply. Comp Biochem Physiol 58C:143–151

Stevenson SV, Grove DJ (1978) The extrinsic innervation of the stomach of the plaice. *Pleuronectes platessa* L. – II. The splanchnic nerve supply. Comp Biochem Physiol 60C:45–50

Stiemens MJ (1934) Anatomische Untersuchungen über die vago-sympathische Innervation der Baucheingeweide bei den Vertebraten. Verh K Akad Wet Amsterdam Sect 2 33:1–356

Stjärne L (1972) The synthesis, uptake and storage of catecholamines in the adrenal medulla. The effect of drugs: In: Blaschko M, Muscholl E (eds) Handbuch der experimentellen Pharmakologie, vol 33. Catecholamines. Springer, Berlin Heidelberg New York, pp 231–269

Stjärne L (1979) Role of prostaglandins and cyclic adenosine monophosphate in release. In: Paton DM (ed) The release of catecholamines from adrenergic neurons. Pergamon Press, Oxford, pp 111–142

Stoll A (1950) Recent investigations on ergot alkaloids. Chem Rev 47:197–218

Stone CA, Torchiana ML, Navarro A, Beyer KH (1956) Ganglionic blocking properties of 3-methylamino-isocamphane hydrochloride (Mecamylamine): a secondary amine. J Pharmacol 117:169–183

Strasser A, Wolf H (1905) Über die Blutversorgung der Milz. Pflügers Arch Gesamte Physiol Menschen Tiere 108:590–626

Stray-Pedersen S (1970) Vascular responses induced by drugs and by vagal stimulation in the swimbladder of the eel, *Anguilla vulgaris.* Comp Gen Pharmacol 1:358–364

Su C, Bevan JA, Burnstock G (1971) ^{3}H-adenosine triphosphate: Release during stimulation of enteric nerves. Science 173:337–339

Sundler F, Aluments J, Fahrenkrug J, Håkanson R, Schaffalitzky de Muckadell OB (1979) Cellular localization and ontogeny of immunoreactive vasoactive intestinal polypeptide (VIP) in the chicken gut. Cell Tissue Res 196:193–201

Sundler F, Håkanson R, Leander S (1980) Peptidergic nervous system in the gut. Clin Gastroenterol 3:517–543

Szerb JC (1975) Endogenous acetylcholine release and labelled acetylcholine formation from [^{3}H]-choline in the myenteric plexus of the guinea-pig ileum. Can J Physiol Pharmacol 53:566–574

Szerb JC (1976) Storage and release of labelled acetylcholine in the myenteric plexus of the guinea-pig ileum. Can J Physiol Pharmacol 54:12–22

Takamine J (1901) Adrenaline the active principle of the supra-renal glands and its mode of preparation. Am J Pharmacol 73:523–531

Taxi J (1976) Morphology of the autonomic nervous system. In: Llinas R, Precht W (eds) Frog neurobiology. Springer, Berlin Heidelberg New York, pp 93–150

Taylor EW, Short S, Butler PJ (1977) The role of the cardiac vagus in the response of the dogfish (*Scyliorhinus canicula* L) to hypoxia. J Exp Biol 70:57–75

Tazawa H, Mochizuki M, Piiper J (1979) Respiratory gas transport by the incompletely separated double circulation in the bullfrog, *Rana catesbeiana*. Respir Physiol 36:77–95

Thoenen H (1970) Induction of tyrosine hydroxylase in peripheral and central adrenergic neurons by cold-exposure in rats. Nature (London) 228:861–862

Thoenen H (1972) Surgical, immunological and chemical sympathectomy. Their application in the investigation of the physiology and pharmacology of the sympathetic nervous system. In: Blaschko H, Muscholl E (eds) Catecholamines. Springer, Berlin Heidelberg New York, pp 813–844

Thoenen H, Tranzer JP (1968) Chemical sympathectomy by selective destruction of adrenergic nerve endings with 6-hydroxydopamine. Naunyn-Schmiedebergs Arch Pharmakol Exp Pathol 261:271–288

Thoenen H, Mueller RA, Axelrod J (1969a) Increased tyrosine hydroxylase activity and drug-induced alteration of sympathetic transmission. Nature (London) 221:1264

Thoenen H, Mueller RA, Axelrod J (1969b) Trans-synaptic induction of adrenal tyrosine hydroxylase. J Pharmacol Exp Ther 169:249–254

Thomas JE, Baldwin MV (1968) Pathways and mechanisms of regulation of gastric motility. In: Code CF (ed) Handbook of physiology, sect 6: Alimentary canal. Am Physiol Soc, Washington DC, pp 1937–1968

Thompson JW (1961) The nerve supply to the nictitating membrane. J Anat 95:371–385

Tischendorf F (1969) Blutgefäß und Lymphgefäßapparat innersekretorische Drüsen. In: Möllendorf W v, Bargmann W (eds) Handbuch der mikroskopischen Anatomie des Menschen, vol VI. Springer, Berlin Heidelberg New York

Tosaka T, Kobayashi H (1977) The SIF-cell as a functional modulator of ganglionic transmission through the release of dopamine. Arch Histol Jpn 40:187–196

Tranzer JP, Richards JG (1976) Ultrastructural cytochemistry of biogenic amines in nervous tissue: methodological improvements. J Histochem Cytochem 24:1178–1193

Tranzer JP, Thoenen H (1967) Electronmicroscopic localization of 5-hydroxytryptamine (3,4,5-trihydroxyphenyl-ethylamine) a new "false" sympathetic transmitter. Experientia 23:743–745

Tranzer JP, Thoenen H (1968) An electronmicroscopic study of selective acute degeneration of sympathetic nerve terminals after administration of 6-hydroxydopamine. Experientia 24:155–156

Trendelenburg P (1917) Physiologische und pharmakologische Versuche über die Dünndarmperistaltik. Arch Exp Pathol Pharmakol 81:55–129

Trendelenburg U (1959) The supersensitivity caused by cocaine. J Pharmacol Exp Ther 125:55–65

Trendelenburg U (1963) Supersensitivity and subsensitivity to sympathomimetic amines. Pharmacol Rev 15:225–276

Trendelenburg U (1966) Mechanisms of supersensitivity and subsensitivity to sympathicomimetic amines. Pharmacol Rev 18:629–640

Trendelenburg U (1972) Factors influencing the concentrations of catecholamines at the receptors. In: Blaschko H, Muscholl E (eds) Handbuch der experimentellen Pharmakologie, vol XXXIII. Catecholamines. Springer, Berlin Heidelberg New York, pp 726–761

Trendelenburg U, Maxwell RA, Pluchino S (1970) The importance of the intraneuronal uptake in the cat's nictitating membrane as assessed with methoxamine, a sympathomimetic amine which is not taken up by adrenergic nerves. J Pharmacol Exp Ther 172:91–99

Tretjakoff D (1927) Das peripherische Nervensystem des Flußneunauges. Z Wiss Biol 129:359–452

Trifaró JM, Cubeddu XL (1979) Exocytosis as a mechanism of noradrenergic transmitter release. In: Kalsner S (ed) Trends in autonomic pharmacology, vol I. Urban & Schwarzenberg, Baltimore München, pp 195–249

Triggle DJ, Moran JF (1979) The identification and isolation of adrenergic receptors. In: Kalsner S (ed) Trends in autonomic pharmacology, vol I. Urban & Schwarzenberg, Baltimore München, pp 387–419

Tuček S (1975) Transport of choline acetyltransferase and acetylcholinesterase in the central stump and isolated segments of a peripheral nerve. Brain Res 86:259–270

Tummons JL, Sturkie PD (1968) Cardio-accelerator nerve stimulation in chickens. Life Sci 7:377–380

Tummons JL, Sturkie PD (1969) Nervous control of heart rate during excitement in the adult White Leghorn cock. Am J Physiol 216:1437–1440

Tummons JL, Sturkie PD (1970) Beta adrenergic and cholinergic stimulants from the cardio-accelerator nerve in the domestic fowl. Z Vergl Physiol 68:268–271

Tytler P, Blaxter JHS (1977) The effect of swimbladder deflation on pressure sensitivity in the saithe *Pollachius virens*. J Mar Biol Assoc UK 57:1057–1064

Uddman R, Fahrenkrug J, Malm L, Aluments J, Håkanson R, Sundler F (1980) Neuronal VIP in salivary glands: Distribution and release. Acta Physiol Scand 110:31–38

Udenfriend S (1959) Biochemistry of serotonin and other indoleamines. Vitam Horm (NY) 17:133–154

Udenfriend S (1966) Tyrosine hydroxylase. Pharmacol Rev 18:43–51

Udenfriend S, Zaltzman-Nirenberg P, Nagatsu T (1965) Inhibitors of purified beef adrenal tyrosine hydroxylase. Biochem Pharmacol 14:837–845

Ungell A-L, Nilsson S (1979) Metabolic degradation of (^{3}H)-adrenaline in the Atlantic cod, *Gadus morhua*. Comp Biochem Physiol 64C:137–141

Ungell A-L, Nilsson S (1982a) Adrenaline and noradrenaline content in blood plasma and tissues of the toad, *Bufo marinus*. Mol Physiol (in press)

Ungell A-L, Nilsson S (1982b) Catabolism of ^{3}H-adrenaline in the toad, *Bufo marinus*. Mol Physiol (in press)

Unsicker K (1976a) Chromaffin small granule-containing and ganglion cells in the adrenal gland of reptiles: A comparative ultrastructural study. Cell Tissue Res 165:477–508

Unsicker K (1976b) Comparative ultrastructural aspects of adrenal chromaffin cells in reptiles. In: Coupland RE, Fujita T (eds) Chromaffin, enterochromaffin and related cells. Elsevier, Amsterdam New York, pp 13–23

Ursillo RC, Clark BB (1965) The action of atropine on the urinary bladder of the dog and on the isolated nerve bladder strip preparation of the rabbit. J Pharmacol Exp Ther 148:338–347

Ursillo RC, Jacobson J (1965) Potentiation of norepinephrine in the isolated vas deferens of the rat by some CNS stimulants and antidepressants. J Pharmacol Exp Ther 148:247–251

Utterback RA (1944) The innervation of the spleen. J Comp Neurol 81:55–66

Vaillant C, Dimaline R, Dockray GJ (1980) The distribution and cellular origin of vasoactive intestinal polypeptide in the avian gastrointestinal tract and pancreas. Cell Tissue Res 211:511–523

Vairel J (1933) Action de l'adrénaline et de l'acétylcholine sur la rate. J Physiol Pathol Gen 31:42–52

Vanhoutte PM (1977) Cholinergic inhibition of adrenergic transmission. Fed Proc 36:2444–2449

Vanov S (1965) Responses of the rat urinary bladder *in situ* to drugs and to nerve stimulation. Br J Pharmacol Chemother 24:591–600

Veach HO (1925) Studies on the innervation of smooth muscle. I. Vagus effects on the lower end of the oesophagus, cardia and stomach of the cat, and the stomach and lung of the turtle in relation to Wedensky inhibition. Am J Physiol 71:229–264

Viveros OH (1975) Mechanism of secretion of catecholamines from adrenal medulla. In: Blaschko H (ed) Handbook of Physiology, sect 7, vol VI. Am Physiol Soc, Washington DC, pp 389–426

Vladimirova IA, Shuba MF (1978) Strychnine, hydrastine and apamin effect on synaptic transmission in smooth muscle cells. Neurofizologija 10:295–299

Vogel V, Vogel W, Pfautsch M (1976) Arterio-venous anastomoses in rainbow trout gill filaments. Cell Tissue Res 167:373–385

Volle RL, Koelle GB (1970) Ganglionic stimulating and blocking agents. In: Goodman LS, Gilman A (eds) The pharmacological basis of therapeutics, 4th edn. Macmillan, New York, pp 585–600

Wahlqvist I (1980) Effects of catecholamines on isolated systemic and branchial vascular beds of the cod, *Gadus morhua*. J Comp Physiol 137:139–143

Wahlqvist I (1981) Branchial vascular effects of catecholamines released from the head kidney of the Atlantic cod, *Gadus morhua*. Mol Physiol 1:235–241

Wahlqvist I, Nilsson S (1977) The role of sympathetic fibres and circulating catecholamines in controlling the blood pressure in the cod, *Gadus morhua*. Comp Biochem Physiol 57C:65–67

Wahlqvist I, Nilsson S (1980) Adrenergic control of the cardio-vascular system of the Atlantic cod, *Gadus morhua*, during "stress". J Comp Physiol 137:145–150

Wahlqvist I, Nilsson S (1981) Sympathetic nervous control of the vasculature in the tail of the Atlantic cod, *Gadus morhua*. J Comp Physiol 144:153–156

Wasserman G, Tramezzani JH (1963) Separate distribution of adrenaline and noradrenaline-secreting cells in the adrenal of snakes. Gen Comp Endocrinol 3:480–489

Watanabe H (1971) Adrenergic nerve elements in the hypogastric ganglion of the guinea-pig. Am J Anat 130:305

Watson AHD (1979) Fluorescent histochemistry of the teleost gut: Evidence for the presence of serotonergic neurones. Cell Tissue Res 197:155–164

Watson AHD (1980) The structure of the coeliac ganglion of a teleost fish, *Myoxocephalus scorpius*. Cell Tissue Res 210:155–165

Webb G, Heatwole H, Bavay de J (1971) Comparative cardiac anatomy of the reptilia. – I. The chambers and septa of the varanid ventricle. J Morphol 34:335–350

Weight FF, Weitsen HA (1977) Identification of small intensely fluorescent (SIF) cells as chromaffin cells in bullfrog sympathetic ganglia. Brain Res 128:213–226

Weiner N (1970) Regulation of norepinephrine biosynthesis. Annu Rev Pharmacol 10:273–290

Wells JW, Wight PAL (1971) The adrenal glands. In: Bell DJ, Freeman BM (eds) Physiology and biochemistry of the domestic fowl. Academic Press, London New York, pp 489–520

Westfall TC (1977) Local regulation of adrenergic neurotransmission. Physiol Rev 57:659–728

Wharton J, Polak JM, McGregor GP, Bishop AE, Bloom SR (1981) The distribution of substances P-like immunoreactive nerves in the guinea-pig heart. Neuroscience 6:2193–2204

Whitby LG, Hertting G, Axelrod J (1960) Effect of cocaine on the disposition of noradrenaline labelled with tritium. Nature (London) 187:604–605

White FN (1970) Central vascular shunts and their control in reptiles. Fed Proc 29:1149–1153

White FN, Ross G (1966) Circulatory changes during experimental diving in the turtle. Am J Physiol 211:15–18

Wikberg J (1977) Release of ^{3}H-acetylcholine from isolated guinea pig ileum. A radiochemical method for studying the release of the cholinergic neurotransmitter in the intestine. Acta Physiol Scand 101:302–317

Wikberg JES (1979) The pharmacological classification of adrenergic α_1 and α_2 receptors and their mechanisms of action. Acta Physiol Scand 106 Suppl 468:1–99

Williams JT, North RA (1978) Inhibition of firing of myenteric neurones by somatostatin. Brain Res 155:165–168

Williams JT, North RA (1979a) Effects of endorphins on single myenteric neurons. Brain Res 165:57–65

Williams JT, North RA (1979b) Vasoactive intestinal polypeptide excites neurones of the myenteric plexus. Brain Res 175:174–177

Williams TH (1967) Electron microscopic evidence for an autonomic interneuron. Nature (London) 214:309–310

Wilson AJ, Furness JB, Costa M (1981a) The fine structure of the submucous plexus of the guinea pig ileum: I. The ganglia, neurons, Schwann cells and neuropil. J Neurocytol 10:759–784

Wilson AJ, Furness JB, Costa M (1981b) The fine structure of the submucous plexus in the guinea pig ileum: II. Description and analysis of vesiculated nerve profiles. J Neurocytol 10:785–804

Winberg M, Holmgren S, Nilsson S (1981) Effects of denervation and 6-hydroxydopamine on the activity of choline acetyltransferase in the spleen of the cod, *Gadus morhua*. Comp Biochem Physiol 69C:141–143

Winslow J-B (1732) Exposition anatomique de la structure du corps humain. Traité de nerfs. Desprez-Desessartz, Paris, pp 424–468

Wood CM (1974) A critical examination of the physical and adrenergic factors affecting blood flow through the gills of the rainbow trout. J Exp Biol 60:241–265

Wood CM (1975) A pharmacological analysis of the adrenergic and cholinergic mechanisms regulating branchial vascular resistance in the rainbow trout (*Salmo gairdneri*). Can J Zool 53:1569–1577

Wood CM (1976) Pharmacological properties of the adrenergic receptors regulating systemic vascular resistance in the rainbow trout. J Comp Physiol 107:211–228

Wood CM (1977) Cholinergic mechanisms and the response to ATP in the systemic vasculature of the rainbow trout. J Comp Physiol 122:325–345

Wood CM, Shelton G (1975) Physical and adrenergic factors affecting systemic vascular resistance in the rainbow trout: A comparison with branchial vascular resistance. J Exp Biol 63:505–523

Wood CM, Shelton G (1980a) Cardiovascular dynamics and adrenergic responses of the rainbow trout in vivo. J Exp Biol 87:247–270

Wood CM, Shelton G (1980b) The reflex control of heart rate and cardiac output in the rainbow trout: interactive influences of hypoxia, haemorrhage, and systemic vasomotor tone. J Exp Biol 87:271–284

Wood JD (1981) Synaptic interactions in the enteric plexuses. J Auton Nerv Syst 4:121–133

Wood JD, Mayer CJ (1978) Slow synaptic excitation mediated by serotonin in Auerbach's plexus. Nature (London) 276:836–837

Wood JD, Mayer CJ (1979) Serotonergic activation of tonic-type enteric neurons in guinea pig small bowel. J Neurophysiol 42:582–593

Wood JD, Grafe P, Mayer CJ (1980) Comparison of the action of 5-hydroxytryptamine and substance P on intracellularly recorded electrical activity of myenteric neurons. In: Christensen J (ed) Gastrointestinal motility. Raven Press, New York, pp 131–138

Wood ME, Burnstock G (1967) Innervation of the lung of the toad (*Bufo marinus*). I. Physiology and pharmacology. Comp Biochem Physiol 22:755–766

Wood WG, Britton BJ, Irving MH (1979) Effects of adrenergic blockade on plasma catecholamine levels during adrenaline infusion. Horm Metab Res 11:52–57

Woodland WNF (1910) Notes on the structure and mode of action of the "oval" in the pollack (*Gadus pollachius*) and mullet (*Mugil chelo*). J Mar Biol Assoc UK 2:561–565

Woods RI (1970a) The innervation of the frogs heart – I. An examination of the autonomic postganglionic nerve fibres and a comparison of autonomic and sensory ganglion cells. Proc R Soc London Ser B 176:43–54

Woods RI (1970b) The innervation of the frogs heart – III. Electronmicroscopy of the autonomic nerve fibres and their vesicles. Proc R Soc London Ser B 176:63–68

Wooten GF, Coyle JT (1973) Axonal transport of catecholamine synthesizing and metabolizing enzymes. J Neurochem 20:1361–1371

Wooten GF, Saavedra JM (1974) Axonal transport of phenylethanolamine-N-methyltransferase in toad sciatic nerve. J Neurochem 22:1059–1064

Wurtman RJ (1966) Control of epinephrine synthesis in the adrenal medulla by the adrenal cortex: Hormonal specificity and dose-response characteristics. Endocrinology 79:608–614

Wurtman RJ, Axelrod J (1965) Adrenaline synthesis: Control by the pituitary gland and adrenal glucocorticoids. Science 150:1464–1465

Wurtman RJ, Watkins CJ (1977) Suppression of noradrenaline synthesis in sympathetic nerves by carbidopa, an inhibitor of peripheral dopa decarboxylase. Nature (London) 265:79–80

Wurtman RJ, Axelrod J, Tramezzani J (1967) Distribution of the adrenaline-forming enzyme in the adrenal gland of the snake-*Xenodon merremii*. Nature (London) 215:879–880

Wurtman RJ, Axelrod J, Vesell ES, Ross GT (1968) Species differences in inducibility of phenylethanolamine-N-methyl transferase. Endocrinology 82:584–590

Wyman LC, Lutz BR (1932) The effect of adrenaline on the blood pressure of the elasmobranch *Squalus acanthias*. Biol Bull Mar Biol Lab 62:17–22

Yamamoto KI, Itazawa Y, Kobagashi H (1980) Supply of erythrocytes into the circulating blood from the spleen of exercised fish. Comp Biochem Physiol 65A:5–11

Yamauchi A (1969) Innervation of the vertebrate heart as studied with the electron microscope. Arch Histol Jpn 31:83–117

Yamauchi A, Burnstock G (1968) An electronmicroscope study of the innervation of the trout heart. J Comp Neurol 132:567–588

Yamauchi A, Chiba T (1973) Adrenergic and cholinergic innervation of the turtle heart ventricle. Z Zellforsch Mikrosk Anat 143:485–493

Yntema CL, Hammond WS (1945) Depletions and abnormalities in the cervical sympathetic system of the chick following extirpation of neural crest. J Exp Zool 100:237–259

Young JZ (1931a) The pupillary mechanisms of the teleostean fish *Uranoscopus scaber*. Proc R Soc London Ser B 107:464–485

Young JZ (1931b) On the autonomic nervous system of the teleostean fish *Uranoscopus scaber*. Q J Microsc Sci 74:491–535

Young JZ (1933a) Comparative studies on the physiology of the iris – I. selachians. Proc R Soc London Ser B 112:228–241

Young JZ (1933 b) Comparative studies on the physiology of the iris. – II. *Uranoscopus* and *Lophius*. Proc R Soc London Ser B 112:242–249
Young JZ (1933 c) The autonomic nervous system of selachians. Q J Microsc Sci 75:571–624
Young JZ (1936) The innervation and reactions to drugs of the viscera of teleostean fish. Proc R Soc London Ser B 120:303–318
Young JZ (1939) Partial degeneration of the nerve supply of the adrenal. A study in autonomic innervation. J Anat 73:540–550
Young JZ (1980a) Nervous control of stomach movements in dogfishes and rays. J Mar Biol Assoc UK 60:1–17
Young JZ (1980b) Nervous control of gut movements in *Lophius*. J Mar Biol Assoc UK 60:19–30
Youson JH (1976) Fine structure of granulated cells in the posterior cardinal and renal veins of *Amia calva* L. Can J Zool 54:843–851
Yüh L (1931) On the innervation of the stomach of the Japanese frog. Jpn J Med Sci III Biophys 2:25–33
Yung IK, Woo KC, Doo WK (1965) Peristaltic movement of the tortoise intestine. Experientia 21:540–541
Zachariasen RD, Newcomer WS (1974) Phenylethanolamine-N-methyl transferase activity in the avian adrenal following immobilization or adrenocorticotropin. Gen Comp Endocrinol 23:193–198
Zeglinski N (1885) Experimentelle Untersuchungen über die Irisbewegung. Arch Anat Physiol 1–37
Zuckerkandl E (1901) Über Nebenorgane des Sympathicus im Retroperitonealraum des Menschen. Anat Anz 19:95–107
Zwaardemaker H (1924) Action du nerf vague et radio activite. Arch Neerl Physiol 9:213–228
Zwillenberg HHL (1964) Bau und Funktion der Forellenmilz. Hans Huber, Bern Stuttgart

Subject Index

AAD, see aromatic-amino-acid decarboxylase
Acceleransstoff   41
Acetylcholine   58, 94, 155
Acetylcholinesterase   58, 94
Acetylthiocholine   59
ACTH-like peptide   65
Actinopterygians   3, 4
Adenosine   60, 83
Adenosine triphosphate, see ATP
ADP   83
Adrenal
  gland   100
  medulla   100, 103
Adenaline
  in blood plasma   106
  in chromaffin tissue   101
  degradation   55, 90
  in spleen   88
  synthesis   51, 86
Adrenergic neurons   41, 49
Adrenoceptor
  agonists   91
  antagonists   91
  interconversion   82
Adrenoceptors   50, 79, 80, 81
Affinity   74
Agonist   68, 71
Alimentary canal   146
Amidephrine   82
AMP   83
c-AMP   77, 78
Amphetamine   90
Amphibians   3, 4, 23, 109, 126, 168, 183, 189, 194
Angiotensin   65
Antagonist   68, 71
Antazoline   98
Anurans   3, 4
Apamin   86, 98
Apodans   3, 4
APUD-cells   62, 100
Arecholine   95
Aromatic-amino-acid decarboxylase   51, 52, 63
Ascending excitatory reflex   149
ATP   59–61, 83, 155
Atropine   83, 93, 96
"a-type" nerve profile   33, 44
Auerbach's plexus, see myenteric plexus
Automodulation   47

Autonomic
  ground plexus   43
  neurons   43
Autoreceptors   47
Axillary body   17, 18, 39, 107
Axonal transport   46
Axon-hillock   28, 43
Azygos vein   23, 39, 109

Baroreceptor reflex   113, 114
Benserazid   87
Bethanecol   95
Bidder's ganglion   126
Birds   3, 4, 30, 110, 133, 171, 185, 196
Bladder, see urinary bladder and swimbladder respectively
"Blaschko pathway"   50
Bombesin   65, 159
Botulinum toxin   93
Brain-gut peptide   157
Branchial vasculature   116, 118, 120
Bretylium   89, 90
2-Bromolysergic acid   99
Bronchi   185, 186
Butoxamine   82
BW 284 c51   94

Caffeine   84
Carbachol   95
Carbidopa   87
Carotid
  body   100
  labyrinth   129
  plexus   12
Catecholamines
  in blood plasma   106
  in chromaffin tissue   101
  degradation   54, 90
  in spleen   88
  synthesis   49–51, 86
Catechol-O-methyl transferase   55
Cephalic sympathetic chains   20, 21, 27, 29
Cervical ganglia   11, 12
ChAT, see choline acetyltransferase
"Chemical tools"   85
Chemoreceptors   113
Chimaeroids   3, 4
Chlorimipramine   98

Chlorisondamine 93, 96
Choline 58, 95
Choline acetylase, see choline acetyltransferase
Choline acetyltransferase 57, 58
"Cholinergic link hypothesis" 48, 93
Cholinergic neurons 41, 57
Cholinesterase 58, 94
Cholinoceptor
  agonists 94
  antagonists 96
Cholinoceptors 58, 79, 83
Chondrosteans 4
Chorda tympani 12, 13
Chromaffin tissue 100, 101
Chromatic tract 201
Chromatophores 198
Chromogranin 45, 50
Chronotropic effect 112
Ciliary
  ganglion 12, 13, 18, 21, 24, 27, 192
  nerve 13, 192
  root 13, 192
Clonidine 82
Cocaine 89, 90
Coeliac ganglion 10–12, 21, 24, 25, 27
Collateral ganglion, see prevertebral ganglion
Competitive antagonism 71, 74, 75
COMT, see catechol-O-methyl transferase
Concentration-response curve 69
Corticosterone 90
Co-transmitter 43
Cranial autonomic system 6, 7, 9, 13
Crop 171
"c-type" nerve profile 33, 44
Cyclostomes 3, 4, 15, 105, 115, 159

"Dale's principle" 48
DBH, see dopamine-β-hydroxylase
Decamethonium 96
Depolarization 78
Descending inhibitory reflex 149
Dibenamine 91
Dibenzyline 91
Dichloroisoprenaline 80
Dihydroergotamine 91
3,4-Dihydroxymandelic acid, see DOMA
3,4-Dihydroxyphenyl glycol, see DOPEG
3,4-Dihydroxyphenylalanine, see DOPA
5,6- and 5,7-Dihydroxytryptamine 98, 160
3,5-Diiodo-L-tyrosine 86
Dipnoans 3, 4, 23, 109, 125, 182
Dipyridamole 97
Dissociation constant 72
Disulfiram 87
Diving, in birds 135
DMPP 83, 96
DOMA 55

DOPA 51, 52
DOPA decarboxylase, see aromatic-amino-acid decarboxylase
DOPEG 55
Dopamine 51, 52, 81, 100
Dopamine-β-hydroxylase 51, 52
Dose-response curve 69
Drug 68

Edinger-Westphal nucleus 13, 34
Edrophonium 94
Efficacy 73
Elasmobranchs 3, 4, 17, 106, 117, 161, 193
Endocrine cell 100
Enkephalin 61, 65, 159
Enteramine 62
Enteric nervous system 6, 7, 9, 153
Enterochromaffin cell 62, 64, 100, 156
Ephedrine 90
Epinephrine 91, 103 (see also adrenaline)
EPSP 78, 104
Ergot alkaloids 91
Eserine 94
Excitatory postsynaptic potential, see EPSP
Exocytosis 45
Extra-adrenal chromaffin cells 104
Extraneuronal uptake 54

Facial nerve 12, 13
Falck-Hillarp technique 56
Fluoxetine 98
Förderungssubstanz 41

Ganglion impar 12, 20, 27
Ganoids 4, 22, 108, 124, 182
Gas gland 178, 179
Gastric receptive relaxation 148, 149
Gastrin/cholecystokinin 65, 159
Gills 116, 118, 120
Gizzard 171
Glossopharyngeal nerve 12, 13
Glyoxylic acid 57
Granular uptake 50, 61, 63, 88
Guanethidine 89, 90

HC-3, see hemicholinium
Head kidney 107
Heart 114, 115, 117, 119, 126, 129, 133, 136, 137
Hemicholinium 93
Heteromodulation 47
Hexamethonium 83, 93, 96
Holocephali, see chimaeroids
Holosteans 4
Homeostasis 1
Homo-ATP 97
5-HT, see 5-hydroxytryptamine
6-Hydroxydopamine 87, 88

5-Hydroxyindole acetic acid   63, 64
5-Hydroxytryptamine   60, 62, 63, 84, 98, 100, 156
5-Hydroxytryptamine receptors, see serotonergic receptors
Hyomandibular nerve   13
Hyoscine   83, 96
Hyoscyamine   96
Hyperpolarization   79
Hypogastric nerve   11, 12

Imidazolines   84
Inhibitory postsynaptic potential, see IPSP
Inotropic effect   112, 131
Interrenal tissue   106
Intestine   149, 159, 162, 167, 169, 170, 172, 174
Intrinsic activity   73, 74
Intrinsic amine-handling neurons   62, 156
Involuntary nervous system   1
3-Iodo-α-methyl-L-tyrosine   86
3-Iodotyrosine   86, 87
IPSP   79, 104
Iris   192, 195
Isoprenaline   80, 87
Isoproterenol, see isoprenaline

Jugular ganglion   14
Junctional adrenoceptors   80, 127

Lacrimal glands   12, 13
Lampetroids   3, 4
Leu-enkephalin   61, 159
LGV, large granular vesicles   44
Lobeline   96
Ludvig's ganglion   126
Lung   182, 184
Lungfish, see dipnoans

Mammals   3, 4, 7, 103, 137, 147, 186, 190, 197
Mandibular nerve   13
MAO, see monoaminoxidase
Mecamylamine   97
Meissner's plexus, see submucous plexus
Melanocyte stimulating hormone   198
Melanophore   198
Melatonin   198
Membrane potential   78
Mesenteric ganglion   11, 12
Metactoid antagonism   75
Metaffinoid antagonism   76
Metanephrine   55, 56, 90
Met-enkephalin   159
Methacholine   95
Methoxamine   82
3-Methoxy-4-hydroxyphenyl glycol, see MOPEG
α-Methyl-DOPA   87, 88
α-Methyl-dopamine   88

α-Methyl-noradrenaline   82, 88
α-Methyl-tyrosine   87
Methylxanthines   84, 98
Methysergide   84, 98
Mipafox   94
MN, see metanephrine
Monoaminoxidase   55, 63, 64
α-Monofluoromethyldopa   87
3-Monoiodo-L-tyrosine, see 3-iodotyrosine
MOPEG   55, 56
Muscarine   83, 95
Muscarinic cholinoceptor   79, 83
Myelin   8
Myenteric plexus   14, 15, 147
Myxinoids   3, 4

NANC   42
Neostigmine   94
Nervi accelerantes   11, 138
Nervi erigentes   11
Nervus intestinalis impar, see *ramus intestinalis impar*
Neurocrine cell   100
Neuromodulator   43
Neuronal uptake   50, 53, 58, 61, 63, 90
Neurotensin   61, 65, 159
Neurotransmission   41
Nialamide   90
Nicotine   83, 95
Nicotinic cholinoceptor   79, 83
Nodose ganglion   14, 21, 27
Non-adrenergic, non-cholinergic, see NANC
Non-competitive antagonism   71, 75, 76
Noradrenaline
  in blood plasma   106
  in chromaffin tissue   101
  degradation   55, 90
  in gut   154
  in spleen   88
  synthesis   51, 52, 86
Noradrenergic neurons   42 (see also adrenergic neurons)
Norepinephrine   91 (see also noradrenaline)
Normetanephrine   55, 56

Oculomotor nerve   12, 13, 192
Opiate receptors   159
Otic ganglion   12–14
Oxygen receptors   119, 123

Palatine nerve   13
Pancreatic polypeptide   159
Paracrine cell   100
Paraganglia   100, 104
Parasympathetic   6, 7
Parasympathicolytics   93
Parasympathicomimetics   92

251

Paravertebral ganglia 7–9
Pargyline 90
Partial agonist 73
Pelvic ganglia, nerves and plexus 6, 10–12
Pempidine 97
Peptidergic neurons 42, 64, 99
Peptidoceptors 65, 79
Peristaltic reflex 149, 150
Petromyzontids, see lampetroids
Petrous ganglion 14, 21, 27
Phenoxybenzamine 91
Phentolamine 84, 91, 92, 98
Phenylbiguanide 84, 99
Phenylephrine 82, 87
Phenylethanolamine-N-methyl transferase 51, 53, 103
Physoclist 176
Physostigmine 94, 95
Physostome 176
Pilocarpine 83, 95
Pneumatic duct 176
PNMT, see phenylethanolamine-N-methyl transferase
Postganglionic neuron 8–10
Practolol 82
Prazosin 82
Preganglionic neuron 8–10
Prenalterol 82
Presynaptic
  receptors 47
  supersensitivity 57, 74
Prevertebral ganglia 8
Pronethalol 134
Propranolol 92
Prostaglandins 72
Prostigmine, see neostigmine
Proventriculus 171
Pseudobranch 124
Pseudo-cholinesterase 58, 94
"p-type" nerve profile 31, 33, 44
Pulmonary vasculature 127, 132
Purinergic neurons 42, 59, 97
Purinoceptor
  agonists 97
  antagonists 97
Purinoceptors 61, 79, 83

Quinacrine 62
Quinidine 84

Radioligand binding to receptors 77
Radix brevis 13, 192
Radix longa 13, 192
Radix sympathica 13, 192
Ramus communicans 8, 10, 21
Ramus intestinalis impar 16, 160
Rauwolscine 82

"Rebound excitation" 151, 164, 166
Receptor
  reserve 73
  for transmitters 68, 79
Release of transmitter 45
Remak's ganglion 126
Remak's nerve 32, 34, 172
Reptiles 3, 4, 26, 110, 129, 170, 185, 189, 194, 201
Reserpine 88, 89, 98
Rete mirabile 177, 179
Retractor lentis 19

Salbutamol 82
Salivary glands 12–14, 66
Sarcopterygians 3, 4
Sciatic nerve 24
Scopolamine 96
SCV, small clear vesicles 44
Secondary messengers 77
Selachians 3, 4
Selectivity of drugs 69
Serotonergic
  agonists 98
  antagonists 98
  neurons 42, 62
  receptors 63, 79, 84, 98
Serotonin 62, 156
Serotonin binding protein 63, 64
SG-cell 104, 105
SGV, small granular vesicles 44
SIF-cell 105
SK&F 29661 87
SK&F 64139 87, 88
Somatostatin 61, 65, 158
Sotalol 92
Specificity of drugs 68
Sphenopalatine ganglion 12, 13
Splanchnic nerve 11, 12, 18, 21, 24, 25
Spleen 141
Spinal
  ganglia 10
  nerves 10, 16
  roots 10
Spinal autonomic system 7–10
Stellate ganglion and complex 11, 12, 27
Stomach 148, 161, 164, 168, 170, 173
Striated muscle
  in gut 163
  in iris 194, 196
Submandibular ganglion 12, 13
Submucous plexus 14, 15, 147
Substance P 61, 65, 99, 157
Superior cervical ganglion 11, 12
Suprarenal tissue 106
Swimbladder 176, 177
"Sympathectomy" 54, 74, 88

Sympathetic   6, 7
Sympathetic chains   7, 8, 12
Sympathicolytic   86
Sympathicomimetic   86
Systemic vasculature   116, 118, 122, 128, 133, 134, 138

Taenia coli   147
Teleosts   3, 4, 20, 107, 119, 163, 176, 188, 193, 198
Terbutaline   82
Tetraethylammonium   96
Tetramethylammonium   96
Tetrapods   4
Tetrodotoxin   86
TH, see tyrosine hydroxylase
Theophylline   84
Threshold phenomenon   73
Tolazoline   91
Tonus, nervous etc.   113
Tramazoline   82
Transmitter substance   42
Transsynaptic modulation   47
Tranylcypromine   90
Trigeminal nerve   192
Tryptamine   84
Tryptophan   63, 64, 98
Tryptophan hydroxylase   63
TTX, see tetrodotoxin

d-Tubocurarine   97
Tyramine   54, 87, 90
Tyrosine   49–51
Tyrosine hydroxylase   50, 51

"Uptake$_1$,"   54, 89, 90
"Uptake$_2$,"   54, 90
Urinary bladder   187
Urodeles   3, 4

"Vagosympathetic trunk"   20, 26
Vagus nerve   12, 13
Vagusstoff   41
Vanillylmandelic acid, see VMA
Varicosity   29, 43
Vasculature, see branchial v., pulmonary v. and systemic v.
Vasoactive intestinal polypeptide, see VIP
Vasodilator
   nerves   138, 151
   reflex in gut   151
Vegetative nervous system   1
Vidian's nerve   34
VIP   60, 61, 65, 66, 156
VMA   55, 56

Yohimbine   82, 91, 92

Zuckerkandl, organ of   104

Zoophysiology (formerly Zoophysiology and Ecology)

Coordinating Editor: **D. S. Farner**
Editors: **W. S. Hoar, B. Hoelldobler, K. Johansen, H. Langer, G. Somero**

Volume 1
P. J. Bentley

Endocrines and Osmoregulation

A Comparative Account of the Regulation of Water and Salt in Vertebrates
1971. 29 figures. XVI, 300 pages. ISBN 3-540-05273-9

"The author ... has, with competence and insight, succeeded in the difficult task of covering two fields ... Bentley presents a thoroughly competent synthesis, and the result is a wellintegrated and balanced book. The book follows the zoological point of view, not only in outline, but also in the integration of physiological function with the natural life of the animal. The coherent viewpoint makes the text **readable and interesting**, and a large number of clear tables makes the material **easily accessible**. The adequate coverage can serve as an introduction to the research literature in both fields treated in this book. If future volumes are of equal quality and value, **the series will be a significant contribution.**"
Quarterly Review of Biology

Volume 2
L. Irving

Arctic Life of Birds and Mammals

Including Man
1972. 59 figures. XI, 192 pages. ISBN 3-540-05801-X

"The author's intense and unabated interest in arctic biology over the last three decades is reflected in the content and perspective of this volume. His unusually keen insight into the life of arctic birds and mammals (including man), which has led to this competent synthesis, is based on a familiarity conceivable only in a person who has experienced arctic life..."
Quarterly Review of Biology

Volume 3
A. E. Needham

The Significance of Zoochromes

1974. 54 figures. XX, 429 pages. ISBN 3-540-06331-5

"Dr. Needham's book is doubly welcome, for it not only considers animal pigments from the points of view of structure and function, but it also presents information and concepts which have never been assembled in one volume before ...
The format of the book is very pleasing, with particularly high quality typeface and paper. It was a good idea to preface each chapter with a brief synopsis of its subject matter and to include a conclusion section at the end. There are many tables which collect data not easily found elsewhere: the book is **a valuable and unique contribution** to the literature on pigments."
Nature

Volume 4/5
A. C. Neville

Biology of the Arthropod Cuticle

1975. 233 figures. XVI, 448 pages. ISBN 3-540-07081-8

"... The layout is clear and orderly throughout ... As the text has a clear and economical style, as there are numerous figures and electron micrographs and an extensive but selective bibliography, this is an essential work of reference. But the treatment throughout is of a critical review ... the book is admirably produced and printed ..."
Quarterly J. Exp. Physiology

Volume 6
K. Schmidt-Koenig

Migration and Homing in Animals

1975. 64 figures, 2 tables. XII, 99 pages. ISBN 3-540-07433-3

"The author ... has provided a valuable service in collecting together examples of homing and migration in a diversity of animal groups. The plan is an excellent one: each chapter, devoted to a single taxonomic group, is subdivided into examples of field performance in orientation and its experimental analysis ..."
The IBIS

Springer-Verlag Berlin Heidelberg New York

Volume 7
E. Curio
The Ethology of Predation
1976. 70 figures, 16 tables. X, 250 pages. ISBN 3-540-07720-0

"... It is good because it is stimulating, exhaustive and logical. No important aspect of the subject is missed. The author illustrates all his main points with a multiplicity of examples drawn from recent research. The reference list of nearly 700 items is evidence of the thoroughness of the treatment and the marshalling of examples used in explanation. Curio is an enthusiast and conveys the excitement to be found in much of the research on this subject; he also draws pointed attention to the gaps in our knowledge. For all these reasons **the book is a must for ethologists, ecologists, experimental psychologists and university libraries.** As a first treatment of seminal quality the book could well become a reference classic and inspire numerous research projects.
... The illustrations are clear and relevant ..."
The Quart. Review Biology

Volume 8
W. Leuthold
African Ungulates
A Comparative Review of Their Ethology and Behavioral Ecology
1977. 55 figures, 7 tables. XIII, 307 pages. ISBN 3-540-07951-3

"... Dr. Leuthold displays a masterly command of his subject ... The work is basically a review of published knowledge with an original approach, enlivened by the author's interpretations and based on his intimate first-hand knowledge of the subject. The first chapter, on the application of ethological knowledge to wildlife management, covers an important area ... a wealth of references is given so that the chapter provides a useful guide to the literature. The illustrations are good and well chosen to demonstrate points made verbally and not first to embellish the text. The book will provide **excellent background reading for undergraduates and research students as well as for anyone seriously interested in African wildlife.** On the whole, **it can be thoroughly recommended.**"
J. Applied Ecology

Volume 9
E. B. Edney
Water Balance in Land Arthropods
1977. 109 figures, 36 tables. XII, 282 pages. ISBN 3-540-08084-8

"... Dr. Edney has provided a wealth of organized information on prior work and ideas for needed research, **all of which make the book a bargain.** The volume should prove useful, not only to those who work in arthropod water relations (it is a must for them), but also to those of us interested in invertebrate and general ecology, entomology, comparative physiology, and biophysics."
AWRA Water Resources Bull.

Volume 10
H.-U. Thiele
Carabid Beetles in Their Environments
A Study on Habitat Selection by Adaptations in Physiology and Behaviour
Translated from the German by J. Wieser
1977. 152 figures, 58 tables. XVII, 369 pages. ISBN 3-540-08306-5

"... Because the book is comparative both in method and interpretation, it is a contribution to systematics as well as to ecology ... **a fine synthesis of current knowledge** of homeostatic aspects of ecological relationships of carabids, and it is a fitting tribute to the man to whom it is dedicated: Carl H. Lindroth, who was instrumental in formulating the approaches and techniques that are commonly used in ecological research on these fine beetles. The material is **well organized** and the text is **easily readable, thanks to the clarity of thought and expression** of the author and to the skill of an able translator." *Science*

Volume 11
M. H. A. Keenleyside
Diversity and Adaptation in Fish Behaviour
1979. 67 figures, 15 tables, XIII, 208 pages. ISBN 3-540-09587-X

"... it is important as the first serious attempt by a senior researcher to produce an overview of the discipline. Previous works have all been symposium volumes or collections of papers haphazardly assembled, and Keenleyside has produced a volume that is of substantially greater value than these. In clearly perceiving that the unique and valuable features of fish behavior are its diversity of form and circumstance, he has charted a course that future authors would be wise to follow.
The book is well produced, well written, and easy to read. The illustrations are clear and straightforward." *Science*

Volume 12
E. Skadhauge
Osmoregulation in Birds
1981. 42 figures. X, 203 pages. ISBN 3-540-10546-8

Contents: Introduction. - Intake of Water and Sodium Chloride. - Uptake Through the Gut. - Evaporation. - Function of the Kidney. - Function of the Cloaca.- Function of the Salt Gland. - Interaction Among the Excretory Organs. - A Brief Survey of Hormones and Osmoregulation. - Problems of Life in the Desert, of Migration, and of Egg-Laying. - References. - Systematic and Species Index. - Subject Index.

Springer-Verlag Berlin Heidelberg New York

UNIVERSITY OF DELAWARE

Please return this book as soon
with it. In order to
 test date stamp